Smart Materials-Based Actuators at the Micro/Nano-Scale

AF781199

Micky Rakotondrabe
Editor

Smart Materials-Based Actuators at the Micro/Nano-Scale

Characterization, Control, and Applications

Springer

Editor
Micky Rakotondrabe
Automatic & MicroMechatronic
Systems Department
University of Franche-Comté at Besançon
Besançon, France

ISBN 978-1-4939-4560-3 ISBN 978-1-4614-6684-0 (eBook)
DOI 10.1007/978-1-4614-6684-0
Springer New York Heidelberg Dordrecht London

© Springer Science+Business Media New York 2013
Softcover reprint of the hardcover 1st edition 2013
This work is subject to copyright. All rights are reserved by the Publisher, whether the whole or part of the material is concerned, specifically the rights of translation, reprinting, reuse of illustrations, recitation, broadcasting, reproduction on microfilms or in any other physical way, and transmission or information storage and retrieval, electronic adaptation, computer software, or by similar or dissimilar methodology now known or hereafter developed. Exempted from this legal reservation are brief excerpts in connection with reviews or scholarly analysis or material supplied specifically for the purpose of being entered and executed on a computer system, for exclusive use by the purchaser of the work. Duplication of this publication or parts thereof is permitted only under the provisions of the Copyright Law of the Publisher's location, in its current version, and permission for use must always be obtained from Springer. Permissions for use may be obtained through RightsLink at the Copyright Clearance Center. Violations are liable to prosecution under the respective Copyright Law.
The use of general descriptive names, registered names, trademarks, service marks, etc. in this publication does not imply, even in the absence of a specific statement, that such names are exempt from the relevant protective laws and regulations and therefore free for general use.
While the advice and information in this book are believed to be true and accurate at the date of publication, neither the authors nor the editors nor the publisher can accept any legal responsibility for any errors or omissions that may be made. The publisher makes no warranty, express or implied, with respect to the material contained herein.

Printed on acid-free paper

Springer is part of Springer Science+Business Media (www.springer.com)

Preface

Micro/nanorobots and microactuators find their applications in various domains: microUAV in military, microrobots for in-body exploration in medicine, microactuators for microassembly and micromanipulation and for surface characterization with nanometric resolution, etc. In order to reach the severe performances required for these "micro/nano" applications—such as very high resolution, micrometric or submicrometric accuracy, and high bandwidth—convenient design of the actuators and convenient control of them are necessary. Smart materials like piezoelectric and electroactive polymers and flexible structures are among the best candidates to design the actuators, but their characteristics (nonlinearities, badly damped vibrations, etc.) require the use of efficient control techniques. In addition to these characteristics, the particularity of working at the micro/nano-scale (lack of embeddable sensors, high sensitivity to the environment, difficulty to directly sense, general uncertainties on the model used) makes their control even more challenging.

Several researches have been carried out since many years with tremendous success; however, limitations still exist in terms of precision, operating speed, and reliability. Furthermore, emerging and new requirements in micro/nano-scale positioning bring additional challenges for the design and the control of the systems. These challenges include the high axiscoupling in multi-degrees of freedom precise positioning systems, the limitations of the existing sensors to measure the signals in them, the high environmental (thermal, vibration...) sensitivity and the high noise-to-signal ratio as they are more and more small. During these last years, in order to tackle these challenges, advanced design concepts, where some are jointly coupled with control theory to include the performances at the design level, novel sensing and actuation combined technologies, noise measurement and resolution estimation techniques, and advanced control techniques with or without sensors have been developed. Several projects have been launched at the international level for that. Finally, many technical meetings such as workshops and tutorials dealing with the design or with the control of actuators based on smart materials were organized in different international conferences.

This book gives a state of the art of emerging techniques to the characterization and control of actuators based on smart materials working at the micro/nano-scale.

The case of piezoelectric and electroactive polymeric actuators is focused. The book was initiated after a scientific tutorial held during the IEEE—International Conference on Robotics and Automation (ICRA) in May 2011 at Shanghaï, China and organized by the book editor. The tutorial has brought researchers and engineers together to present, discuss, and exchange ideas on the challenging topic: "Dynamics, characterization and control at the micro/nano scale." The exciting discussions and exchanges between the speakers of the workshop and the audience, composed of engineers, researchers, and students, have resulted in the necessity to make a perennial archive available for a large public of the interesting presentations and discussions. This is the motivation of this book which contains a potential both for industrial applications and for research. The writing of the book is also such that novice academic level (undergrads, masters and Ph.D. students) can start with the domain of the micro/nano-scale and related actuators without difficulty.

This book is composed of twelve chapters that organized into four main parts.

The first part, made of three chapters, concerns the introduction to piezoelectric materials and polymeric materials and their use for the design of actuators working at the micro/nano-scale:

- Chapter 1 deals with the main motivations of using smart materials as a fundamental component in micro/nano-positioning applications. The authors show that actuators based on smart materials are chosen as an alternative to classical actuators, since the design of classical actuated systems is not suitable for the design of very small ones. The chapter particularly gives an emphasis to piezoelectric materials which are widely used in the microworld. Over the chapter, the authors provided the favorable properties of piezoelectric materials which make them very interesting for the design and development of microsystems working at micro/nano-scale.
- Chapter 2 deals with the case of newer polymeric materials as base for actuators. Synthesis of the new materials, modeling, and experimental characterization on these are detailed in the chapter.
- In Chapter 3, a new method to design actuators based on piezoelectric materials is presented. The method uses interval techniques combined with the geometrical and physical model of the actuator. The new technique has an advantage to provide guaranteed performances, thanks to the properties of interval tools.

The second part of the book deals with the closed-loop (or feedback) control of smart materials-based actuators. This part includes four chapters.

- Chapter 4 proposes a decoupling method to model the behavior of a nonlinear and oscillating piezoelectric actuator that has 2-degrees of freedom (2-dof). A robust H-inf technique is afterward employed to control the actuator. Experimental results demonstrate the efficiency of the proposed technique.
- In Chapter 5, a model-based control system to enhance the performances of nano-positioning systems is proposed. Different control methods are applied and they can be classified into: (a) inverse-based control schemes and (b) model-based control schemes.

- Chapter 6 combines interval tools and related techniques with the classical control theory to model a piezoelectric actuator and to synthesize a robust controller for this. The main advantages of the approach are the natural way to characterize parametric uncertainties in the actuator's model, these uncertainties being due to the difficulties of identification and to the high sensitivity of systems working at the micro/nano-scale. The approach also derives low-order controllers which are well appreciated because of their ease of implementation. Finally, the chapter proposes a new way to analyze the performance robustness *a posteriori* by still using interval techniques but combined with the H-inf tool.

- In Chapter 7, a state-feedback control with integral action is introduced to improve the performances of a nonlinear and noisy piezoelectric microsystem. As the state-feedback control requires a linear system, a feedforward controller to compensate the nonlinearity (hysteresis) is first utilized. Furthermore, in order to make the state of the actuator available for the feedback, a Kalman filtering is also proposed. This filtering permits at the same time to reduce the noises seen at the sensor output. The experimental results demonstrated the efficiency of these techniques combined and their interest for micro/nano-positioning.

The third part of the book, made of three chapters, treats the feedforward control of smart materials-based actuators. Feedforward control techniques are very appreciated in systems where the use of sensors is impossible or difficult. In particular, they are of great interest in systems working at the micro/nano-scale due to the lack of embeddable and convenient sensors usable at this scale.

- Chapter 8 presents a hysteresis model based on least squares support vector machine (LSSVM) and proposes feedforward compensators by neglecting the inverse hysteresis. It presents a comparative experimental study to present the advantage of LSSVM over Bouc–Wen Model.
- In Chapter 9, the rate-dependent Prandtl–Ishlinskii hysteresis modeling and control are proposed. The techniques allow to reduce the hysteresis that depends on the rate or the frequency of the input control and that is found in many hysteretic dynamical systems.
- Chapter 10 treats the modeling and the simultaneous feedforward control of the hysteresis, the creep nonlinearity, and the badly damped vibration found in nonlinear and flexible piezoelectric actuators. The design of the three controllers (compensators) for these three behaviors, put in cascade, is based on precise models and on their inversion. The chapter includes experimental results which demonstrate the efficiency of the approach.

The last part presents two of the most emerging topics and applications at the micro/nano-scale: nanorobotics and biological cells micro/nano-manipulation. This part is composed of two chapters.

- Chapter 11 presents the fabrication with nanorobotic techniques and the characterization of piezoresistive force sensors based on helical nanobelts. The process

of fabrication and assembly of the sensors are well detailed and experimental characterization provides their interesting performances.
- Chapter 12 describes computer vision-based sperm analyses and manipulation methods. The chapter also introduces recent progress in automating sperm manipulation procedures, including sperm immobilization, aspiration, and positioning inside a micropipette.

I would like to thank all the contributors of this book who describe new results in a very didactic way in these chapters. Most of the contributors participated in the above-mentioned tutorial. I am also very grateful to Merry Stuber from Springer Verlag for her assistance and encouragement along the preparation of this book. It was a great pleasure to work with her. Finally, I give my thanks to Alison Waldron (from Springer Verlag) to whom I initially contacted for the idea of this book.

Besançon, France Micky Rakotondrabe

Contents

Contributors

Mohammad Al Janaideh Department of Aerospace Engineering, The University of Michigan, Ann Arbor, MI, USA

Department of Mechatronics Engineering, The University of Jordan, Amman, Jordan

Alex Bienaimé AS2M Department, FEMTO-ST Institute, UMR CNRS 6174 – UFC/ENSMM/UTBM, Besançon, France

Juan-Antonio Escareno AS2M Department, FEMTO-ST Institute, UMR CNRS 6174 – UFC/ENSMM/UTBM, Besançon, France

Didace Habineza University of Franche-Comté at Besançon

AS2M Department, FEMTO-ST Institute, UMR CNRS 6174 – UFC/ENSMM/UTBM, Besançon, France

Hideki Hashimoto Department of Electrical, Electronic, and Communication Engineering, Faculty of Science and Engineering, Chuo University, Tokyo, Japan

Gilgueng Hwang Laboratory for Photonics and Nanostructures (LPN), Centre National de la Recherche Scientifique (CNRS), CNRS-LPN, Marcoussis, France

Sofiane Khadraoui Electrical and Computer Engineering Department, Texas A&M University at Qatar, Doha, Qatar

Sergio Lescano University of Franche-Comté at Besançon

AS2M Department, FEMTO-ST Institute, UMR CNRS 6174 – UFC/ENSMM/UTBM, Besançon, France

Clement Leung Department of Electrical and Computer Engineering, University of Toronto, Toronto, ON, Canada

Jun Liu Department of Electrical and Computer Engineering, University of Toronto, Toronto, ON, Canada

Beatriz Cristina López-Walle Programa Doctoral en Ingeniería de Materiales, División de estudios de Posgrado, Facultad de Ingeniería Mecánica y Eléctrica (FIME), Centro de Investigación, Innovacióny Desarrollo en Ingeniería y Tecnología (CIIDIT), Universidad Autonoma de Nuevo Leon (UANL), Km. 10 Autopista al Aeropuerto Internacional de Monterrey, Apodaca, México

Zhe Lu Department of Electrical and Computer Engineering, University of Toronto, Toronto, ON, Canada

Philippe Lutz University of Franche-Comté at Besançon

AS2M Department, FEMTO-ST Institute, UMR CNRS 6174 – UFC/ENSMM/UTBM, Besançon, France

John Lygeros Department of Information Technology and Electrical Engineering, Automatic Control Laboratory, ETH Zurich, Zürich, Switzerland

Angeliki Pantazi Storage Technologies, IBM Research - Zurich, Rüschlikon, Switzerland

Micky Rakotondrabe University of Franche-Comté at Besançon

AS2M Department, FEMTO-ST Institute, UMR CNRS 6174 – UFC/ENSMM/UTBM, Besançon, France

Edgar Reyes-Melo División de estudios de Posgrado, Facultad de Ingeniería Mecánica y Eléctrica (FIME), Centro de Investigación, Innovacióny Desarrollo en Ingeniería y Tecnología (CIIDIT), Universidad Autonoma de Nuevo Leon (UANL), Km. 10 Autopista al Aeropuerto Internacional de Monterrey, Apodaca, México

Abu Sebastian Storage Technologies, IBM Research - Zurich, Rüschlikon, Switzerland

Yu Sun Departments of Mechanical and Industrial Engineering and Electrical and Computer Engineering, Institute of Biomaterials and Biomedical Engineering, University of Toronto, Toronto, ON, Canada

Tomas Tuma Storage Technologies, IBM Research - Zurich, Rüschlikon, Switzerland

Department of Information Technology and Electrical Engineering, Automatic Control Laboratory, ETH Zurich, Zürich, Switzerland

Qingsong Xu Department of Electromechanical Engineering, Faculty of Science and Technology, University of Macau, Taipa, Macao SAR, China

Chapter 1
Introduction: Smart Materials as Essential Base for Actuators in Micro/Nanopositioning

Micky Rakotondrabe, Mohammad Al Janaideh, Alex Bienaimé, and Qingsong Xu

Abstract The main motivations of using smart materials as fundamental in micro/nanopositioning systems are presented in this chapter. It is shown that the design of classical (or macro) actuated systems cannot directly be used to design small ones, particularly those used for micro/nanopositioning. While in macro, many components are assembled to form the actuated systems, in micro one attempts to reduce the number of elements in order to ensure some resolution and accuracy of positioning and in order to make easy their fabrication. Smart and active materials are therefore seen as the principal and essential component in microsystems and systems working at the micro/nano-scale. Their advantages are detailed in the chapter and some of the behaviors (hysteresis and creep) that are often encountered are explained. A particular attention is given to piezoelectric materials since nine chapters of the book treat them.

M. Rakotondrabe (✉) • A. Bienaimé
Automatic Control and Micro-Mechatronic Systems Department, AS2M, FEMTO-ST Institute, 24 rue Alain Savary, Besançon 25000, France
e-mail: mrakoton@femto-st.fr; alex.bienaime@femto-st.fr

M. Al Janaideh
Department of Mechatronics Engineering, The University of Jordan, Amman 11942, Jordan

Department of Aerospace Engineering, The University of Michigan, Ann Arbor, MI 48109, USA
e-mail: aljanaideh@gmail.com

Q. Xu
Department of Electromechanical Engineering, University of Macau, Taipa, Macau, China
e-mail: qsxu@umac.mo

M. Rakotondrabe (ed.), *Smart Materials-Based Actuators at the Micro/Nano-Scale: Characterization, Control, and Applications*, DOI 10.1007/978-1-4614-6684-0_1,
© Springer Science+Business Media New York 2013

1.1 Introduction

This chapter is an introduction to the principal motivations of why using smart materials in the design and development of systems working at the micro and nano-scale, in particular for micro/nanopositioning applications. More focus will be given to piezoelectric materials as nine chapters of the book use such materials. The chapter is organized as follows. In Sect. 1.2, the principal difficulties encountered when using classical design to develop systems for micro/nano-scale are presented. Some smart materials are cited as possible alternative to the classical design (DC motors and articulations) and among the well-recognized smart materials for micro/nanopositioning, we find piezoelectric materials. Section 1.3 is therefore devoted to present the basic principle of piezoelectricity and Sect. 1.4 deals with their particular advantageous. In Sect. 1.5 we present some of characteristics (hysteresis, creep) that may limit the utilization of smart materials in general such that the readers could have an idea of the motivations of the modeling and control presented in the different chapters of the book.

1.2 Why Using Smart Materials in Micro/Nanopositioning

In positioning systems with classical dimensions (robots, manipulators, etc.), we often use several components to compose them. First, among the most used actuators, we find: DC motors, pneumatic actuators, magnetic actuators, etc. These actuators are themselves made up of several subcomponents such as stator, rotor, and movable part. In addition to the actuators, other mechanical components are utilized to transform the angular (resp. linear) motion into linear (resp. angular) motion to amplify the displacement stroke or to reduce the speed. In general, these mechanical amplifiers and transformers contain themselves several subcomponents such as passive articulations. At the macro-scale, the assembly of the different components is evident. Remind that many actuators and systems at this scale can perform theoretically infinite stroke, for instance the angle obtained with an electrical motor can be unlimited. Although the large stroke that can be obtained, the methods and the technologies used to design systems working at the macro- are not suitable for the design and the development of systems working at the micro-. The principal reasons of such incompatibility are as follows:

- The sizes of systems used in macro are relatively bulky face to the available space when working at the micro-nano. For instance, the available space in an atomic force microscopy (AFM) is very reduced and could not welcome an actuator based on DC motor to position the sample of material to be scanned.
- With "macro" systems, the consumed energy is huge and is not justified face to the small objects to be positioned. As an example, using a "macro" robot to precisely position a biological cell would require many electrical power and

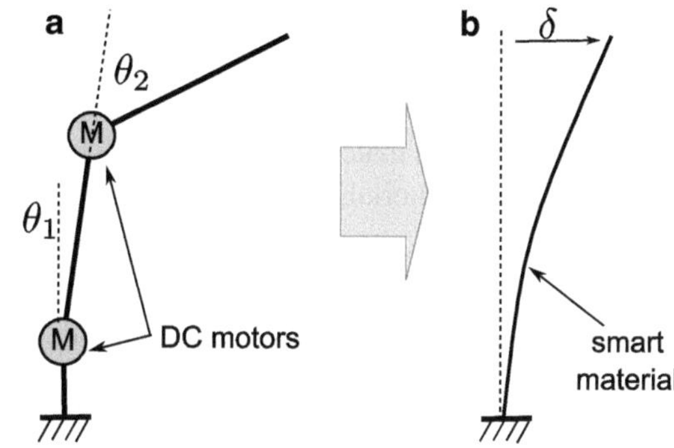

Fig. 1.1 Replacement of a classical system (articulation with DC motors, see (**a**)) by utilizing a smart material (see (**b**))

it would be more convenient to position the same object with lower power consumption systems such as smaller robot.

- The mechanical clearances found in the articulations yield a limited resolution of positioning. This limited resolution is not often adapted to the resolution required in nanopositioning.
- The fact that "macro" systems are based on the assembly of different components, they have minimal sizes that are irreducible to be convenient with the available space in micro. Remind that if we assumed the existence of similar but smaller components, the assembly itself would be very difficult at small scale and then still maintains the difficulty to fabricate assembled micro-components. In addition to that, some components are mobile relative to themselves. This yields some friction and therefore implies a loss of precision of the whole system.

The above reasons lead researchers and engineers to use new design of the systems devoted to micro/nanopositioning. The main idea is that, instead of using several assembled components (actuators and articulations), one employs smart materials, i.e. materials that react and that can generate motion when excited electrically, magnetically, thermally, etc. Indeed, it is possible to replace an actuator and related articulations by utilizing the same bulk of smart material which consequently removes many of the above cited limitations substantially. Figure 1.1a depicts a 2-degrees of freedom (2dof) classical manipulator based on two motors and bars, while Fig. 1.1b depicts its replacement with a cantilever structured smart material. Both systems can provide a displacement at their extremities. As we can see, the articulations are removed in the smart material-based actuator and the use and assembly of several components are bypassed. The resolution is highly increased as the friction and the mechanical clearances do not anymore exist. This resolution is only dependent on the minimal deformation that can perform the smart material, which should be theoretically infinite. Finally, the fabrication of smaller systems are possible since we can use the same bulk material to integrate different functions (actuation, mechanical amplification, etc.).

Among the smart materials that are commonly used to develop micro/nanopositioning systems, we find:

- Shape memory alloys (SMA) and thermal shape memory alloys (TSMA) materials
- Electropolymeric and magnetopolymeric materials
- Magnetostrictive materials and fluid mechanics
- Piezoelectric materials

To resume, the main benefit from smart materials is their ability to deform themselves under the electrical or magnetic excitation. However, there also exist other principles to deform materials that are not necessarily smart. They include:

- The electrostatic principle which is based on the Coulomb force to deform the material
- The magnetic principle, for instance by fixing a ferromagnetic small object at the tip of the cantilever of Fig. 1.1b which, under a magnetic field, is attracted/repulsed and then results a bending of this latter
- The thermal principle based on a bilayered cantilever where each layer has different thermal coefficients

In these smart and active materials, piezoelectric one are among the well recognized thanks to several advantages that they can offer. In this book, nine chapters (Chaps. 2, 3, and 5–11) are devoted to piezoelectric materials used in nanopositioning systems and related applications. In order to give some preliminaries to the readers, the remaining sections will therefore be consecrated to these materials.

1.3 Basics on Piezoelectric Materials

Piezoelectricity is the ability of certain materials to create electrical charges in response to a mechanical stress. This phenomenon is called the direct piezoelectric effect. The piezoelectric effect is a reversible process and piezoelectric materials are able to generate internal mechanical strain when an electrical field is applied. A wide range of applications uses these two phenomena especially for sensors (pressure), actuators, energy harvesting or resonance applications (ultrasonic applications, filters, high sensitive mass sensors). We present here the principle at the microscopic and macroscopic scale before reminding the constitutive equations of a piezoelectric material. We end the section by classifying the different kinds of piezoelectric materials.

1.3.1 Microscopical Principle

Piezoelectric materials are composed of different kinds of atoms, having different electrical charges. At equilibrium (Fig. 1.2a), the electric charges of each compound are compensated, for example in quartz, oxygen atoms charged negatively sharing

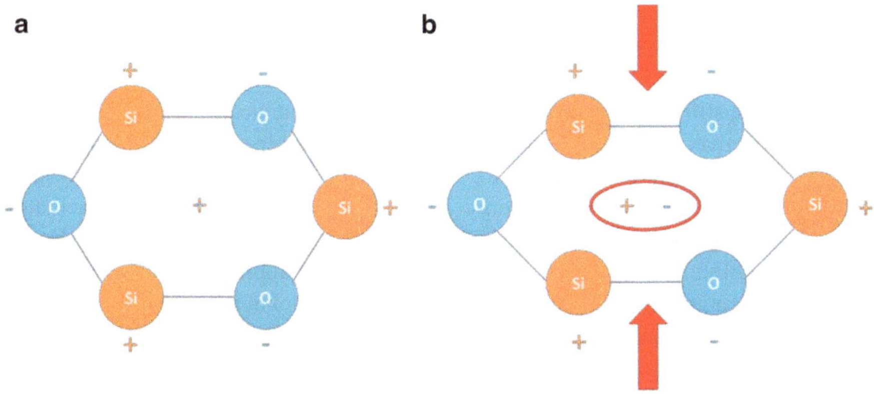

Fig. 1.2 Example of direct/converse piezoelectric effect on an elementary mesh of quartz. (**a**) The mesh at its equilibrium state (no mechanical force or electrical voltage is applied). (**b**) The mesh when a mechanical force or an electrical voltage is applied

electrons with silicon atoms charged positively. However, under a mechanical stress, the material is deformed (Fig. 1.2b) and the electric equilibrium is broken. A polar moment appears in the solid which creates charges in the material. On the other hand, the separation of electrical charges, due to the application of electric potential on the material, induces a displacement of atoms which causes deformations in the solid.

1.3.2 Macroscopical Principle and Equations

At macroscopical scale, the direct or inverse piezoelectric effect is visible by measuring, respectively, the electrical potential difference in function of the stress applied or the deformation generated by the application of an electrical potential difference. For an ideal piezoelectric material, these two relations are proportional and can be written as:

$$\begin{cases} S = dE \\ D = dT \end{cases} \tag{1.1}$$

where D is the electric displacement, T the stress, S the strain, E the electric field, and d the proportional coefficient, called the piezoelectric constant. If we add the relations that link stress with strain (Hooke's law) and the electric displacement with the electric field which traduce, respectively, the mechanical and electrical behavior of a material, we obtain the piezoelectric constitutive equations:

$$\begin{cases} S = s^E T + dE \\ D = dT + \varepsilon^T E \end{cases} \tag{1.2}$$

where s is the stiffness and ε the permittivity for constant electric field and constant stress, respectively. These equations are generally combined with the equilibrium equations to determine the behavior of a device. The direct effect is usually used in sensor applications or energy harvesting and inverse effect for actuator applications (quasi-static case: expansion/compression or shear motions). Some devices combine both effects, especially for resonance applications using different modes of vibration: compression/expansion, flexion, thickness shear, or face shear.

A piezoelectric material can be viewed as an energy converter and its ability to convert electrical (resp. mechanical) energy into mechanical (resp. electrical) energy is given by the piezoelectric coupling factor:

$$k^2 = \frac{\text{mechanical (or electrical) energy converted to electrical (or mechanical) energy}}{\text{input mechanical (or electrical) energy}} \tag{1.3}$$

and can be expressed as:

$$k^2 = \frac{U_{\mathrm{C}}^2}{U_{\mathrm{D}} U_{\mathrm{E}}} \tag{1.4}$$

where U_{C} is the coupling energy, U_{D} the deformation energy, and U_{E} the electrical energy.

1.3.3 *Piezoelectric Materials*

Two materials are widely used in devices: the PZT (lead zirconate titanate, $Pb(Z_x Ti_{1-x})O_3$) which are generally used for actuators, and quartz crystals used for resonators. But various materials could be adapted depending on the applications.

Crystals, such as quartz, present generally a high stability, especially face to temperature variation, but they have a low piezoelectric module and high acoustic impedance. Monocrystals based on langasite (langasite, langatate, langanite) or lithium (lithium niobate or lithium tantalite) show higher piezoelectric coefficients although they are still lower than those of ceramics. Then, their costs of production and their fragility limit their applications. New monocrystals such as PMN-PT or PZN-PT, with the same composition as ceramics, offer high coupling properties and are under numerous research investigations nowadays, in the aim to substitute PZT ceramics in sensors or actuators applications.

Ceramics, in opposition to crystals, have a lower stability but a higher piezoelectric coefficient and a low cost of production. Ceramics are generally made of PZT with various concentration of titanium and it is possible to add some dopants to modify the material properties and make easier electrical exchanges. Two kinds of

PZT are developed: soft and hard PZT. Hard PZT are doped with acceptor atoms which reduce loss in material but decrease piezoelectric constant. Soft PZT are doped with donator atoms which confer better piezoelectric constants but increase losses in material due to internal friction. Finally other piezoelectric materials are also found in small systems applications such as small sensors and small actuators. They include GaAs, AlN, ZnO, and piezoelectric polymers such as PVDF.

1.4 Gains Obtained with Piezoelectric Materials

Piezoelectric materials are commercially available from a number of companies, such as Noliac, Physik Instrumente (PI), and NEC/Tokin. Acting as an actuator, the piezoelectric material converts the electrical energy into motion. They are electromechanical devices for generating movements in the micrometer range. Piezoelectric actuators possess some attractive properties, such as compact size, high resolution, high bandwidth, ease of fabrication of small systems, avoidance of mechanical plays/clearances, and no electromagnetic interference. They can be operated over billions of cycles without wear or deterioration. Their bandwidth is very high, which is only limited by the inertia of the object being moved and the output capability of the electronic driver. In addition, virtually no power is consumed to maintain a piezoelectric actuator in an energized state. Moreover, they enable the capability of self-sensing [1]. Taking PI actuators as example, specific advantages of the piezoelectric actuators are enumerated as follows [2]:

- *Ultrahigh resolution:* A piezoelectric actuator can produce extremely fine position changes down to the subnanometer range. The smallest changes in operating voltage are converted into smooth movements. Motion is not influenced by friction effect.
- *High bandwidth:* Piezoelectric actuators offer the highest bandwidth available. Microsecond time characteristics can be easily obtained.
- *Large force generation:* Piezoelectric actuators can generate a force of from several kN to several tons within a typical range of tens micrometer.
- *No magnetic fields:* Piezoelectric actuators are especially well suited for applications where magnetic fields cannot be tolerated.
- *Low power consumption:* The piezoelectric material absorbs electrical energy during movement only. Static operation, even holding heavy loads, does not consume power.
- *No wear and tear:* A piezoelectric actuator has neither gears nor rotating shafts. Its displacement is based on pure solid-state effects and exhibits no wear and tear.
- *Vacuum and clean-room compatible:* Piezoelectric actuators employ ceramic elements that do not need any lubricants and exhibit no wear or abrasion. This makes them clean-room compatible and ideally suited for ultra-high-vacuum applications.

- *Operation at cryogenic temperatures:* The piezoelectric effect is based on electric fields, and it functions down to almost 0 K although at reduced specifications.

In view of the obtained gains mentioned above, piezoelectric actuators have been applied extensively in engineering applications. The most popular types of piezoelectric actuators are piezoelectric stacks, bending actuators, and shear actuators. Piezoelectric stack actuators are constructed by stacking multiple layers of piezoelectric materials together. They are usually adopted in micro-/nanopositioning stages [3, 4], auto focusing of cell phone camera, vibration sources, vibration controls, mirror/prism positioning, AFM [5], etc. Alternatively, piezoelectric bender actuators consist of one to several layers of piezoelectric materials. With one end fixed, the free-end of the bender delivers motion once powered. Such kind of cantilever-based actuators have been widely used in the scenarios of smart microgrippers [6], valves, active vibration damping, AFM [7], energy harvesting [8], etc. Besides, piezoelectric shear actuators present electrodes on top and bottom surfaces. They have been widely employed in the applications of active vibration control [9], structural health monitoring [10], microscopy, switches, etc.

1.5 Some Problems Encountered When Using Smart Materials

1.5.1 Background

Smart actuators invariably exhibit hysteresis, which is a path-dependent memory effect where the output relies not only on the current state but also on the past output history [11]. The presence of the hysteresis in smart actuators, such as piezoceramic, magnetostrictive, and SMA actuators has been widely associated with various performance limitations [12]. These include the oscillations in the responses of the open-as well as closed-loop systems, and poor tracking performance and potential instabilities in the closed-loop system [13]. Smart actuators have also shown strong creep effects in the output displacement during slow and fast operations [14, 15]. A number of studies are calling the creep effects at high excitation frequencies as rate-dependent hysteresis [15, 16]. These creep effects yield significant loss in precision when positioning is required over extended periods of time and high oscillations at high excitation frequencies [17].

Considerable continuing efforts are thus being made to seek methods for effective compensation of hysteresis and creep effects in order to enhance the tracking performance of smart actuators, particularly for closed-loop micro-positioning systems. The characterization and modeling of the hysteresis and creep properties of smart actuators, however, is vital for designing efficient compensation algorithms. Considering that the hysteresis properties of such actuators are strongly dependent upon the type of materials, magnitude of input and the rate of input in a highly non-

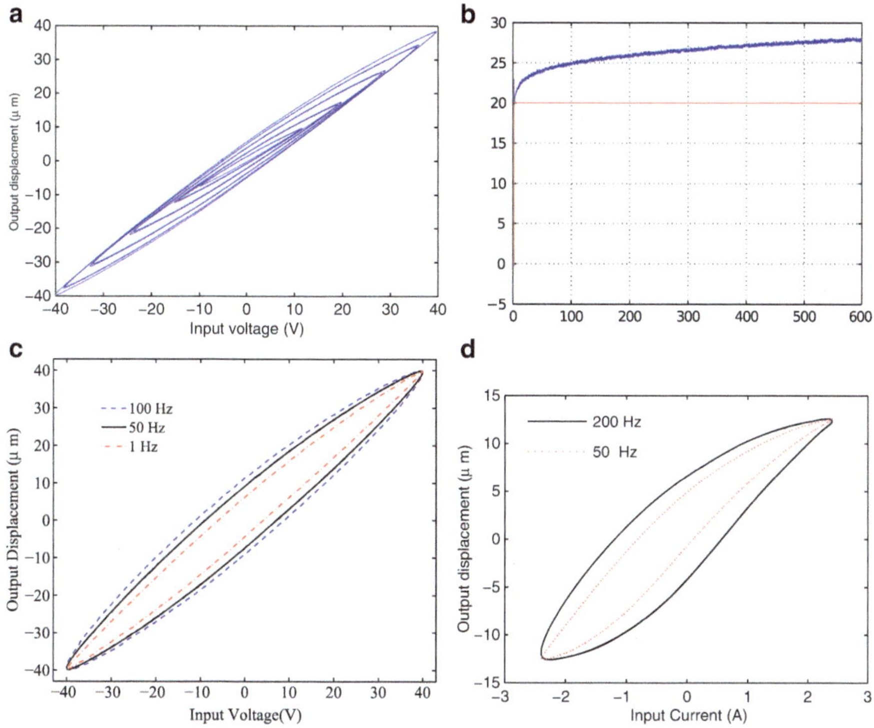

Fig. 1.3 (**a**) Measured displacement of a piezo micropositioning actuator under sinusoidal input voltage, (**b**) measured displacement of a piezocantilever when a step input reference is applied, (**c**) measured hysteresis loops of a piezo micropositioning actuator when a sinusoidal input voltage is applied at different excitation frequencies, and (**d**) measured hysteresis loops of a magnetostrictive actuator when a sinusoidal input current is applied at different excitation frequencies

linear manner, the characterizations as well as modeling of the phenomenon pose considerable challenges. For instance, piezoceramic actuators generally exhibits symmetric convex minor and major hysteresis loops [18], while magnetostrictive and SMA actuators yield highly asymmetric concave hysteresis loops [12, 19], which further depend upon the rate of the applied input. Smart actuators also exhibit output saturation, which further contributes to the modeling challenge. Figure 1.3 shows measured hysteresis and creep nonlinearities in smart actuators.

1.5.2 Hysteresis Models

A number of hysteresis models have been proposed in the literature for characterizing the hysteresis properties of smart actuators [13]. These models could be broadly classified into phenomenological models and physics-based models [11, 20]. Different dynamic models have been proposed to model the creep effects in smart actuators. These models could be classified into linear creep models and nonlinear creep models. Linear creep models characterize the creep effects using series connection between springs and dampers, while nonlinear creep models apply nonlinear equations.

The physics-based models are generally derived on the basis of a physical measure, such as energy, displacement, or stress–strain relationship. These hysteresis models generally require comprehensive knowledge of the physical phenomenon for the hysteretic system. Alternatively, the phenomenological models describe the hysteresis properties without attention to the physical properties of the hysteretic system [13]. Many of these models were initially proposed for specific physical systems and were later generalized for applications to other systems. The primary goal of these models is to accurately predict the hysteresis in order to study the hysteresis effects and to facilitate the design of controllers for compensating the hysteresis effects.

The most widely cited models based on the input and output behaviors include: the operator-based hysteresis models such as Preisach model [11] and Prandtl–Ishlinskii model [20] and differential equation-based hysteresis models such as Duhem model and Bouc–Wen model [21]. These models generally constitute a nonlinear differential equation relating the output to the magnitude and direction of the input. Unlike the differential equation-based model, the operator-based models are considered to be better suited for the design of control algorithms for compensating hysteresis effects due to their invertibility. Such models have been widely applied for modeling hysteresis nonlinearities in smart actuators, and are briefly described below.

1.5.2.1 The Preisach Model

The Preisach model has been most widely applied for characterizing the hysteresis properties of smart actuators, see, for example, [13, 18]. The Preisach model can be presented analytically as [11]:

$$\Gamma[v](t) = \iint_{\alpha \geq \beta} p(\alpha,\beta)\gamma_{\alpha\beta}[v](t)\mathrm{d}\alpha\mathrm{d}\beta \tag{1.5}$$

where $\gamma_{\alpha\beta}[v](t)$ is the output of the relay operator, α and β are thresholds, and $p(\alpha,\beta)$ is a positive integrable density function identified from the measured data for a particular smart actuator. The Preisach model is rate-independent hysteresis model.

The Preisach model is completely characterized by two properties [11]: wiping-out and congruent minor-loop properties. The wiping out property means that the output is affected only by the current input and the history of the output, while the effect of all other inputs is wiped out. The congruent minor-loop property requires that all equivalent minor loops be similar. Two minor loops are said to be equivalent if they are generated under monotonically varying inputs of identical amplitudes. Different forms of the classical Preisach model have thus evolved to model hysteresis in various materials and smart actuators [11].

1.5.2.2 The Prandtl–Ishlinskii Model

The Prandtl–Ishlinskii model is constructed using the play hysteresis operator. Unlike the relay operators in the Preisach model, the play operator is continuous hysteresis operator characterized by the input v and the threshold r. A detailed discussion about these operators can be found in [20]. The play operator has been described by the motion of a piston within a cylinder of length $2r$ [20]. Analytically the output of the Prandtl–Ishlinskii model is expressed as [20]:

$$\Psi[v](t) = \int_0^R p(r)F_r[v](t)\mathrm{d}r, \tag{1.6}$$

where $F_r[v](t)$ is the output of the play operator and $p(r)$ a positive integrable density function identified from the measured data for a particular smart actuator. The Prandtl–Ishlinskii model is a rate-independent hysteresis model, attributed to the time independent play operator that the model employs.

The Prandtl–Ishlinskii model has been applied to characterize hysteresis effects of different smart actuators. The model, however, is limited to symmetric hysteresis loops, such as those observed in many piezoceramic actuators, which is attributed to the play operator. The model, thus, cannot be applied for predicting asymmetric input–output hysteresis, which is invariably observed in SMA and magnetostrictive actuators. Furthermore, unbounded nature of the play operator does not permit the Prandtl–Ishlinskii model applications for saturation property, which is widely observed in SMA actuators. Different developments have been carried out in a number of studies to enhance the performance of the Prandtl–Ishlinskii model. These studies include the generalized Prandtl–Ishlinskii model [13], the classical Prandtl–Ishlinskii model [22], the modified Prandtl–Ishlinskii model [23], the rate-dependent Prandtl–Ishlinskii model [16].

1.5.3 Hysteresis Compensation

The hysteresis in smart actuators has been associated with oscillations and poor tracking performance of the closed-loop system [24]. Consequently, considerable efforts have been made towards the design of controllers for compensation of hysteresis. A vast number of controllers have been proposed to reduce the error due to hysteresis effects. The proposed control algorithms could be classified into two broad categories, namely non-inverse-based control methods and inverse-based control methods.

1.5.3.1 Model-Based Control Methods

Compensation of hysteresis nonlinearities have been carried out in many studies without considering the inverse of the hysteresis models. Model-based hysteresis compensation methods employ the phenomenological hysteresis models to construct controllers to compensate for the actuator hysteresis. A number of control methods have been proposed to compensate for smart actuators such as adaptive control [25, 26], energy-based control methods [12], and sliding model control systems [27], which employ the hysteresis model of the actuator for constructing the controller.

1.5.3.2 Inverse-Based Control Methods

Control algorithms based on inverse compensators have been suggested to be more effective in compensating the hysteresis and creep effects [28–30]. The inverse model-based hysteresis compensation methods generally employ a cascade of a hysteresis model and its inverse together with a controller to compensate for the hysteresis effects. These methods, however, necessitate the formulation of the hysteresis model inverse, which is often a challenging task.

Some reported hysteresis models have thus been employed for deriving the inverse hysteresis models to serve as a compensator for the hysteresis effects, particularly these based on the Preisach model and Prandtl–Ishlinskii models [13]. The Preisach model is not analytically invertible; numerical methods are thus employed to obtain approximate inversions of the model. The effectiveness of the approximate inversions in conjunction with different controllers in hysteresis compensation has been demonstrated in a number of studies, see, for example, [31].

1.6 Conclusion

This chapter presented the main motivations of using smart materials as core components in systems working at the micro/nanoscale, in particular systems for micro/nanopositioning. Piezoelectric materials are considered as one of the well recognized among the existing smart materials in these applications. Indeed nine chapters of the book are devoted to the use of these materials. Hence, this chapter provided a remind of the basics of piezoelectricity and of their main advantages. Finally, the chapter presented some of the main behaviors (hysteresis and creep) that limit the performances of smart materials and in particular of piezoelectric materials. Most of the chapters in the book will treat these behaviors.

References

1. I. Ivan, M. Rakotondrabe, P. Lutz, N. Chaillet, Quasistatic displacement self-sensing method for cantilevered piezoelectric actuators. Rev. Sci. Instrum. **80**(6), 065102-1–065102-8 (2009)
2. Physik Instrumente, Piezo Actuator Tutorial (2012), http://www.piceramic.com/piezotutorial1.php
3. M. Al Janaideh, S. Rakheja, C.-Y. Su, An analytical generalized Prandtl–Ishlinskii model inversion for hysteresis compensation in micropositioning control. IEEE/ASME Trans. Mechatron. **16**(4), 734–744 (2011)
4. Q. Xu, Design and development of a flexure-based dual-stage nanopositioning system with minimum interference behavior. IEEE Trans. Autom. Sci. Eng. **9**(3), 554–563 (2012)
5. K. Leang, A. Fleming, High-speed serial-kinematic SPM scanner: design and drive considerations. Asian J. Contr. **11**(2), 144–153 (2009)
6. M. Rakotondrabe, I. Ivan, Development and force/position control of a new hybrid thermopiezoelectric microgripper dedicated to micromanipulation tasks. IEEE Trans. Autom. Sci. Eng. **8**(4), 824–834 (2011)
7. Y. Yong, A. Fleming, S. Moheimani, A novel piezoelectric strain sensor for simultaneous damping and tracking control of a high-speed nanopositioner. IEEE/ASME Trans. Mechatronics **18**(3), 1113–1121 (2013)
8. S. Lee, B. Youn, A new piezoelectric energy harvesting design concept: multimodal energy harvesting skin. IEEE Trans. Ultrason. Ferroelectrics Freq. Contr. **58**(3), 629–645 (2011)
9. A. Wills, D. Bates, A. Fleming, B. Ninness, S. Moheimani, Model predictive control applied to constraint handling in active noise and vibration control. IEEE Trans. Contr. Syst. Technol. **16**(1), 3–12 (2008)
10. M. Viscardi, L. Lecce, An integrated system for active vibro-acoustic control and damage detection on a typical aeronautical structure, in *Proceedings of IEEE International Conference on Control Applications*, Glasgow, Scotland, 2002, pp. 477–482
11. Mayergoyz, *Mathematical Models of Hysteresis* (Elsevier, New York, 2003)
12. R. Gorbet, Control of hysteresis systems with Preisach representations, Ph.D. Dissertation, University of Waterloo, ON, 1997
13. C. Visone, Hysteresis modelling and compensation for smart sensors and actuators. J. Phys.: Conf. Ser. **138**, 1–25 (2008)
14. M. Rakotondrabe, C. Clevy, P. Lutz, Complete open loop control of hysteretic, creeped, and oscillating piezoelectric cantilevers. IEEE Trans. Autom. Sci. Eng. **7**(3), 440–450 (2010)
15. M. Al Janaideh, S. Rakheja, C.-Y. Su, Experimental characterization and modeling of rate-dependent hysteresis of a piezoceramic actuator. Mechatronics **17**(5), 656–670 (2009)
16. P. Krejčí, M. Al Janaideh, F. Deasy, Inversion of hysteresis and creep operators. Physica B **407**(9), 1354–1356 (2012)
17. S. Devasia, E. Eleftheriou, R. Moheimani, A survey of control issues in nanopositioning. IEEE Trans. Contr. Syst. Technol. **15**(5), 802–823 (2007)
18. P. Ge, M. Jouaneh, Tracking control of a piezoceramic actuator. IEEE Trans. Contr. Syst. Technol. **4**(3), 209–216 (1996)
19. D. Hughes, J. Wen, Preisach modelling of piezoceramic and shape memory alloy hysteresis. Smart Mater. Struct. **6**(3), 287–300 (1997)
20. M. Brokate, J. Sprekels, *Hysteresis and Phase Transitions* (Springer, New York, 1996)
21. M. Rakotondrabe, Bouc–Wen modeling and inverse multiplicative structure to compensate hysteresis nonlinearity in piezoelectric actuators. IEEE Trans. Autom. Sci. Eng. **8**(2), 428–431 (2011)
22. M. Rakotondrabe, Classical Prandtl–Ishlinskii modeling and inverse multiplicative structure to compensate hysteresis in piezoactuators, in *Proceedings of American Control Conference*, Montreal, Canada, 2012, pp. 1646–1651
23. K. Kuhnen, Modeling, identification and compensation of complex hysteretic nonlinearities—a modified Prandtl–Ishlinskii approach. Eur. J. Contr. **9**(4), 407–418 (2003)

24. G. Tao, P. Kokotovic, Adaptive control of plants with unknown hysteresis. IEEE Trans. Automat. Contr. **40**(2), 200–212 (1995)
25. M. Al Janaideh, J. Yan, A. D'Amato, D.S. Bernstein, Retrospective-cost adaptive control of uncertain Hammerstein-Wiener systems with memoryless and hysteretic nonlinearities, in *Proceedings of AIAA Guidance, Navigation, and Control Conference*, Minneapolis, USA, AIAA-2012-4449-671, 2012, pp. 1–26
26. M. Al Janaideh, E. Sumer, J. Yan, D. Anthony, B. Drincic, K. Aljanaideh, D. Bernstein, Adaptive control of uncertain linear systems with uncertain hysteretic input nonlinearities, in *Proceedings of ASME Dynamic Systems and Control Conference*, Lauderdale, FL, USA, 2012, pp. 1–10
27. M. Al Janaideh, Y. Feng, S. Rakheja, Y. Tan, C.-Y. Su, Generalized Prandtl–Ishlinskii hysteresis: modeling and robust control, in *Proceedings of IEEE Conference on Decision and Control*, Shanghai, China, 2009, pp. 7279–7284
28. K. Leang, Q. Zou, S. Devasia, Feedforward control of piezoactuators in atomic force microscope systems: inversion-based compensation for dynamics and hysteresis. IEEE Contr. Syst. Mag. **19**(1), 70–82 (2009)
29. K. Kuhnen, P. Krejčí, Compensation of complex hysteresis and creep effects in piezoelectrically actuated systems: a new Preisach modeling approach. IEEE Trans. Automat. Contr. **54**(3), 537–550 (2009)
30. P. Krejčí, K. Kuhnen, Inverse control of systems with hysteresis and creep. IEE Proc. Contr. Theor. Appl. **148**(3), 185–192 (2001)
31. R. Iyer, X. Tan, P. Krishnaprasad, Approximate inversion of the Preisach hysteresis operator with application to control of smart actuators. IEEE Trans. Automat. Contr. **50**(6), 798–810 (2005)

Chapter 2
Characterization and Dynamics of Polymer Microactuators

Beatriz Cristina López-Walle and Edgar Reyes-Melo

Abstract A magnetic hybrid material consisting of iron oxide nanoparticles (4 nm) embedded in a polymer matrix of Na-CMC was synthesized. The synthesis was done from a chemical treatment on a precursor hybrid material previously synthesized. The structure, morphology, and magnetic properties for this magnetic hybrid material were studied by X-ray diffraction, IR spectroscopy, transmission electron microscopy, and magnetometry. Additionally, the dynamic response was analyzed in order to probe the feasibility to use this magnetic hybrid material as a bending-type actuator. The experimental results show that the responses of the deflection have a linear trend over a reasonable range, suggesting that the magnetic hybrid material can be used as bending-type actuators in small mechanical systems and devices. First simulations have also been done considering the two components of the magnetic hybrid material: the oxide iron nanoparticles and Na-CMC. The displacement response takes in account the viscoelastic properties of the polymeric matrix and the magnetization of the nanoparticles.

2.1 Introduction

Recent advances in polymer science have allowed a better utilization of the properties of polymeric materials and their composite and/or hybrid materials [3, 10, 11, 17, 25, 27–30, 33, 39]. Polymers offer an attractive basis for microsystems [24]. Some of the characteristics of the polymer materials that benefit from both the field of microsystems and the electroactive polymer actuation technologies are: (1) elasticity, they can absorb impact energy and tolerate large degree of deformation;

B.C. López-Walle (✉) • E. Reyes-Melo
Programa Doctoral en Ingeniería de Materiales-FIME and Centro de Innovación, Investigación y Desarrollo en Ingeniería y Tecnología, Universidad Autónoma de Nuevo León, Av. Universidad s/n, Cuidad Universitaria, San Nicolás de los Garza, Nuevo León, México
e-mail: beatriz.lopezwl@uanl.edu.mx; mreyes@gama.fime.uanl.mx

M. Rakotondrabe (ed.), *Smart Materials-Based Actuators at the Micro/Nano-Scale: Characterization, Control, and Applications*, DOI 10.1007/978-1-4614-6684-0_2,

© Springer Science+Business Media New York 2013

(2) cost, they are relatively inexpensive compared to silicon; and (3) manufacturing, they can be formed into three-dimensional structures. In this chapter, a new hybrid material with potential applications as an electroactive microactuator is presented.

The synthesis of new hybrid materials has acquired great importance due to the development of experimental techniques allowing that two inherently incompatible components (e.g., organic polymers and inorganic oxides) can be combined by a pre-mix of both at the molecular level, before its conversion into a new hybrid material. This has allowed the design and synthesis of hybrid advanced materials with very specific physicochemical properties, including a multi-functional character, essential in modern organic electronics and microsystems. From the point of view of structure and morphology, a hybrid material is one phase or multi-phases dispersed into a matrix, which is typically a polymeric material. Additionally, the synthesis of hybrid materials with a polymeric matrix compatible to biological systems, as carboxymethyl cellulose or Na-CMC, represents an area that has paid a lot of attention of many research groups, due to the biological or surgical applications that this kind of materials could offer [30, 39]. The hybrid material described through the next sections is composed of a polymeric matrix of Na-CMC and inorganic nanoparticles of iron oxide (Fe_2O_3). Polymeric matrices are also suitable to encapsulate the metallic nanoparticles during their synthesis, avoiding the formation of agglomerates [3, 11, 26, 27, 29]. For these reasons, the development of hybrid magnetic materials using biocompatible polymeric matrices as Na-CMC is of great scientific and technological interest.

This work deals with a magnetic hybrid material based on Na-CMC and Fe_2O_3 nanoparticles. Section 2.2 describes the synthesis of this innovative material; Sect. 2.3 presents its structural, morphological, and magnetic characterization; and Sect. 2.4 shows the experimental results and finite element simulation (FES) of this material considering that it could work as a bending-type microactuator.

2.2 Synthesis of the Magnetic Hybrid Material

The synthesis of new hybrid materials by the scientific community has led to the improvement of the process of combining two or more materials with the purpose of obtaining multifunctional materials. In polymer technology, it is a common practice to strengthen conventional polymeric materials with inorganic fibers or fillers to improve their mechanical properties. This kind of materials have a great application today mainly for the construction of light vehicles and utensils for sports [8, 16]. In this sense, the development of new microactuators for the electronic industry, enabling it to extend their applications to biological systems, requires the synthesis of new hybrid materials with polymer matrix that respond to the application of a magnetic field [12, 18, 27, 29, 37], either, which can also carry electricity, without losing their viscoelastic properties and biocompatibility. To achieve these objectives, an important alternative is the development of nanostructured hybrid materials. Na-CMC due its low cost and its functional groups, it is a polymer matrix which

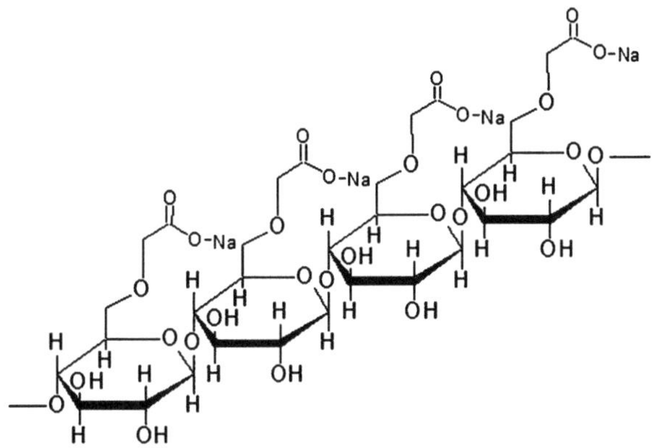

Fig. 2.1 Chemical structure of a chain segment of Na-CMC

can be used as a stabilizing agent of metallic nanoparticles. The aim of this section is the synthesis of magnetic nanoparticles that can be dispersed and stabilized in a polymeric matrix of Na-CMC, resulting in a hybrid magnetic material.

2.2.1 The Polymer Matrix or Na-CMC

Sodium carboxymethyl cellulose or Na-CMC is an ether derivative of cellulose with the carboxymethyl groups bounded to the hydroxyl groups of the β-anhydroglucose units (see Fig. 2.1) of cellulose macromolecules. The role of Na-CMC in the industry is related to its hydrophilic character and its high viscosity, allowing it to be used as base material to form thin films with good rheological properties. In addition Na-CMC also can be used as a thickener or even as adhesive in many industrial processes. In the process of Na-CMC synthesis, chemical groups of "sodium carboxymethyl" (CH_2COONa) are introduced in the repetitive units (β-anhydroglucose) of cellulose. These CH_2COONa groups give Na-CMC a certain degree of solubility in water; this property does not have cellulose macromolecules. Figure 2.1 shows a segment of Na-CMC chain, same hydroxyl groups have been replaced by sodium carboxymethyl groups into the repetitive unit: β-anhydroglucose. Because of its chemical structure, Na-CMC can form coordinate bonds with divalent ions without loss of its process ability [14, 15, 26, 28, 29, 36].

A very important aspect that defines the functional properties of Na-CMC is its degree of substitution or *DS*. The value of *DS* is a function of the average number of OH-groups replaced by CH_2COONa groups into the β-anhydroglucose units [26, 29]. Figure 2.1 shows that each β-anhydroglucose unit has 3 OH-groups available for a maximum of 3 *DS*. For example, in a sample of Na-CMC with a 1.5 *DS*, it means that on average 50% of the hydroxyl groups were replaced by CH_2COONa groups, and the 50% of remaining OH-groups remain free. The ability of the Na-CMC as a thickening agent for the control of flow of fluids (rheology control) depends largely on its *DS* value. At the same time the magnitude of *DS* depends on reaction kinetics for the synthesis of Na-CMC. For samples with low values of *DS*, for example $DS = 0.2$, most of the reaction was carried out in amorphous cellulose regions or on the surface, and consequently Na-CMC sample is relatively insoluble in water, but it absorbs significantly more liquid than the starting (cellulose) material. For values of $DS = 0.5$, Na-CMC shows partial solubility, due to little replaced regions that appear very swollen and opaque, as an opaque gel. Samples with a $DS = 0.7$ or greater are the result of sufficient replacement of OH-groups producing a higher solubility of Na-CMC in water and minimize the physical interactions among chains that produce agglomerates, these interactions can be interrupted by the inherent stress in the fluid and produce thixotropic rheology. The chemical reactions necessary for the synthesis of Na-CMC, which reach a value of $DS = 1$, produce polymeric chains that have a very low concentration of OH-groups; these OH-groups have not been replaced by carboxymethyl groups. Consequently this specimen has a little tendency to the formation of agglomerates. These last aspects lead to a rheology behavior of Na-CMC which is known as pseudoplastic (or non-Newtonian rheology). In pseudoplastic behavior viscosity decreases as shear rate increases. It is important to remark here that when OH-groups into chains of Na-CMC are uniformly replaced by CH_2COONa groups, these groups interact with water molecules producing a uniform flow of Na-CMC solution. On the other hand, when OH-groups are replaced by CH_2COONa groups at a random way, the less substituted regions or hydrophobic chains, they tend to swell, since these regions tend to be associated through hydrogen bonds and form three-dimensional networks that give rise to a structure with thixotropic fluid characteristics. Na-CMC compatibility is another important property that is also affected by *DS*-magnitude; this is because the less soluble regions of Na-CMC tend to precipitate easily. Na-CMC samples with high values of *DS* which correspond to Na-CMC uniformly replaced are more stable to acid pH, because of that acid hydrolysis takes place in the union ether between two β-anhydroglucoses units. An important technological information of commercial Na-CMC is as follows, it is solid white, odorless, tasteless, and without toxicity. Viscosity in aqueous solutions to 2% ranges between 10 and 50,000 mPa s [26].

2.2.2 In Situ Synthesis of the Magnetic Hybrid Material

Today, a wide variety of methodologies that allow a combination at the molecular level of different materials have been developed. Many of these methodologies are based on some mechanism of precipitation, controlling the size of particle produced in this way [3, 26, 28, 29]. These methodologies for the synthesis of hybrid materials are classified as in situ and ex situ methodologies. For methodologies classified as in situ growth of nanoparticles is given into the polymer matrix, instead on methodologies ex situ nanoparticles are synthesized independently and then they are dispersed into polymer matrix. In this work, we used an in situ methodology to obtain a magnetic hybrid material (Fe_2O_3/Na-CMC).

For in situ methodologies, it is essential that, at a first stage must be mixed inorganic precursors with the polymer matrix, and then inorganic precursors into polymer matrix are transformed to a new phase. The pre-mix can be carried out in several ways:

- Absorption of metal ions in ion-exchange resin or polymeric gels
- Dissolution of precursors in a polymer solution
- Deposition of precursors in a porous polymer, chemical, electrochemical methods or steam

A prerequisite for the development of the in situ methodology is that the polymer matrix must have functional groups, which should be establishing chemical or physical interactions with the metallic ions of precursor salts, favoring a homogeneous dispersion of the precursor salt into the matrix. Once the metallic ions interact with the polymer matrix, a chemical or physical treatment is carried out with the purpose of obtaining the desired structure of the dispersed phase. In a previous work, it has been reported the synthesis of a material hybrid in which were dispersed in a homogeneous way nanoparticles of Fe_2O_3 into a polymeric matrix of chitosan [27].

For the synthesis of Fe_2O_3/Na-CMC hybrid material, the process was carried out in two stages. In the first stage, the main objective is that the precursor salt and Na-CMC should be combined in a way such that the precursor salt ions are dispersed evenly in the Na-CMC, to obtain a precursor hybrid material. In the second stage, the precursor hybrid material is subjected to a certain chemical treatment for obtaining Fe_2O_3/Na-CMC.

Figure 2.2 is a schema of the first stage; Na-CMC and $FeCl_2$–$4H_2O$ were dissolved in distilled water obtaining two solutions Na-CMC/H_2O and $FeCl_2/H_2O$. Both solutions were subjected to an agitation process for 5 h, and then mix both solutions; the product obtained is subjected to another agitation process during 4 h. After that, this last solution is poured in a Petri dish and is subjected to a process of drying by natural convection in a hot plate at an average temperature of 40 °C for 24 h. The resulting product is the precursor hybrid material. This material is stored in a desiccator where samples are taken for analysis and also to be subjected to chemical treatment that constitutes the second stage.

In the second stage, the precursor hybrid material was subjected to a chemical treatment with an aqueous solution of NaOH 6.7 M at 40 °C. Consequently the precursor hybrid material undergoes a change of color, dark coffee to black. After that, the "black" material is immersed in the alkaline solution and then 30 mL of hydrogen peroxide (H_2O_2) was added, this produced a new change of color, black to reddish brown. After 15 min the "reddish brown" material was removed from the alkaline solution and it was subjected to a process of washing with distilled water and ethyl alcohol, with the aim of eliminating waste substances. This new material is the final product that presumably is a hybrid magnetic material composed of Fe_2O_3 nanoparticles dispersed in a polymer matrix of Na-CMC.

The magnetic hybrid material obtained has been characterized structurally, morphologically, and magnetically. The techniques applied and the results obtained are described in the next section.

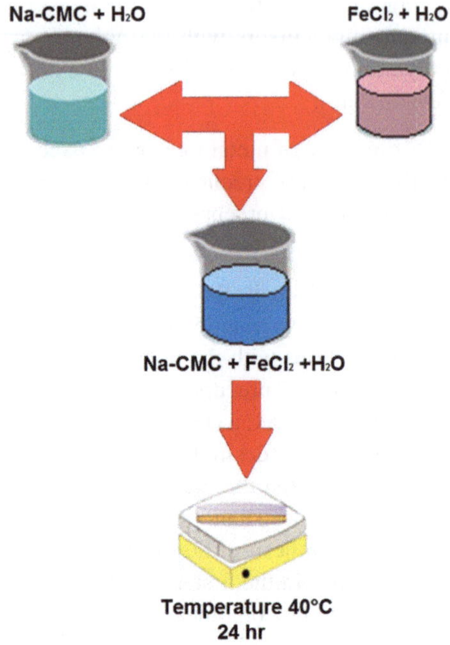

Fig. 2.2 First step of the synthesis of the magnetic hybrid material: the precursor hybrid material

2.3 Characterization of the Magnetic Hybrid Material

In this section, the experimental results obtained from the structural, morphological, and magnetic characterizations are presented and discussed. In the first part, we deal with the structure and morphology for both the precursor hybrid material and the magnetic hybrid material. In the second part, the magnetic properties of the last one are analyzed.

2.3.1 Morphology and Structure of the Precursor Hybrid Material and the Magnetic Hybrid Material

For the synthesis of the precursor hybrid material two aqueous solutions were mixed, one of $FeCl_2$ and another of Na-CMC, this with the purpose of obtaining an aqueous solution containing $FeCl_2$ and Na-CMC. After that, it has removed the solvent by natural convection; it allowed us to obtain a thin film in which presumably the Fe^{2+} ions are bound to carboxymethyl groups of the Na-CMC. Figure 2.3 shows three images, the first one (Fig. 2.3a) is a thin film of Na-CMC, the second one (Fig. 2.3b) corresponds to a film of the precursor hybrid material, and Fig. 2.3c is a powder of the magnetic hybrid material obtained.

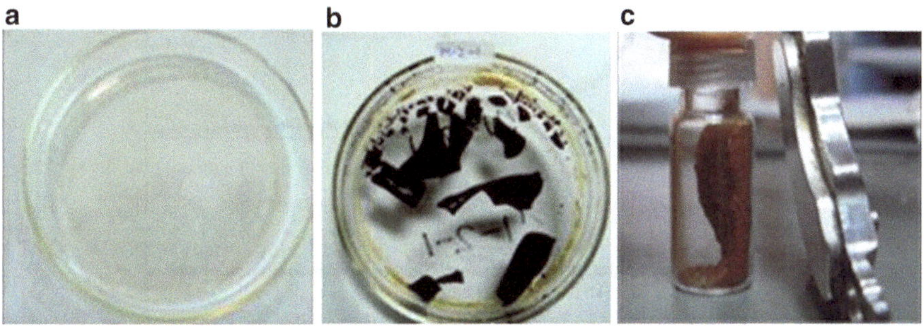

Fig. 2.3 The magnetic hybrid material at different steps: (**a**) Na-CMC sample, (**b**) the precursor hybrid material, and (**c**) powder of the magnetic hybrid material

In order to verify that the Fe^{2+} ions have some kind of chemical interaction with carboxymethyl groups of Na-CMC, the precursor hybrid material (Fig. 2.3b) was analyzed by X-ray spectroscopy and IR-spectroscopy. On the one hand, the X-ray diffraction technique was used mainly to corroborate the complete dissolution of precursor salt into the polymer matrix of Na-CMC. The device used was a SIEMENS D5000 diffractometer with a radiation source of CuK. On the other hand, by IR-spectroscopy, the different vibration modes of the chemical groups for both the precursor hybrid material and Na-CMC samples were analyzed. Using these results, it was possible to study the interactions between functional groups of the Na-CMC and the ions of precursor salt. The device used to obtain these experimental measurements was a Nicolet FTIR spectrometer.

Figure 2.4 shows the patterns of X-ray diffraction which were obtained for the samples of Na-CMC (Fig. 2.4a), for the precursor hybrid material in question (Fig. 2.4b), and for salt precursor $FeCl_2{\cdot}4H_2O$ (Fig. 2.4c). The diffractogram obtained for Na-CMC sample (Fig. 2.4a) is a typical curve corresponding to materials with amorphous structure. While the diffraction pattern that corresponds to the precursor hybrid material is displayed in Fig. 2.4b. It is clear that the diffraction peaks in Fig. 2.4b do not correspond to the precursor salt, $FeCl_2{\cdot}4H_2O$, see Fig. 2.4c, these diffraction peaks correspond to NaCl crystals embedded into Na-CMC matrix. These results suggest that the precursor salt ($FeCl_2{\cdot}4H_2O$) was completely dissolved into the polymer matrix, probably forming chemical bonds between Fe^{2+} ions and carboxymethyl groups of the Na-CMC. On the other hand, the presence of crystals of NaCl into the precursor material is a consequence that in the first stage of the synthesis process, there are conditions to carry out a process of crystallization, due to the ionic interactions that occur between the Cl^- of the precursor salt and the Na^+ from the Na-CMC [26, 29]. In order to corroborate the chemical interaction of ions of Fe^{2+} with the Na-CMC, the different modes of vibration which have functional groups both the precursor hybrid material and Na-CMC samples were determined by IR-spectroscopy. The comparison of the spectra obtained for these samples (see Fig. 2.5) go allowed clarifying the way in which Fe^{2+} ions interact with the polymer matrix for the formation of the precursor hybrid material.

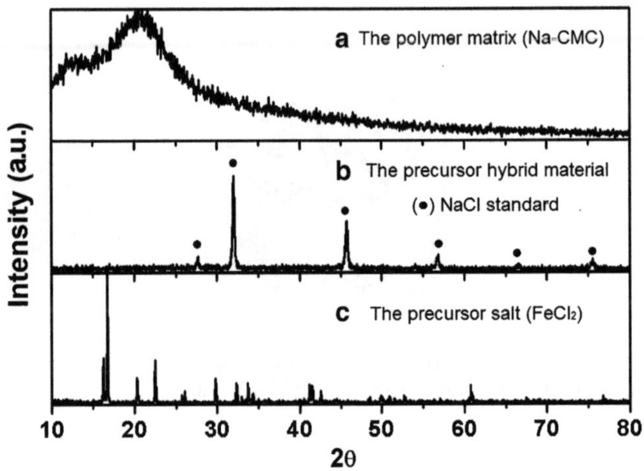

Fig. 2.4 X-ray diffraction patterns for samples: (**a**) Na-CMC, (**b**) the precursor hybrid material, and (**c**) precursor salt ($FeCl_2{\cdot}4H_2O$)

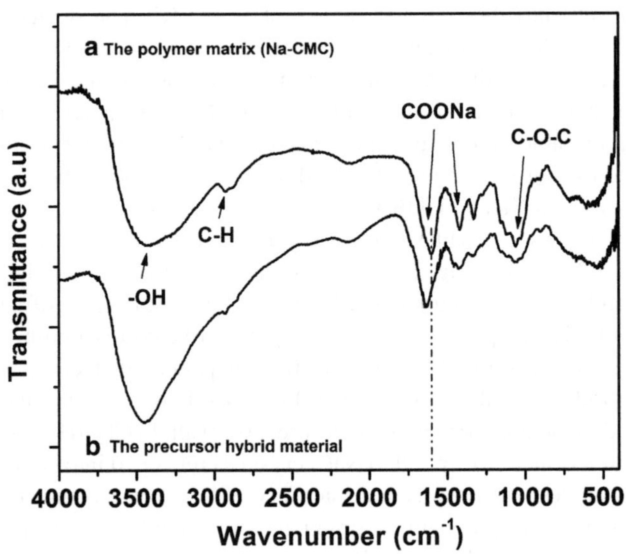

Fig. 2.5 Infrared spectra for: (**a**) NA-CMC and (**b**) the precursor hybrid material

Figure 2.5a corresponds to the infrared (IR) spectrum obtained for Na-CMC sample. The characteristic vibration mode associated with chemical groups –OH is observed at 3,500 cm^{-1}. Another vibration mode is stretching of C–H, and its corresponding band is observed at 2,925 cm^{-1}. At 1,417, 1,600, and 1,058 cm^{-1} are identified another bands related to stretching of the asymmetric ether group of the carboxymethyl groups. On the other hand, Fig. 2.5b corresponds to infrared

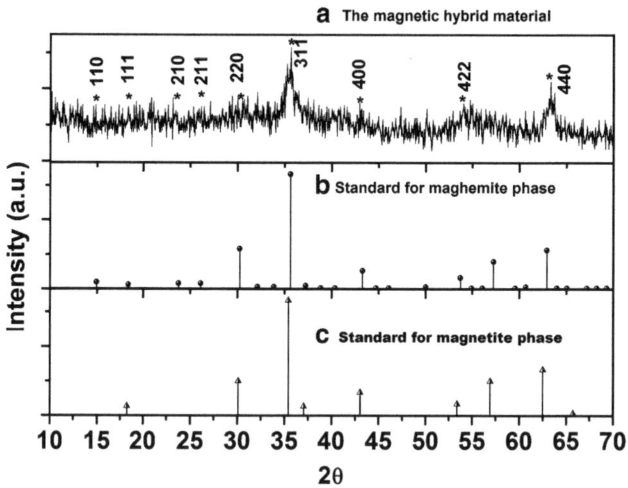

Fig. 2.6 Diffraction patterns for: (**a**) the magnetic hybrid material, (**b**) standard maghemite phase, and (**c**) standard magnetite phase

spectrum of the precursor hybrid material; in this figure the band associated with the carboxymethyl groups has a shift towards high values of the wavenumber, about 1,633 cm^{-1}. The shift of this band is related to chemical interactions that occur between Fe^{2+} ions and carboxymethyl groups of Na-CMC. The results obtained by X-ray diffraction and IR-spectroscopy allow us to ensure that Fe^{2+} ions are distributed evenly in the polymer matrix of the precursor hybrid material. This condition is necessary and indispensable for the precursor hybrid material to be used as raw material base for the synthesis of Fe_2O_3 nanoparticles homogeneously dispersed into Na-CMC.

The hybrid magnetic material samples were also analyzed by X-ray diffraction with the aim of determining the crystal structure and the size of nanoparticles dispersed into polymer matrix. The Scherrer equation [26, 29] was used for the calculus of particle size [26]. Figure 2.6a shows the diffraction pattern obtained for the magnetic hybrid material. The diffraction peaks which are marked with an asterisk (*) 18.35°, 30.25°, 33.65°, 42.25°, 53.7°, and 62.95° correspond to crystallographic planes (1 1 1), (2 2 0), (3 1 1), (4 0 0), (4 2 2), and (4 4 0), respectively, and they are consistent with the bars shown in Fig. 2.6b coming from a standard diffraction phase of maghemite [1 0 0], and also with standard diffraction phase (Fig. 2.6c) of magnetite [1 0 1]. In addition, it also displayed diffraction peaks associated with the crystallographic planes (2 1 0), (2 1 1), and (1 1 0) belonging to the phase of maghemite [1 0 0]. These results are important since they corroborate us the existence of a phase Fe_2O_3 into the polymer matrix of Na-CMC [26, 29].

From experimental results of Fig. 2.6a, inter-planar distances (*d*) were calculated using Bragg law. These calculated values of *d* are consistent with those reported in the literature for the crystal structure of the [1 0 0] maghemite. Table 2.1 shows

Table 2.1 Computed inter-planar distances $d = d(d_{exp})$ from experimental results

$2\theta_{exp}$	d	$d(\gamma Fe_2O_3)$	$d(Fe_3O_4)$
14.95	0.5921	0.5918	
18.35	0.4800	0.4822	0.4852
23.75	0.3740	0.3740	
26.10	0.3411	0.3411	
30.25	0.2952	0.2953	0.2967
35.65	0.2516	0.2517	0.2532
43.25	0.2090	0.2088	0.2099
53.70	0.1705	0.1704	0.1714
62.95	0.1475	0.1475	0.1484

d-computed values and the reported values in ICDD database for maghemite $d(\gamma Fe_2O_3)$ and magnetite $d(Fe_3O_4)$ phases. The computed d values are very close to the ICDD values for maghemite crystalline structure [26, 29].

On the other hand, from the experimental data of Fig. 2.6a, it is also possible to compute the crystal size $\langle L \rangle$ by Scherrer equation:

$$\langle L \rangle = \frac{0.89\lambda}{\beta \cos \theta} \tag{2.1}$$

where λ is the wavelength of the incident X-rays, θ is the half of the diffraction angle 2θ in degrees, and β is the full width at half maximum of the diffraction peak. The computed value $\langle L \rangle$ obtained was 5.6 nm. These results suggest that iron oxide nanoparticles are embedded into the polymer matrix.

To obtain additional information about the chemical structure of the magnetic hybrid material, an IR-spectroscopy analysis was performed. Results are displayed in Fig. 2.7. The IR-spectrum obtained in Fig. 2.7b shows the characteristic bands of Na-CMC at 3,500, 2,925, and 1,058 cm^{-1} already mentioned. In this case, unlike the IR-spectrum of the precursor hybrid material (Fig. 2.5b), the bands associated with the carboxymethyl groups remain at 1,600 and 1,417 cm^{-1} for the magnetic hybrid material.

This result can be interpreted as a weak-chemical interaction between iron oxide nanoparticles and the carboxymethyl groups of the Na-CMC. In addition, the band at 1,600 cm^{-1} also suggests that during the iron oxide precipitation, the sodium-carboxymethyl groups were again formed through the reaction of carboxymethyl groups with Na^+ of NaOH added to obtain alkaline conditions (see Fig. 2.8). It is important to remark here the presence of two peaks in the IR-spectrum for magnetic hybrid material (Fig. 2.7b), they are located at 570 and 437 cm^{-1} wavenumber. In [26, 29], these bands were associated with the Fe–O bending vibrations of the maghemite iron oxide phase.

The size and morphology of Fe_2O_3 nanoparticles into magnetic hybrid material were studied by Scanning Transmission Electron Microscope (STEM) and High-Resolution Transmission Electron Microscope (HRTEM) techniques. A scanning electron microscope JEOL 2010 transmission with acceleration 200 kV voltage was used for this purpose. For this analysis, samples were prepared in the form of powder as the specimen shown in Fig. 2.3c, and they were subsequently dispersed

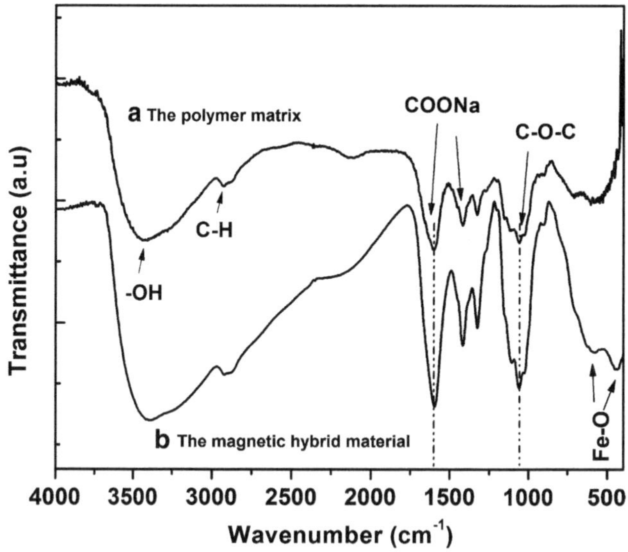

Fig. 2.7 IR spectra for: (**a**) Na-CMC and (**b**) the magnetic hybrid material

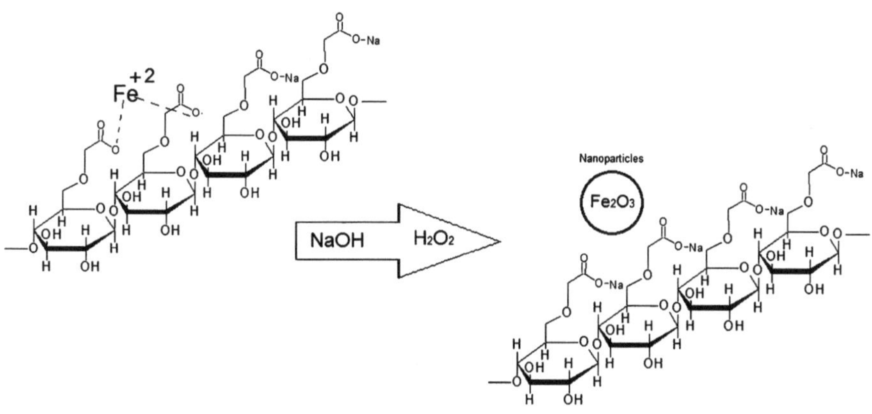

Fig. 2.8 Schematic representation of the synthesis process of magnetic hybrid material from a precursor hybrid material

in acetone, using for this purpose a sonificator (ultrasound). After that, it is taken an aliquot and settles on a grid of copper-coated with a film of carbon.

The STEM images (Fig. 2.9) show a higher number of iron oxide nanoparticles with sphere-like morphology. The average particle size of the iron oxide was measured using image analysis, and the frequency histogram shows a size of 4 nm (inset Fig. 2.9), very similar to the previously computed value $\langle L \rangle$ obtained using the Scherrer equation.

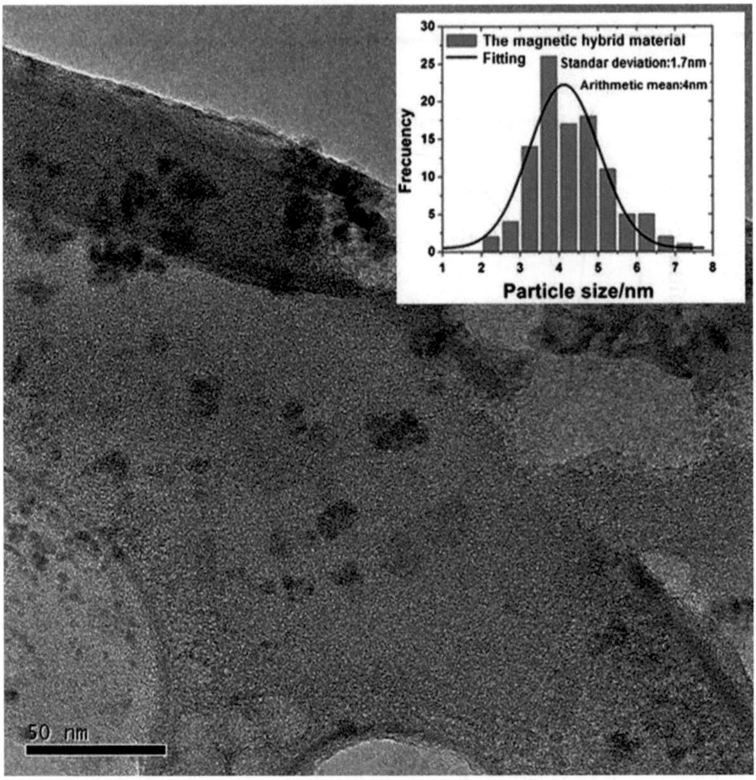

Fig. 2.9 STEM image of the magnetic hybrid material and frequency histogram

HRTEM image of Fe_2O_3 nanoparticles is shown in Fig. 2.10. The well-defined lattice fringes correspond to crystallographic planes of Fe_2O_3 nanoparticles and they are identified in this HRTEM image. In addition, inter-planar distances: 0.477, 0.332, 0.297, and 0.249 nm were also computed from this figure. These computed values correspond to inter-planar distances of (1 1 1), (2 1 1), (2 2 0), and (3 1 1) planes. These crystallographic planes are of maghemite crystalline structure, which is consistent with the X-ray diffraction analysis. These inter-planar distances are very close to the computed values, shown in Table 2.1.

The study of the structure and morphology of the magnetic hybrid material indicates that iron oxide nanoparticles are evenly dispersed in the polymer matrix. The following section presents the results obtained from the analysis of the magnetic behavior of the magnetic hybrid material.

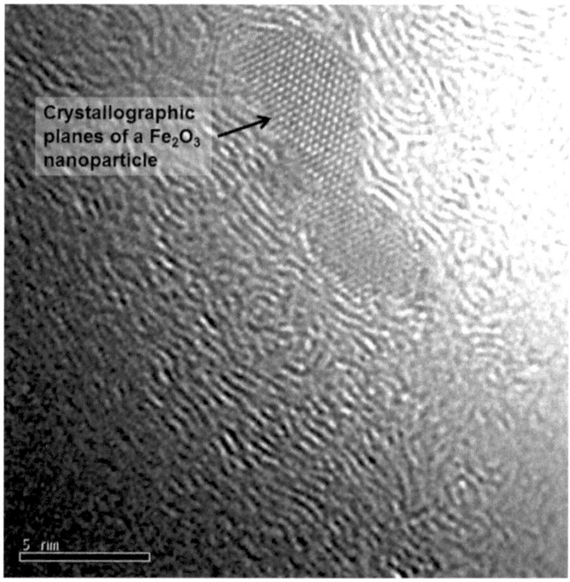

Fig. 2.10 HRTEM images of Fe_2O_3 nanocrystals into the magnetic hybrid material

2.3.2 Magnetic Properties of Magnetic Hybrid Materials

The magnetic hybrid material (Fe_2O_3/Na-CMC) was synthesized by subjecting the precursor hybrid material to a chemical treatment in an alkaline solution and hydrogen peroxide. At this stage it changed its color to black when the precursor hybrid material is in contact with the alkaline solution, then to reddish brown by gradually adding the peroxide. This color change is considered a macroscopic evidence of the formation of nanoparticles of iron oxide (in situ) in the polymer matrix of the CMC, obtaining with this a magnetic hybrid material, Fe_2O_3/Na-CMC. For the study of magnetic properties, the samples of the magnetic hybrid material were powdered as the specimen shown in Fig. 2.3c.

The study of the magnetic properties of the synthesized hybrid magnetic material was carried out through a Quantum Design MPMS magnetometer of the SQUID VSM type by magnetization measurements. Figure 2.11 shows the magnetization curve at room temperature (300 K). The saturation magnetization value (Ms) is 13.9 emu/g whereas both, remnant magnetization and coercivity (Hc) are undetectable. These last results are characteristics of a magnetic hybrid material with superparamagnetic behavior.

Conversely, Fig. 2.12 presents the magnetization curve at temperature 2 K. Under these conditions both coercivity field and remanent magnetization values are different to zero. It can be seen that the corresponding values for the different magnetic parameters are 19.9 emu/g for saturation magnetization, 357 Oe for coercive field (Hc), and 6.14 emu/g for remanence value (Mr) (inset Fig. 2.12); a ferromagnetic behavior appears when the magnetic analysis is performed below the blocking temperature (TB).

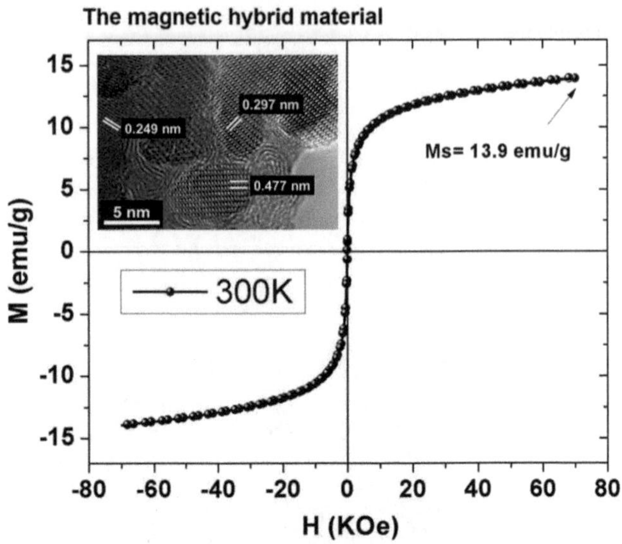

Fig. 2.11 Magnetization curve for the magnetic hybrid material at room temperature

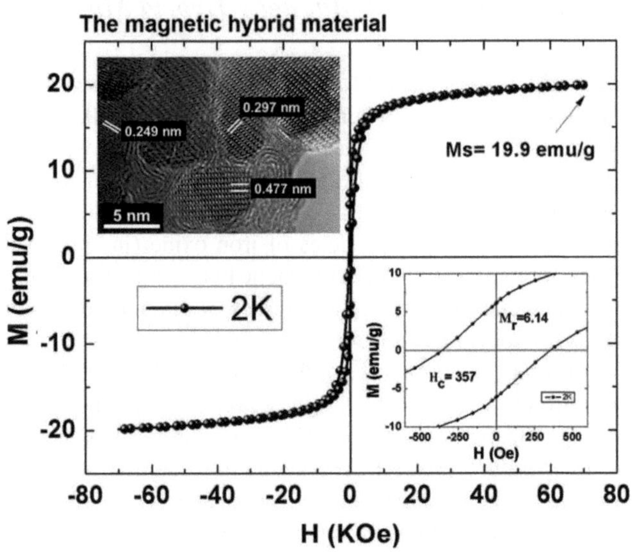

Fig. 2.12 Magnetization curve for the magnetic hybrid material at 2 K

To estimate the TB for the magnetic hybrid material, cooling at zero magnetic field (ZFC) and field cooling (FC) analyses was performed at a magnetic field of 100 Oe (Fig. 2.13). When the magnetic hybrid material is cooled at ZFC, Fig. 2.13 shows that the total magnetization is small, but not zero (20% of the maximum),

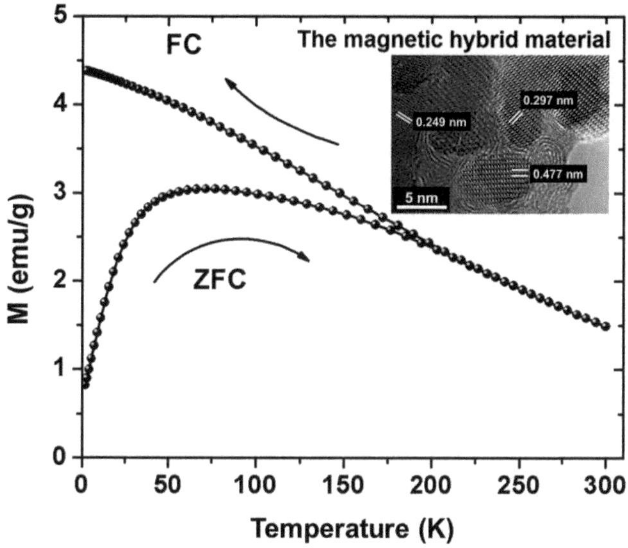

Fig. 2.13 ZFC and FC magnetization curves at 100 Oe

as the magnetic particles are not fully random. When temperature increases, the nanoparticle magnetic moment is oriented with the external field increasing the total magnetization until it reaches a maximum at 71 K which is the value of the blocking temperature (TB). At this temperature, the thermal energy becomes comparable to the energy gained by aligning the nanoparticle magnetic vector in the weak field. At this point, the transition from ferromagnetic to superparamagnetic behavior is observed. When all nanoparticles are at the superparamagnetic relaxation state, above TB, their magnetization follows Curie's law decreasing with increasing temperature. In the case of field cooling (FC), magnetization monotonically increases as the temperature decreases because the nanoparticles are cooled from room temperature under a magnetic field and the magnetization direction of all the nanoparticles is frozen in the field direction. The magnetization shows the maximum at 2 K in the FC process (Fig. 2.13).

In conclusion, the analysis of the magnetic properties of the magnetic hybrid material shows a superparamagnetic behavior at room temperature changing to ferromagnetic below 71 K, the blocking temperature.

The next section presents the experimental and FES response of the magnetic hybrid material for working as a bending-type actuator.

2.4 Actuation Properties of the Magnetic Hybrid Material

Electroactive materials with low driving voltage and large displacement are of critical importance for advancing the technology of microactuation. The major drawbacks of magnetic actuation are the by-effects arising from the relatively high currents involved in conventional magnetic actuation: the Joule losses in conductors imply overheating which may call for cooling techniques but also energy wastage [7]. The innovative material described above mixes the magnetic properties of the Fe_2O_3 and the insulating characteristics of the polymeric Na-CMC. Hence, it presents a good response face to an external magnetic field, but it does not heat because of the Joule effect.

Even if electromagnetic systems typically are rather complex, not only does magnetism already dominate the macroworld, but it also scales down very well to the microworld. Electromagnetic interactions that deserve a larger interest from the microsystems community, magnetic fields, and gradients can be effective over long distances relatively to the size of the microsystems. Additionally, magnetic microsystems offer large forces, large strokes, remote or distance control, bi-stability, robustness, high energy conversion efficiency, levitation, etc., all with great potential for new devices in many domains of applications.

The global market for magnetic sensors has been growing in terms of technology and applications. Position sensors, speed sensors, and record heads in hard disk drives in computers are the most commonly used magnetic sensor types currently in vogue. One of the rapidly expanding application areas includes e-compassing used in passenger cars, GPS-enabled handheld devices, cell phones and dead reckoning (DR) in personal vehicle, aircraft and marine navigation [32]. Contactless magnetic interaction allows remote actuation making magnetic actuators very well suited to harsh environment or for medical applications, through the skin. Laboratory-developed prototypes include RF microswitches for mobile phones, read/write heads and microposition systems, optical microcross-connect for fiber optic networks, microscanners, micromotors for less-invasive surgery or microrobotics, micropumps or microvalves for lab-on-chip and microfluidic devices, electrical microgenerators for autonomous power supplies, micromirrors for adaptive optics, microscanners for retinal scanning displays, magnetic suspensions for hard disk drives, etc. [7,20,22,35,38].

For these innovative applications, the insertion of smart functions was not previously possible with macroscopic devices. Performing research in the field of magnetic microactuators becomes thus a necessity. This section deals with the actuation properties of the magnetic hybrid material (Na-CMC/Fe_2O_3). As it has been explained, the hybrid material reacts to external magnetic fields. The resulting motion and force could be used as a magnetic field-controlled actuator. The focus of the macroscopic study of the magnetic behavior of the magnetic hybrid material is to investigate its dynamic response, specifically the displacement curves as a function of the applied magnetic field. This study has been performed experimentally and via a FES.

Table 2.2 Geometries of the analyzed samples of the magnetic hybrid material

Geometry	Length (mm)	Width (mm)	Thickness (mm)
I	17.94	2.81	0.3
II	18.00	1.06	0.02

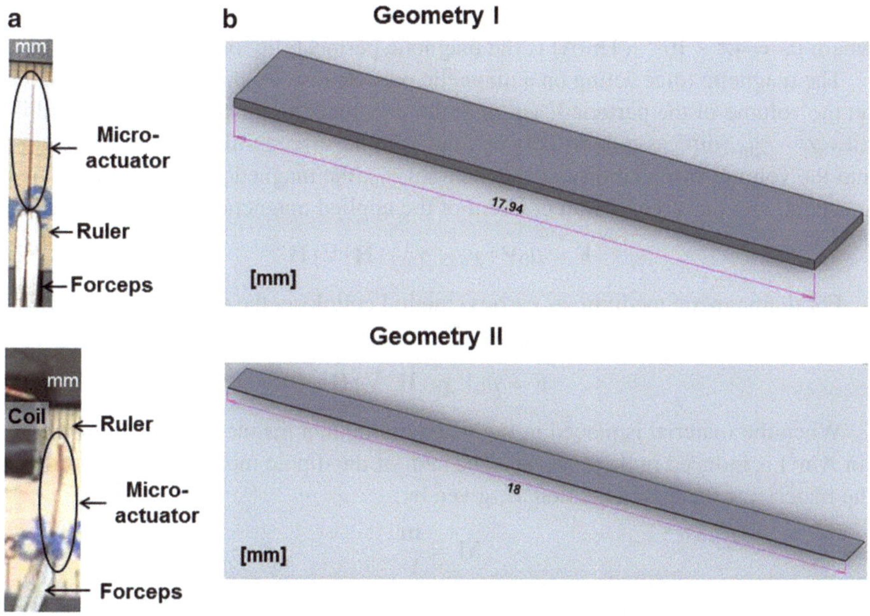

Fig. 2.14 Geometries of the magnetic hybrid actuator samples (see Table 2.2): (**a**) lateral view photograph and (**b**) schematic representation

The range of variation of the magnetic field for both analyses, experimental and FES, is based on previous results, which indicates a magnetic field around 100 Oe necessary to excite the magnetic hybrid material. The geometries of the two samples analyzed are described and illustrated in Table 2.2 and Fig. 2.14. Both geometries seam to a long beam.

Next sections will describe the basic theory of magnetics and bending beams involved in the actuation study of the magnetic hybrid material, as well as the procedures and results concerning the experimental and simulated actuation response of the material.

2.4.1 Basic Theory

The following section shows a summary of the relevant theory in magnetics and bending of beams involved in the actuation response of the magnetic hybrid material.

Concerning magnetic equations, two vectorial variables are used to describe magnetic fields: the magnetic field strength **H**, in Ampere per meter (A/m), and the magnetic induction **B**, in Tesla (T). When an external field **H** is applied to the material, the magnetic field **B** is induced in the material. The two variables are the same only in free space; otherwise, they are connected by the equation $\mathbf{B} = \mu_0\mathbf{H}$, where $\mu_0 = 4\pi \times 10^{-7}$ (Tm/A) is the magnetic permeability of the free space.

The magnetic force acting on a magnetic particle inside a magnetic field depends on the volume of the particle V (in m^3), the difference in the magnetic susceptibilities $\chi_p - \chi_m$ (dimensionless) between the volume susceptibility of the particle χ_p and the volume susceptibility of the medium χ_m, the magnetic permeability of free space μ_0, and the strength and gradient of the applied magnetic field **H** [19, 31, 34]:

$$\mathbf{F} = \mu_0 V \left(\chi_p - \chi_m\right) (\mathbf{H} \cdot \nabla) \mathbf{H} \tag{2.2}$$

For diamagnetic medium, as carboxymethyl cellulose, the volume susceptibility χ_m can be neglected and (2.2) becomes:

$$\mathbf{F} = \mu_0 V \chi_p (\mathbf{H} \cdot \nabla) \mathbf{H} \tag{2.3}$$

When the material is placed in the magnetic field, a magnetic dipole moment **m** (in Am^2) is induced in the material. The sum of the dipole moments in a volume of the material is the magnetization M, given by:

$$\mathbf{M} = \frac{\mathbf{m}}{V} \tag{2.4}$$

In a uniform magnetic field **B**, a magnetic dipole experiences a magnetic torque defined by:

$$\mathbf{T} = \mathbf{m} \times \mathbf{B} \tag{2.5}$$

This torque tends to line up the magnetic moment with the magnetic field, so this represents its lowest energy configuration.

Considering bending equations, simplifying approximations are employed to study problems of beam bending. Euler–Bernoulli beam theory is one of them [5]. It provides a means of calculating the load-carrying and deflection characteristics of beams and covers the cases for small deflections of a beam which is subjected only to lateral loads. The principal hypothesis of the Euler–Bernoulli theory is that plane sections remain plane and normal to the axis of the beam.

Considering this hypothesis, the displacement δ (in m) at the free end of a cantilever beam (rectangular cross-section) with a single concentrated load F (in N) at the same free end, is done by the expression:

$$\delta = \frac{FL^3}{3EI} \tag{2.6}$$

where E is the Young's modulus of the cantilever's material (in Pa) and $I = bh3/12$ is its area moment of inertia (in m4). The geometry of the cantilever is defined by its length L, width b, and thickness h (all in meters, m).

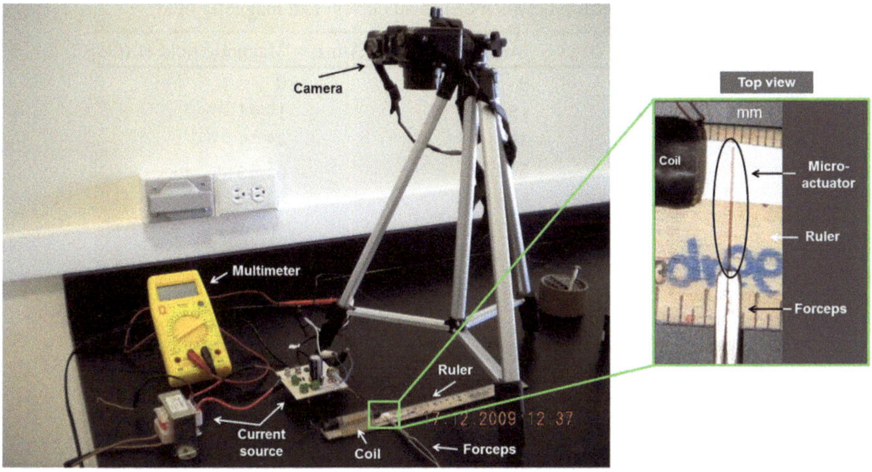

Fig. 2.15 Experimental setup to quantify the bending of the magnetic hybrid material samples

2.4.2 *Experimental Actuation Response*

The next study will focus on experimental evaluation of the magnetic hybrid material in an effort to determine the feasibility of utilizing these materials as bending-type actuators. This study has been realized for the two samples of the magnetic hybrid material whose geometries are described in Table 2.2. The experimental setup is shown in Fig. 2.15. After mounting the magnetic samples in a fixed-free cantilever configuration, uniform magnetic fields are applied to the samples; it causes the samples to bend, similar to a cantilever beam. The samples were oriented such that the length of the sample was perpendicular to the direction of the magnetic field. A wood ruler is placed below the sample to quantify the displacement of the bending. A digital camera Canon Powershot G10 records the experiments. In order to produce a varying magnetic field, an electrical current was supplied to a coil. The objective of this experimentation is to measure the displacement response of the magnetic hybrid material due to an external applied magnetic field. Linearity is an important characteristic for both the use and the characterization of actuators [21].

The coil has been designed using (2.7), considering a magnetic field $H = 7957.75$ A/m (100 Oe), and an electrical current $i = 1$ A. The resulting coil consists of 398 turns of 15-AWG wire wounded around a ferrite core with 5 cm of length l and 1 cm of diameter ϕ. For the two samples studied, the relationship between the applied electrical current and the resulting magnetic field using (2.7) is shown in Table 2.3.

$$N = \frac{Hl}{i} \tag{2.7}$$

Table 2.3 Relationship between electrical current and magnetic field

Sample	Current i (A)	Magnetic field H (A/m)	Magnetic field H (Oe)
I	0	0	0
	0.20	1586.69	19.94
	0.44	3441.37	43.25
	0.65	5197.88	65.32
	0.84	6689.05	84.06
	1.05	8373.92	105.23
	1.23	9809.37	123.27
	1.40	11114.81	139.67
II	0	0	0
	0.19	1544.24	19.41
	0.43	3422.80	43.01
	0.61	4855.60	61.02
	0.83	6582.92	82.72
	1.06	8461.48	106.33
	1.21	9639.56	121.13
	1.38	10984.80	138.04

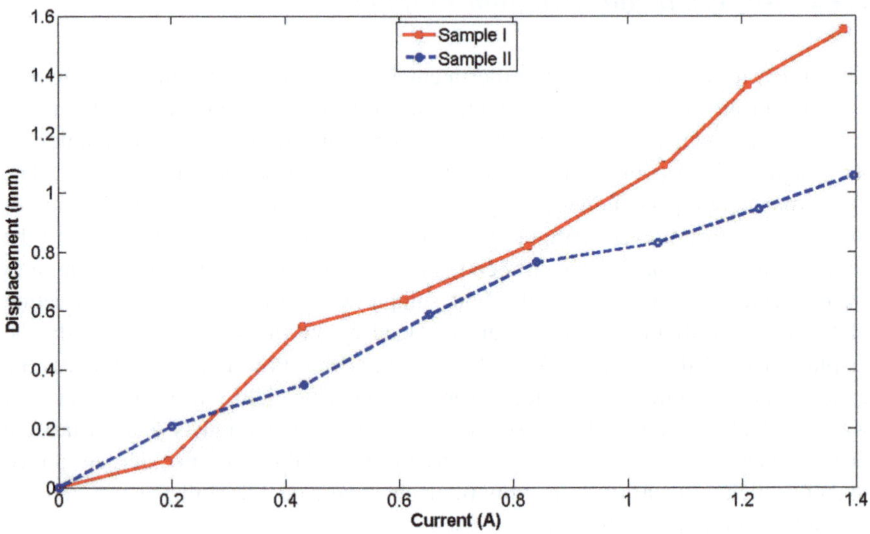

Fig. 2.16 Experimental displacement of the two samples of the magnetic hybrid material as a function of the electrical current applied to the coil

Figure 2.16 gives evidence that the geometry and shape affect the displacement response. The thinner sample, sample II, presents 48% more displacement than the bigger sample (sample I), being the maximal displacement attempted 1.55 mm. Considering an uniform distribution of the nanoparticles in the polymeric matrix, the rotation motion of the samples depends principally on the medium in which they are

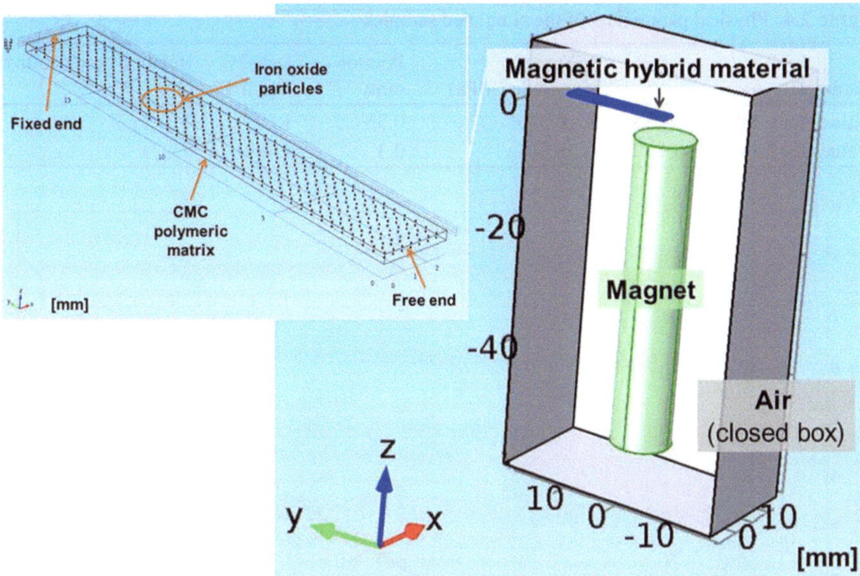

Fig. 2.17 Finite element setup of the magnetic hybrid material (sample II)

embedded. For both cases, the displacement generated by the magnetic field could show a positive linear correlation over the range studied. Hence, the magnetic hybrid material can be considered as a candidate of a flexible actuator.

Next section shows the FES results of the magnetic hybrid material.

2.4.3 Finite Element Simulation of Actuation Dynamics

As improved materials emerging, it becomes necessary to address key issues such as the need of effective magnetomechanical modeling and guiding parameters in scaling the actuators [24]. The dynamic FES of the magnetic hybrid film allows the examination of its behavior in order to extend experimental study and to validate the performance of the hybrid material.

Generally, FES on magnetic hybrid materials is done considering the material as one material [4]. The objective of this section is to present the first simulations of the magnetic hybrid material considering the two components that form it: Na-CMC and Fe_2O_3. These FES of the magnetic hybrid material as a bending-type actuator was done using COMSOL Multiphysics 4.3. We simulate the geometry of the two samples defined in Table 2.2 introducing only a limited number of particles to take into account the magnetic effect produced by the iron oxide particles and the viscoelastic effect inherent to the carboxymethyl cellulose.

Figure 2.17 shows the simulated setup of the samples. The varying magnetic field has been simulated as a permanent magnet with a magnetization varying linearly

Table 2.4 Physical parameters of the simulated samples

Parameter	Young's module (MPa)	Poisson's ratio	Density (kg/m^3)	Relative permeability $(1+\chi)$
Values for CMC [1,4,9,23]	6.37	0.35	1.59	1
Values for Fe_2O_3 [2,6,13,34]	18.82	0.3	1,089	7

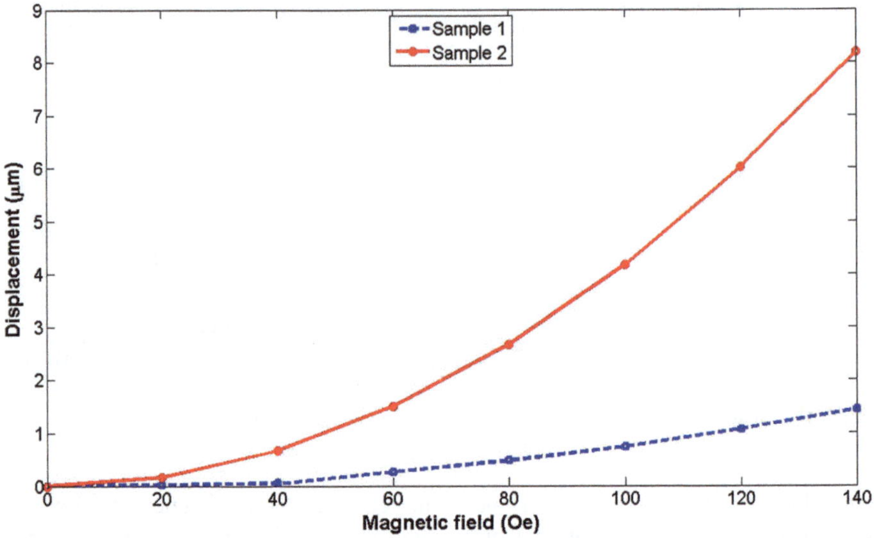

Fig. 2.18 FES displacement of the magnetic hybrid material face to an external magnetic field

from 0 to 140 Oe. The iron oxide (Fe_2O_3) particles, as spheres, are embedded on the polymeric matrix beam. The sample I contains 528 particles whose diameter is 0.1 mm, and the sample II contains 348 particles of 0.015 mm of diameter. Table 2.4 describes the physical parameters for the CMC and the Fe_2O_3 considered for the FES.

The simulation has been developed in two steps. First, we evaluate the magnetic forces acting on the superparamagnetic particles due to the permanent magnet. Second, we simulate the displacement of the hybrid material produced by these magnetic forces. Meshed models consist of triangular elements on the surface and tetrahedral elements in the volume. The automatic meshing available in COMSOL Multiphysics has been used to define the meshing. Depending on the geometry of the sample, during the magnetic simulation, the mesh includes 128,268–447,198 elements which represent between 7,479 and 10,111 nodes. Considering the same conditions, during the second step of the simulation, the mesh includes 134,074–149,293 elements which represent between 7,068 and 10,178 nodes, depending on the geometry of the sample. All the simulations are done in statics conditions. The resulting curves of the displacement as a function of the magnetic field are shown in Fig. 2.18. The nonlinearity of the displacement response is a consequence of the polymeric medium in which the particles are embedded.

Results obtained by FES (Fig. 2.18) could be considered as a first approximation of experimental results (Fig. 2.16). In experimental results, the displacement seems proportional to the electrical current that produces a magnetic field; while in FES results the displacement is exponentially proportional to the simulated magnetic field. These deviations between experimental results and FES results are associated with the limited number of Fe_2O_3 nanoparticles introduced in the FES models to take into account the magnetic effect produced by the iron oxide particles into the viscoelastic effect, inherent to the carboxymethyl cellulose.

The synthesis in situ of iron oxide nanoparticles into the precursor hybrid material to obtain the magnetic hybrid material allows us to driving the viscoelasticity of CMC, because the displacements or deformations in the magnetic hybrid material can be modified as a function of the magnetic field applied.

2.5 Conclusion

On the one hand, a magnetic hybrid material consisting of nanoparticles of iron oxide in a Na-CMC matrix was obtained. The synthesis of magnetic nanoparticles from the precursor hybrid material was confirmed via XRD and IR analysis. The iron oxide nanoparticles are embedded in the Na-CMC and their measured size was around 4 nm, having a nearly spherical morphology. The analysis of the magnetic properties of the magnetic hybrid material shows a superparamagnetic behavior at room temperature changing to ferromagnetic below 71 K, the blocking temperature. Our results suggest that in situ precipitation of nanoparticles in the precursor hybrid material is a promising route to the production of the magnetic hybrid material.

On the other hand, the dynamic response, specifically the displacement of the magnetic hybrid material, has been observed in order to probe the feasibility to use this material as a bending-type actuator. The experimental results show that the responses of the deflection have a linear trend over a reasonable range, suggesting that the magnetic hybrid material can be used as bending-type actuators in small mechanical systems and devices. The thinner sample showed the largest response among the two samples considered in this study. When an electrical current of 1.40 A is applied to the coil, the tip deflected nearly 1.55 mm. This result suggests that the dynamic experimental behavior provide promising implications for their role as magnetically controlled actuator. First simulations have been also done considering the two components of the magnetic hybrid material: the oxide iron nanoparticles and the carboxymethyl cellulose. The displacement response takes into account the viscoelastic effect of the polymeric matrix and the magnetization of the magnetic particles. Further studies must be addressed in this sense in order to describe the macroscopic magnetic behavior of the hybrid material visualizing a further application of the material working as a magnetic microactuator.

References

1. Aqualon sodium carboxymethylcellulose: physical and chemical properties. Technical Report, Hercules Incorporated, 1999
2. A.B.M. Aowlad Hossain, M.H. Cho, S.Y. Lee, Magnetic nanoparticle density mapping from the magnetically induced displacement data: a simulation study. BioMed. Eng. OnLine, 11, p. 13 (2012). doi:10.1186/1475-925X-11-11
3. M.F. Ashby, Y.J.M. Bréchet, Designing hybrid materials. Acta Mater. **51**, 5801–5821 (2003)
4. M. Barham, D. White, Finite element simulation of permanent magnetoelastic thin films. IEEE Trans. Magn. **47**(5), 1402–1405 (2011)
5. A.P. Boresi, P.C. Ken, *Elasticity in Engineering Mechanics* (PTR Prentice Hall, Englewood Cliffs, 1987)
6. J. Chatterjee, Y. Haik, C.-J. Chen, Size dependent magnetic properties of iron oxide nanoparticles. J. Magn. Magn. Mater. **257**(1), 113–118 (2003)
7. O. Cugat, J. Delamare, R. Gilbert, Magnetic microsystems: Mag-MEMS, in *Magnetic Nanostructures in Modern Technology* (Springer, Berlin, 2008), pp. 105–125
8. D. Eder, Carbon nanotube-inorganic hybrids. Chem. Rev. **110**(3), 1348–1385 (2010)
9. T.J. Fiske, H. Gokturk, D.M. Kalyon, Enhancement of the relative magnetic permeability of polymeric composites with hybrid particulate fillers. J. Appl. Polym. Sci. **65**(7), 1371–1377 (1997)
10. M.A. Garza-Navarro, M. Hinojosa-Rivera, V.A. González-González, Desarrollo de nanocompuestos superparamagnticos quitosán/magnetita. Ingenierías **9**(33), 14–20 (2006)
11. M.A. Garza-Navarro, M.E. Reyes-Melo, V. González-González, C. Guerrero-Salazar, U. Ortiz-Mendez, Modeling of isochronal complex magnetic susceptibility of polymer-magnetic nanocomposites using fractional calculus. J. Appl. Polym. Sci. **23**(4), 2154–2161 (2012)
12. S.P. Gubin, Y.A. Koksharov, G.B. Khomutov, G.Y. Yurkov, Magnetic nanoparticles: preparation, structure and properties. Russ. Chem. Rev. **74**(6), 489–520 (2005)
13. Z. Guo, K. Lei, L. Li, H.W. Ng, S. Prikhodko, H.T. Hahn, Fabrication and characterization of iron oxide nanoparticles reinforced vinyl-ester resin nanocomposites. Compos. Sci. Technol. **68**(6), 1513–1520 (2008)
14. V. Hegedusic, Z. Herceg, S. Rimac, Rheological properties of carboxymethylcellulose and whey model solutions before and after freezing. Food Technol. Biotechnol. **38**(1), 19–26 (2000)
15. C.W. Hoogendam, A. de Keizer, S.M.A. Cohen, B.H. Bijsterbosch, Adsorption mechanisms of carboxymethyl cellulose on mineral surfaces. Langmuir **14**(14), 3825–3839 (1998)
16. J.P. Jose, K. Joseph, Advances in polymer composites: macro- and microcomposites. State of the art, new challenges, and opportunities, in *Polymer Composites. Volume 1: Macro- and Microcomposites* (Wiley-VCH, Weinheim, 2012) [Chapter 1]
17. M. Karg, Multifunctional inorganic/organic hybrid microgels. An overview of recent developments in synthesis, characterization, and application. Colloid Polym. Sci. **290**(8), 673–688 (2012)
18. G. Kickelbick, *Hybrid Materials: Synthesis, Characterization and Applications* (Wiley-VCH, Weinheim, 2007)
19. M.-C. Kim, D.-K. Kim, S.-H. Lee, M.S. Amin, I.-H. Park, C.-J. Kim, M. Zahn, Dynamic characteristics of superparamagnetic iron oxide nanoparticles in a viscous fluid under an external magnetic field. IEEE Trans. Magn. **42**(4), 979–982 (2006)
20. S.-H. Kim, S. Hashi, K. Ishiyama, Methodology of dynamic actuation for flexible magnetic actuator and biomimetic robotics application. IEEE Trans. Magn. **46**(6), 1366–1369 (2010)
21. J.-H. Koo, A. Dawson, H.-J. Jung, Characterization of actuation properties of magnetorheological elastomers with embedded hard magnetic particles. J. Intell. Mater. Syst. Struct. **23**(9), 1049–1054 (2012)
22. D. Lee, S. Kim, Y.-L. Park, R.J. Wood, Design of centimeter-scale inchworm robots with bidirectional claws, in *Proceedings of IEEE International Conference on Robotics and Automation (ICRA)*, Shanghai, China, May 2011, pp. 3197–3204

23. L.P. Li, W. Herzog, R.K. Korhonen, J.S. Jurvelin, The role of viscoelasticity of collagen fibers in articular cartilage: axial tension versus compression. Med. Eng. Phys. **27**(1), 51–57 (2005)
24. C. Liu, Y. Bar-Cohen, Scaling laws of microactuators and potential applications of electroactive polymers in MEMS, in *Proceedings of SPIE International Symposium on Smart Structures and Materials*, CA, USA, March 1999, 10 pp.
25. B. López-Walle, E. Reyes-Melo, E. López-Cuellar, Los materiales híbridos en el desarrollo de sistemas mecatrónicos. Ingenierías **14**(53), 35–43 (2011)
26. J.F. Luna-Martínez, Síntesis y caracterización de materiales nanoestructurados a base de una matriz polimérica de carboximetilcelulosa, Ph.D. thesis, Universidad Autnoma de Nuevo León, FIME, 2011
27. J.F. Luna-Martínez, E. Reyes-Melo, V. González-González, A. Torres-Castro, C. Guerrero-Salazar, S. Sepúlveda-Guzmán, Iron oxide nanoparticles obtained from a Fe(II)-chitosan polymer film. Mater. Sci. Forum **644**, 51–55 (2010)
28. J.F. Luna-Martínez, D.B. Hernández-Uresti, M.E. Reyes-Melo, C.A. Guerrero-Salazar, S. Sepúlveda-Guzmán, V.A. González-González, Synthesis and optical characterization of ZnS–sodium carboxymethyl cellulose nanocomposite films. Carbohydr. Polym. **84**, 566–570 (2011)
29. J.F. Luna-Martínez, E. Reyes-Melo, V. González-González, C. Guerrero-Salazar, A. Torres-Castro, S. Sepúlveda-Guzmán, Synthesis and characterization of a magnetic hybrid material consisting of iron oxide in a carboxymethyl cellulose matrix. J. Appl. Polym. Sci. **127**(3), 2325–2331 (2013)
30. C. Michelina, B. Flavia, Synthesis and characterization of amorphous and hybrid materials obtained by sol–gel processing for biomedical applications, in *Biomedical Science, Engineering and Technology*, Rijeka, Croatia (InTech, 2012), pp. 389–416 [Chapter 6]
31. N. Pamme, Magnetism and microfluidics. Lab Chip **6**, 24–38 (2006)
32. PRWeb, Global magnetic sensors market to reach $3.2 billion by 2017, according to a new report by Global Industry Analysts, Inc., August 2011
33. M.E. Reyes-Melo, J.F. Luna-Martínez, V.A. González-González, Materiales híbridos magnéticos. Ingenierías **14**(53), 6–12 (2011)
34. S.S. Shevkoplyas, A.C. Siegel, R.M. Westervelt, M.G. Prentiss, G.M. Whitesides, The force acting on a superparamagnetic bead due to an applied magnetic field. Lab Chip **7**, 1294–1302 (2007)
35. J. Streque, A. Talbi, R. Viard, P. Pernod, V. Preobrazhensky, Elaboration and test of high energy density magnetic micro-actuators for tactile display applications. Procedia Chem. **1**(1), 694–697 (2009)
36. W. Wang, A. Wanga, Nanocomposite of carboxymethyl cellulose and attapulgite as a novel pH-sensitive superabsorbent: synthesis, characterization and properties. Carbohydr. Polym. **82**, 83–91 (2010)
37. Y. Xiaotun, X. Lingge, N.S. Choon, C.S. Hardy, Magnetic and electrical properties of polypyrrole-coated γ-Fe_2O_3 nanocomposite particles. Nanotechnology **14**(6), 624–629 (2003)
38. Y. Yamanishi, S. Sakuma, K. Onda, F. Arai, Powerful actuation of magnetized microtools by focused magnetic field for particle sorting in a chip. Biomed. Devices **12**(4), 745–752 (2010)
39. H. Yan, M. Guo, K. Liu, Multifunctional magnetic hybrid nanoparticles as a nanomedical platform for cancer-targeted imaging and therapy, in *Biomedical Science, Engineering and Technology*, Rijeka, Croatia (InTech, 2012), pp. 283–300 [Chapter 12]

Chapter 3
Design of Piezoelectric Actuators with Guaranteed Performances Using the Performances Inclusion Theorem

Micky Rakotondrabe and Sofiane Khadraoui

Abstract This chapter presents the design of piezoelectric actuators by using the performances inclusion theorem (PIT). The main objective is to seek for the dimensions of a cantilevered actuator such that its performances will lie within some specifications imposed *a priori*. For that, these specifications are transcribed into an interval transfer function, called interval reference model, while an interval model of the actuator is also provided. Then, from the PIT, a problem of finding the dimensions is yielded such that this latter model is enclosed in the reference model. The problem is seen as a set-inversion problem that can be solved with interval tools such as the SIVIA (Set Inversion Via Interval Analysis) algorithms. The designed piezoelectric actuator is afterwards fabricated and characterized. The experimental characterizations demonstrate the efficiency of the proposed technique.

3.1 Introduction

Piezoelectric materials are well recognized for the design and development of actuators in systems working at the micro/nanoscale, in particular for micro/nanopositioning systems. This recognition is due to their high bandwidth, high resolution and high density force. Furthermore, the fact that they can be used for sensing and actuation and the fact that their energy control is electrical make them more attractive than thermal or other active or smart materials.

M. Rakotondrabe (✉)
Automatic Control and Micro-Mechatronic Systems Department, AS2M,
FEMTO-ST Institute, 24 rue Alain Savary, Besançon 25000, France
e-mail: mrakoton@femto-st.fr

S. Khadraoui
Texas A&M University at Qatar, PO Box 23874, Education city, Doha, Qatar
e-mail: khadraoui_s@hotmail.fr

M. Rakotondrabe (ed.), *Smart Materials-Based Actuators at the Micro/Nano-Scale: Characterization, Control, and Applications*, DOI 10.1007/978-1-4614-6684-0_3,
© Springer Science+Business Media New York 2013

In micro/nanopositioning, there are several approaches to use piezoelectric materials: stepper piezoelectric actuators such as stick-slip or inch-worm motion principle [1–4], ultrasonic piezoelectric actuators [5–7], and flexible and continuously deformed piezoelectric cantilevers used in microgrippers or in microscopes [8–10]. In the fields of micromanipulation and microassembly, piezoelectric cantilever actuators with rectangular section are often used to develop microgrippers able to pick, transport, and place precisely small objects. These piezoelectric cantilevered structures offer an ease of control of the force or of the deflection (position of the object) which is essential when performing precise positioning and manipulation at the same time.

In general, the design of piezoelectric cantilevered actuators are done without explicit and *a priori* information on desired performances. Then, the designed actuators often offer any performances to which the applications should adapt instead of the converse, i.e. instead of designing actuators that would fit with the applications. Recently, an optimal design technique was proposed in [11] to design piezoelectric systems. Based on the gramian tools in control theory, the technique can provide optimal locations of cantilevered piezoelectric actuators and sensors. In this sense, the technique is useful when the system is based on several cantilevers such as treillis and then cannot be used for one cantilever-based system that are used in microgrippers. The aim of this chapter is to propose a novel technique to design piezoelectric cantilever-based actuators and systems. The main difference relative to the work in [11] is that we can design systems with single cantilevered actuators. Furthermore, in the proposed technique, we impose *a priori* some desired performances and specifications. For that, we propose to use interval tools, in particular we will use the performances inclusion theorem (PIT) developed in [12] which is effectively an efficient tool to also design actuators.

Interval tools [13, 14] are techniques that were efficiently used in different applications: control theory and control systems, robotics, signals and parameters estimation, etc. The main advantage of interval tools is the *guarantee* aspect that they can offer with the results, i.e. guaranteed solution or guaranteed nonsolution. In this chapter, we propose to benefit from such advantage to find guaranteed dimensions of piezoelectric actuators that would provide some specified performances. The chapter is organized as follows.

In Sect. 3.2, some preliminaries on intervals techniques are given. The PIT is particularly reminded. Section 3.3 is devoted to the modeling of piezoelectric cantilevered actuators. Models of multilayered actuators are treated in both static and dynamics. In Sect. 3.4, the novel technique to design piezoelectric actuators is proposed. We particularly focus on the design of unimorph cantilevered actuators. Finally, fabrication and experimental tests on the fabricated actuators are carried out which demonstrate the efficiency of the proposed approach.

3.2 Preliminaries on Intervals

3.2.1 Definitions

We remind here some basics on intervals that will be used in the rest of the chapter. The readers who are interested to see more in details the techniques of intervals are suggested to read the references [13, 14].

A real interval $[x]$ is a closed interval such that:

$$[x] = [x^-, x^+] \tag{3.1}$$

where x^- and x^+ are called lower bound and upper bound, respectively. We have: $x^- \leq x^+$. Having $x^- = x^+$ means that the interval $[x]$ is degenerate. By convention, a degenerate interval $[a] = [a, a]$ is identified by the real number a. The designation *point number* is similar to the designation *degenerate interval number*. While the set of real point numbers is $\mathbb{R}$, the set of real intervals (or real interval numbers) is $\mathbb{IR}$.

Instead of using the notation in (3.1), one can also identify a real interval number by its midpoint $\text{mid}([x])$ and its radius $\text{rad}([x])$ such that:

$$\begin{cases} \text{mid}([x]) = \frac{(x^+ + x^-)}{2} \\ \text{rad}([x]) = \frac{(x^+ - x^-)}{2} = \frac{w([x])}{2} \end{cases} \tag{3.2}$$

where $w([x])$ is the width of the interval.

3.2.2 Operations on Intervals

In the arithmetics of intervals, the basic operations are extended to interval numbers. Consider two intervals $[x] = [x^-, x^+]$ and $[y = y^-, y^+]$. So we have:

$$[x] + [y] = [x^- + y^-, x^+ + y^+] \tag{3.3}$$

and

$$[x] - [y] = [x^- - y^+, x^+ - y^-] \tag{3.4}$$

Consequently, we have: $[x] - [x] \neq 0$, except for $x^- = x^+$.

The multiplication and division are defined as follows:

$$\begin{aligned} [x] . [y] = [&\min(x^- y^-, x^- y^+, x^+ y^-, x^+ y^+), \\ &\max(x^- y^-, x^- y^+, x^+ y^-, x^+ y^+)] \end{aligned} \tag{3.5}$$

and

$$[x] / [y] = [x] . [1/y^+, 1/y^-], \; 0 \notin [y] \tag{3.6}$$

We say that an interval $[x]$ is included in an interval $[y]$, i.e. $[x] \subset [y]$, if and if only $[x] \cap [y] = [x]$. We have $[x] > [y]$ if $x^- > y^+$. The real interval $[x]$ is said to be positive if $x^- > 0$. The distributive law does not hold in general for interval. However, the following relation, called subdistributivity, holds: $[x]([y]+[z]) \subseteq [x][y]+[x][z]$. In addition, if $[x]+[y]=[x]+[z]$, the cancellation law for addition holds, and $[y] = [z]$. The same property holds for multiplication: if $[x][y]=[x][z]$ and $0 \notin [x]$, thus $[y]=[z]$.

If f is a function $f:\mathbb{R} \to \mathbb{R}$, then its interval counterpart $[f]$ satisfies:

$$[f]([x]) = [\{f(x) : x \in [x]\}] \tag{3.7}$$

The interval function $[f]$ is called inclusion function because $f([x]) \subseteq [f]([x])$, for all $[x] \in \mathbb{IR}$. An inclusion function $[f]$ is thin if for any degenerate interval $[x] = x$, $[f](x) = f(x)$. It is minimal if for any $[x]$, $[f]([x])$ is the smallest interval that contains $f([x])$. The minimal inclusion function for f is unique and is denoted by $[f]^*([x])$.

An easy way to compute an inclusion function for f is to replace each variable x in the expression of f by $[x]$ and all operations on points by their interval counterpart. Thus, one obtains the natural inclusion function.

3.2.3 Interval Systems

An interval system is a transfer function representation, a state space representation or a differential representation where the parameters are intervals. For an interval transfer function, which is the interest of this chapter, the representation is as follows:

$$[G](s) = \frac{[b_m]s^m + \cdots + [b_1]s^1 + [b_0]}{[a_n]s^n + \cdots + [a_1]s^1 + [a_0]} = \frac{\sum\limits_{l=0}^{m}[b_l]s^l}{\sum\limits_{k=0}^{n}[a_k]s^k} \tag{3.8}$$

where s is the Laplace variable and where $m \le n$, n being the order of the interval system $[G](s)$. The parameters $[a_k]$ and $[b_l]$ are considered to be constant real intervals in order to assume linear time invariant (LTI) systems. The notation $[G](s)$ shall be used if the intervals $[a_k]$ and $[b_l]$ are known. Instead, the notation $[G]([a_k],[b_l],s)$ is used when they are unknown and to be sought for.

The notion of inclusion of systems should also be defined. Consider two interval systems having the same polynomials degrees m and n, i.e. having the same structure:

$$[G_1](s) = \frac{\sum\limits_{l=0}^{m}[b_{1l}]\cdot s^l}{\sum\limits_{k=0}^{n}[a_{1k}]\cdot s^k}, \quad [G_2](s) = \frac{\sum\limits_{l=0}^{m}[b_{2l}]\cdot s^l}{\sum\limits_{k=0}^{n}[a_{2k}]\cdot s^k} \tag{3.9}$$

$[G_1](s) \subseteq [G_2](s)$ is equivalent to saying that for any $s \in [0,\infty)$, we have $[G_1] \subseteq [G_2]$.

Lemma 2.1. *If* $[b_{1l}] \subseteq [b_{2l}]$ *and* $[a_{1k}] \subseteq [a_{2k}]$, $\forall k, l$, *then* $[G_1](s) \subseteq [G_2](s)$.

Proof. See [12].

3.2.4 The Performances Inclusion Theorem [12]

Consider two interval systems having the same polynomial degrees m and n:

$$[G_1](s) = \frac{\sum_{l=0}^{m} [b_{1l}] \cdot s^l}{\sum_{k=0}^{n} [a_{1k}] \cdot s^k}, \; [G_2](s) = \frac{\sum_{l=0}^{m} [b_{2l}] \cdot s^l}{\sum_{k=0}^{n} [a_{2k}] \cdot s^k} \tag{3.10}$$

The PIT which will be used to further design actuators is composed of two results.

Theorem 2.1. *The performances inclusion in the frequency domain:*

$$\text{if} \begin{cases} [a_{1k}] \subseteq [a_{2k}], \forall k = 1, \ldots, n \\ \text{and} \\ [b_{1l}] \subseteq [b_{2l}], \forall l = 1, \ldots, m \end{cases}$$

$$\Rightarrow \begin{cases} [\rho]([G_1](j\omega)) \subseteq \rho([G_2](j\omega)) \\ \text{and} \\ [\varphi]([G_1](j\omega)) \subseteq \varphi([G_2](j\omega)) \end{cases}$$

Theorem 2.2. *The performances inclusion in the time domain:*

$$\text{if} \begin{cases} [a_{1k}] \subseteq [a_{2k}], \forall k = 1, \ldots, n \\ \text{and} \\ [b_{1l}] \subseteq [b_{2l}], \forall l = 1, \ldots, m \end{cases}$$

$$\Rightarrow [g_1](t) \subseteq [g_2](t)$$

where:

- $[\rho]([G_i](j\omega))$ is the modulus of the system $[G_i]$.
- $[\varphi]([G_i](j\omega))$ is the argument.
- $[g_i](t)$ is the impulse response.

Proof. See [12].

3.3 Piezoelectric Cantilevered Actuators and Their Modeling

3.3.1 *Presentation of a Piezoelectric Cantilevered Actuator*

A piezoelectric cantilevered actuator, alternately called piezocantilever, is a cantilever having one or several layers, where at least one layer is piezoelectric material. We are interested here by a cantilever with rectangular section. When the cantilever has many layers, it is called multilayered piezoelectric actuator. Often, if there are p piezoelectric layers in a multilayered cantilever, we also call the actuator a p-morph actuator [15]. In the actuator, the non-piezoelectric layers are called passive layers. They can be silicone, nickel, copper, chrome, polymer materials, etc. The objective with the actuator is that, when a voltage is applied to the piezoelectric layers, the whole cantilever bends. Figure 3.1a presents a multilayered piezoelectric actuator with n layers. In Fig. 3.1b, a bilayer unimorph piezoelectric actuator is presented. The application of a voltage U to the piezoelectric layer makes it contract/expand resulting a global deflection of the cantilever. In Fig. 3.1c, a bilayer bimorph actuator is depicted. The application of the voltage U, yielding an electrical field parallel and anti-parallel to the internal polarization P_{ol} of the two layers yields a contraction and a compression of them. This antagonist longitudinal deformation yields a bending of the cantilever.

An unimorph actuator is more simple to develop and to use, in particular in terms of electrical connection. A bimorph actuator (and a multimorph actuator) is more complex to settle. However, relative to an unimorph actuator, it offers higher deformation and bending with the same applied voltage.

3.3.2 *Static Model*

First, the static model of a multilayered piezoelectric cantilever is given. We are particularly interested in the resulting deflection of the actuator when a voltage is applied. Consider the n-layered actuator pictured in Fig. 3.2a where:

- h_i, with $i \in \{1,2,\ldots,n\}$, is the ith layer.
- $\bar{y}$ is the distance between the neutral fiber and the bottom surface of the cantilever.

Each layer of the cantilever is exclusively a piezoelectric material or a passive material. The only rule is that at least one layer among the i-layers is piezoelectric. The application of a voltage U to the piezoelectric layers yields a bending of the whole cantilever as depicted in Fig. 3.2b. The bending $y(x)$ at distance x from the clamp can be written as follows [16]:

$$y(x) = \frac{m_{\text{piezo}} x^2}{2C} U \tag{3.11}$$

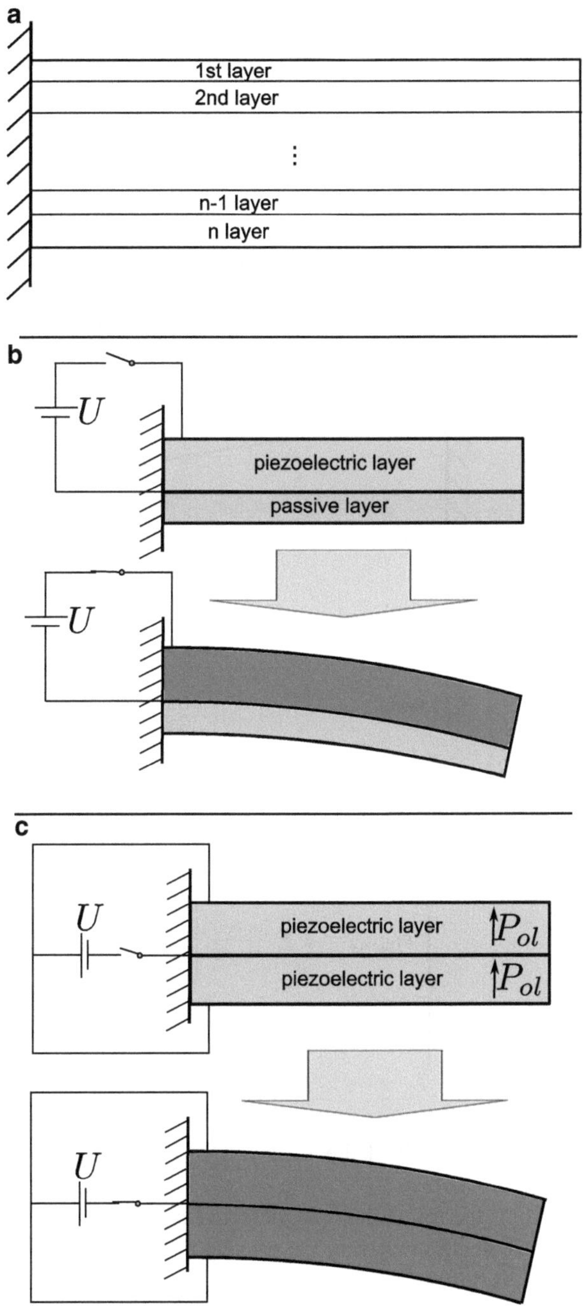

Fig. 3.1 (**a**): a multilayered piezoelectric cantilever. (**b**): a (bilayer) unimorph piezoelectric cantilever. (**c**): a (bilayer) bimorph piezoelectric cantilever

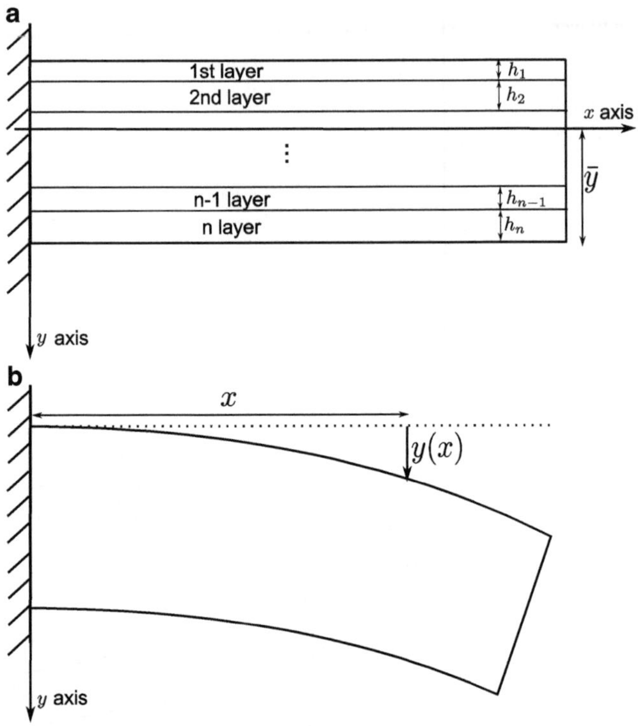

Fig. 3.2 (**a**): a multilayered piezoelectric cantilever. (**b**): bending of a multilayered piezoelectric cantilever

with

$$\begin{cases} m_{\text{piezo}} = \frac{1}{2}\sum_{i=1}^{n} \frac{w_i d_{31,i}}{s_{11,i} h_i}\left[2\bar{y}h_i - 2h_i \sum_{j=1}^{i} h_j + h_i^2\right] \\ C = \frac{1}{3}\sum_{i=1}^{n} \frac{w_i}{s_{11,i}}\left[3h_i\left(\bar{y} - \sum_{j=1}^{i} h_j\right)\left(\bar{y} - \sum_{j=1}^{i-1} h_j\right) + h_i^2\right] \end{cases} \tag{3.12}$$

where

- w_i is the width of the ith layer. In the sequel, it is assumed that $w_i = w$, $\forall i \in \{1, 2, \ldots, n\}$.
- $d_{31,i}$ is the transversal piezoelectric coefficient of the ith layer. If the layer is non-piezoelectric material, i.e. passive material, we have $d_{31,i} = 0$.
- $s_{11,i}$ is the elastic coefficient.

At the tip of the cantilever, this bending is

$$y(x = L) = y = \frac{m_{\text{piezo}} L^2}{2C} U \tag{3.13}$$

where L is the length of the cantilever.

Table 3.1 Numerical value of $k_m L$

m	1	2	3	4	5	...
$k_m L$	1.8751	4.6941	7.8548	10.9955	14.137	...

3.3.3 Dynamic Model

The previous model is static and relates the steady-state deflection $y(x)$ of the cantilever when a voltage U is applied. For performances and control point of view, a static model is insufficient, and it is important to also have an idea of the dynamics of the actuator. An essential parameter that can describe the dynamics of a system is the resonant frequency if the system is oscillating, which is the case for cantilevered actuators in general. A high resonant frequency indicates that it has a large bandwidth. For a multimorph piezoelectric cantilevered actuator, the mth resonant frequency $f_m[Hz]$ is defined as follows [16]:

$$f_m = \frac{(k_m L)^2}{2\pi L^2}\sqrt{\frac{C}{\mu}} \tag{3.14}$$

where C is defined by (3.12) and k_m is called wave number and can be calculated from the Table 3.1. The coefficient μ is given by:

$$\mu = \frac{M}{L} = \sum_{i=1}^{n} \rho_i h_i w_i \tag{3.15}$$

such that M is the mass of the cantilever, L being its length, and ρ_i is the density of the ith layer.

It is noticed, however, that there is no analytical solution to determine the damping ratio associated with each mode. In general, this is provided experimentally.

3.3.4 Equivalent Parametric Model

In order to design a piezoelectric actuator that will satisfy some specified performances, we will use the previous static model augmented with the resonant frequency information. The previous model is general: it provides the deflection at any point x along the cantilever and all resonant frequencies (until to the mth mode) are given. In our applications, we are interested in the deflection at the tip, i.e. for $x = L$, which is the most useful. Indeed, the manipulation of objects is often performed with the extremity of the cantilever. In addition, the first resonant frequency $m = 1$ is sufficient. We therefore provide a model limited to the first mode and where the range of deflection is calculated at the tip of the cantilevered actuator. Such a model is a second order model that is expressed by the following transfer function:

$$G(s)=\frac{Y(s)}{U(s)}=\frac{K}{\frac{1}{(w_n)^2}s^2+\frac{2\xi}{w_n}s+1} \tag{3.16}$$

where:

- s is the Laplace variable.
- $G(s)$ is the name of the model.
- $Y(s)$ and $U(s)$ are the Laplace transforms of $y(t)$ and $U(t)$, respectively.
- K is the statical gain.
- $w_n[\frac{\text{rad}}{s}]$ is the natural frequency.
- ξ is the damping ratio.

The statical gain K is derived from (3.13):

$$K=\frac{Y(s\to 0)}{U(s\to 0)}=\frac{y(t\to\infty)}{U(t\to\infty)}=\frac{m_{\text{piezo}}L^2}{2C} \tag{3.17}$$

The natural frequency is dependent on the damping ratio ξ and on the first resonant frequency $[\frac{\text{rad}}{s}]$ as described as follows:

$$w_n=\frac{w_r}{\sqrt{1-2\xi^2}} \tag{3.18}$$

with $w_r[\frac{\text{rad}}{s}]=2\pi f_1$ and such that f_1 is calculated from (3.14) by letting $m=1$. We have:

$$f_1=\frac{(1.8751)^2}{2\pi L^2}\sqrt{\frac{C}{\mu}} \tag{3.19}$$

Remind that the damping ratio ξ is not defined analytically. Hence, we cannot use this as (extra-)parameter for the design.

In the sequel, we are interested in designing unimorph piezocantilever, i.e. a cantilever made up of two layers: one piezoelectric layer and one passive layer. Hence, the coefficients C and m_{piezo} we have in (3.17) and (3.19) are calculated by setting $n=2$ in (3.12) and by using the coefficients of the materials used. We have:

$$\begin{cases} m_{\text{piezo}}=\frac{-wd_{31}h_{mp}(h_p+h_{mp})}{2(s_{11}^{mp}h_p+s_{11}^{p}h_{mp})} \\ C=\frac{w\left[(s_{11}^{p})^2h_{mp}^4+s_{11}^{mp}s_{11}^{p}(4h_ph_{mp}^3+6h_p^2h_{mp}^2+4h_{mp}h_p^3)+(s_{11}^{mp})^2+h_p^4\right]}{12s_{11}^{mp}s_{11}^{p}(s_{11}^{mp}h_p+s_{11}^{p}h_{mp})} \end{cases} \tag{3.20}$$

where:

- d_{31} is the transversal piezoelectric coefficient, h_p is the thickness, and s_{11}^{p} is the elastic coefficient of the piezoelectric layer.
- h_{mp} is the thickness and s_{11}^{mp} is the elastic coefficient of the passive material.

3.4 Design of a Unimorph Piezoelectric Actuator by Using the PIT

This section presents a methodology for designing piezoelectric cantilevered actuators by combining the analytical modeling presented in the previous section and interval techniques. We are particularly interested in the design of unimorph piezoelectric actuators. This choice is motivated by the ease of their fabrication and of electrical connection with respect to that of the bimorph and multimorph piezocantilevers. The design problem is formulated as a set inversion problem which is then solved using interval techniques. The proposed design technique is novel and very interesting in the sense that the design yields guaranteed performances, if solution exists.

3.4.1 Specifications

We aim to design unimorph piezocantilevers for which some desired performance specifications given either in the time domain or in the frequency domain should be met. These specifications can be transcribed into a model, called reference model, i.e. is a transfer function or a state-space representation. Considering bounded performance measures, the model parameters are also bounded. Taking therefore a reference model with the same structure than the system model in (3.16) (second order model), we consider the following reference (or desired) model where the parameters are intervals:

$$[G_d](s,[p]) = \frac{[a_0]}{[a_2]s^2 + [a_1]s + 1} \tag{3.21}$$

such that $[p] = ([a_0],[a_1],[a_2])^T$ is a vector of interval parameters that can be derived from the specified performances. The unit step response of (3.21) defines upper and lower envelopes of the desired step response. Also, the interval magnitude and phase of (3.21) over a sufficient set of frequencies defines upper and lower bounds of the desired frequency response [12].

3.4.2 Problem Formulation

Reconsider a unimorph piezocantilever having a length L, width w, and thicknesses h_p and h_{mp} for its piezoelectric and passive layers, respectively, as reminded in Fig. 3.3.

As shown in previous section, the static and dynamic models of the unimorph piezocantilever depend on the geometrical sizes and the physical properties of

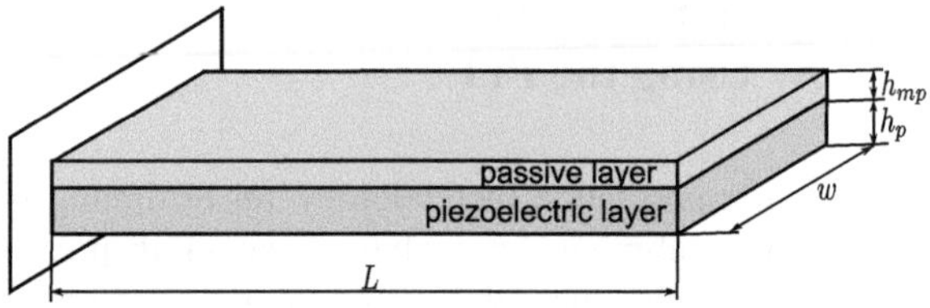

Fig. 3.3 Unimorph piezocantilever

the materials that compose it. The model that relates this fact is given by (3.16). If we admit that, for given piezoelectric and passive materials, there is a set of dimensions (i.e., of the length, of the thicknesses, and of the width) that would provide performances which lie within the specified performances transcribed by (3.21), these geometrical sizes can also be described by intervals. These interval geometrical parameters yield therefore interval model parameters and the initial model in (3.16) becomes an interval model as follows:

$$[G](s,[q]) = \frac{[K]}{\frac{1}{[w_n]^2}s^2 + \frac{2[\xi]}{[w_n]}s + 1} \tag{3.22}$$

where $[q] = ([K],[w_n],[\xi])^T$ is a vector containing the interval static gain $[K]$, the interval natural frequency $[w_n]$ and the interval damping ratio $[\xi]$. $[K]$ and $[w_n]$ can be easily derived using the geometrical sizes and physical properties of the unimorph piezoelectric actuator as presented in Sect. 3.3.

According to (3.17) and (3.19), the resonant frequency of a unimorph piezocantilever is conversely proportional to the square of its length whereas the deflection is directly proportional to the square of its length. Therefore, decreasing the length results in increasing the resonant frequency and therefore enlarging the bandwidth. However, the static gain will be reduced and the range of deflection will be limited. This decrease in the range can be compensated by some setting on the thicknesses of the piezoelectric and passive layers, while holding a large bandwidth. For that, our endeavor consists in finding the best geometrical sizes of the unimorph piezocantilever such that its performance still satisfies the specified performances that are described by the interval reference model (3.21). Such a problem can be formulated as follows: for a pre-selected value of the damping ratio $[\xi] = \xi$, find suitable values of geometrical sizes with which the unimorph piezocantilever model is included in the interval reference model as follows:

$$[G](s,[q]) \subseteq [G_d](s,[p]) \tag{3.23}$$

The statement in (3.23) is from the PIT presented in Theorems 2.1and 2.2. This statement tells us to find a set of geometrical sizes of unimorph cantilevers such that the performances of the corresponding model, denoted $[G]$, are enclosed in that of the reference model $[G_d]$.

In general, the applications (micromanipulation, microassembly, etc.) require that the width $[w]$ is imposed. Indeed, the maximal sizes of the manipulable objects with the actuator depend on this width. Therefore, in the sequel we fix $[w] = w$. Then, the remaining parameters to be sought for are the length $[L]$ and the thicknesses $[h_p]$ and $[h_{mp}]$. Let $[\theta] = ([L], [h_p], [h_{mp}])^T$ be an interval vector containing these design parameters. Now, our design problem consists in finding the possible values for the parameter $[\theta]$ for which the inclusion (3.23) is satisfied. To check the fulfillment of the inclusion $[G](s,[q]) \subseteq [G_d](s,[p])$ one can perform parameter by parameter inclusion as given in Lemma 2.1. Let Θ be the set corresponding to admissible values of the parameter $[\theta]$ for which (3.23) holds. Thus, the design problem to be addressed can be viewed as finding the set Θ so that:

$$\Theta := \left\{ \theta \in \mathscr{D} \;\middle|\; \begin{array}{l} [K]([\theta]) \subseteq [a_0] \\ \dfrac{1}{([w_n]([\theta]))^2} \subseteq [a_2] \\ \dfrac{2\xi}{[w_n]([\theta])} \subseteq [a_1] \end{array} \right\} \tag{3.24}$$

where $\mathscr{D}$ is the definition domain of $[\theta]$. The problem (3.24) is known as a set-inversion problem and a possible way to solve a set-inversion problem is the SIVIA algorithm. In the next subsection, we introduce this algorithm to solve our design problem.

3.4.3 Solution Computation via the SIVIA Algorithm

The SIVIA algorithm [14, 17] is an interval technique that can be used to solve a set-inversion problem such as that of problem (3.24). The set inversion operation consists in computing the reciprocal image of a compact set called subpaving. The set-inversion algorithm SIVIA allows to solve the design problem given in (3.24) and provides an approximation with subpavings of the set solution Θ. This approximation is realized with an inner and outer subpavings, respectively, $\underline{\Theta}$ and $\overline{\Theta}$, such that $\underline{\Theta} \subseteq \Theta \subseteq \overline{\Theta}$. The subpaving Θ corresponds to geometrical sizes of the unimorph piezocantilever for which the problem (3.24) holds. If $\Theta = \emptyset$, then it is guaranteed that no solution exists for the design problem (3.24).

We provide in Table 3.2 the recursive SIVIA algorithm that allows to solve the design problem (3.24) with guaranteed solution. SIVIA algorithm requires a search box $[\theta_0]$ (possibly very large), also called initial box within which $\overline{\Theta}$ is guaranteed to belong. The inner and outer subpavings ($\underline{\Theta}$ and $\overline{\Theta}$) are initially empty. ε represents the wanted accuracy of computation.

Quite often we are interested to compute an inner approximation $\underline{\Theta}$ for which we are sure that $\underline{\Theta}$ is included in the set solution Θ, i.e., $\underline{\Theta} \subseteq \Theta$, but when no inner approximation exists i.e., $\underline{\Theta} = \emptyset$, it is possible to choose parameters inside the outer subpaving, i.e., choose $\theta \in \overline{\Theta}$.

Table 3.2 SIVIA algorithm for solving a set-inversion problem [14, 17]

Step	SIVIA(in: $[K]$, $[w_n]$, $[a_0]$, $[a_1]$ $[a_2]$, $[\theta]$, ε; inout: $\underline{\Theta}$, $\overline{\Theta}$)
1	if $\begin{pmatrix} [K]([\theta]) \\ \frac{1}{([w_n]([\theta]))^2} \\ \frac{2\xi}{[w_n]([\theta])} \end{pmatrix} \cap \begin{pmatrix} [a_0] \\ [a_2] \\ [a_1] \end{pmatrix} = \emptyset$ return;
2	if $\left\{ \begin{array}{c} [K]([\theta]) \subseteq [a_0] \\ \wedge \\ \frac{1}{([w_n]([\theta]))^2} \subseteq [a_2] \\ \wedge \\ \frac{2\xi}{[w_n]([\theta])} \subseteq [a_1] \end{array} \right\}$ then $\underline{\Theta} := \underline{\Theta} \bigcup [\theta]$ and $\overline{\Theta} := \overline{\Theta} \bigcup [\theta]$ return;
4	if $width([\theta]) < \varepsilon$ then $\{\overline{\Theta} := \overline{\Theta} \bigcup [\theta]\}$ return;
5	bisect $[\theta]$ into $L([\theta])$ and $R([\theta])$;
6	SIVIA($[K]$, $[w_n]$, $[a_0]$, $[a_1]$ $[a_2]$, $L([\theta])$, ε; $\underline{\Theta}$, $\overline{\Theta}$);
	SIVIA($[K]$, $[w_n]$, $[a_0]$, $[a_1]$ $[a_2]$, $R([\theta])$, ε; $\underline{\Theta}$, $\overline{\Theta}$).

Remark 1. The number of unknown parameters in (3.24) is 3 and the number of inclusions is also 3. The set solution Θ can be obtained by intersecting the set solution of each inclusion in (3.24), i.e.:

$$\Theta = \bigcap_{i=1}^{3} (set\_sol)_i$$

such as: $(set\_sol)_i$ is the set solution of the ith inclusion.

Remark 2. If the set-inversion problem is not feasible, i.e. $\Theta = \emptyset$, the initial box of parameters must be changed and/or the desired performance specifications (interval reference model) must be relaxed.

3.4.4 Experimental Validation

This part is devoted to a numerical application and an experimental validation of the proposed design technique presented previously.

3.4.4.1 Materials

The layers of the unimorph to be designed are based on materials commercially available: a PZT-PIC151 (lead zirconate titanate) piezoelectric material from *Physike Instrumente* (PI) company for the piezoelectric layer and Nickel material

Table 3.3 Physical characteristics of PZT-PIC151 and of Nickel materials

Materials	Compliance s_{11}	Piezoelectric constant d_{31}	Density ρ
PZT	$15 \times 10^{-12}\,\mathrm{m^2/N}$	$-210 \times 10^{-12}\,\mathrm{m/V}$	$7{,}800\,\mathrm{kg/m^3}$
Nickel	$5 \times 10^{-12}\,\mathrm{m^2/N}$	0	$8{,}900\,\mathrm{kg/m^3}$

from *Goodfellow* company for the passive layer. The thermal glue "EPO-TEK H22" from *PI* is used to glue the piezoelectric and passive layers (PZT-Nickel). Table 3.3 summarizes some useful physical properties of the PZT-PIC151 and Nickel materials. These numerical values will be used during the design process of the unimorph piezocantilever.

3.4.4.2 Interval Reference Model

The reference model (3.21) is a transcription of some specifications. These specifications can be given in the time domain such as the settling time and the maximal overshoot, or in the frequency domain such as the resonant frequency. The relations that link these performances measures in the time domain or in the frequency domain with the coefficients $[a_0]$, $[a_1]$, and $[a_2]$ of a second order model will not be presented here. We directly give the following numerical values instead which, from our experience, corresponds to the performances required in our applications:

$$[G_d](s) = \frac{[0.5, 1]}{[0.156, 2.5] \times 10^{-7} s^2 + [0.25, 1] \times 10^{-5} s + 1} \tag{3.25}$$

3.4.4.3 Unimorph Sizes Computation

Our objective is now to design a unimorph piezocantilever having some performances that are enclosed in the performances of the interval reference model (3.25), i.e. finding the actuator's sizes such that the model $[G]$ in (3.22) is enclosed in the reference model (3.25). The problem will be treated with a pre-fixed value of the damping ratio $[\xi] = \xi = 0.01$. In addition, in order to reduce the number of the design parameters, we set the unimorph piezocantilever length to $L = 16\,\mathrm{mm}$ and its width to $w = 2\,\mathrm{mm}$. At the end, our design problem boils down to find thicknesses $[hp]$ and $[h_{mp}]$ of the piezoelectric and of the passive layers, respectively, so that inclusions (3.24) are satisfied. Using the numerical values defined so far and applying the SIVIA algorithm presented in Table 3.2 with an initial box $[h_p]_0 \times [h_{mp}]_0 = [10, 500] \times [10, 500]$ and an accuracy $\varepsilon = 1\,\mu\mathrm{m}$, we obtain the subpaving shown in Fig. 3.4. In this figure, the area denoted $\underline{\Theta}$ corresponds to the guaranteed solution (inner subpaving), i.e. the set $[h_p] \times [h_{mp}]$ that satisfies the

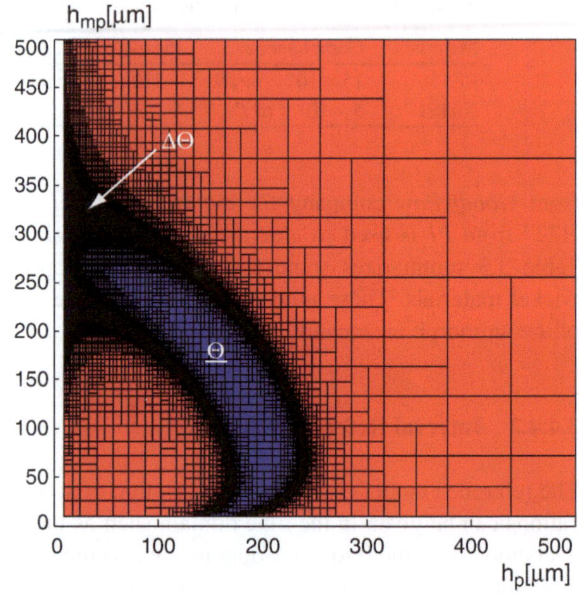

Fig. 3.4 Set solution $\underline{\Theta}$ corresponding to the parameters h_p and h_{mp}

inclusion $[G](s) \subseteq [G_d](s)$. Any choice of the parameters $[h_p]$ and $[h_{mp}]$ within the subpaving $\underline{\Theta}$ ensures the inclusions problem given in (3.24). The area $\Delta\Theta$ contains the boxes for which no decision on the test of inclusion in (3.24) can be taken. Notice that:

$$\overline{\Theta} = \underline{\Theta} \cup \Delta\Theta$$

$\Delta\Theta$ can be minimized by increasing the computation accuracy. The remaining external boxes correspond to the parameters $[h_p]$ and $[h_{mp}]$ for which the inclusions (3.24) do not hold.

3.4.4.4 Fabrication of the Unimorph Piezocantilever and Experimental Verifications

In order to develop and fabricate some prototypes of unimorph piezocantilevers, we choose the following geometrical sizes:

$$\begin{cases} L = 16\,\text{mm} \\ h_p = 200\,\mu\text{m} \\ h_{mp} = 100\,\mu\text{m} \\ w = 2\,\text{mm} \end{cases} \tag{3.26}$$

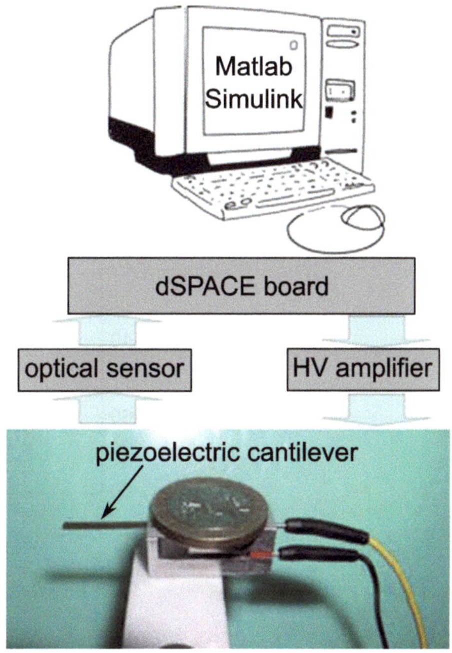

Fig. 3.5 Prototype of a unimorph piezoelectric actuator and experimental setup

where h_p and h_{mp} have been chosen arbitrarily from the solution region $\underline{\Theta}$. Figure 3.5 presents a photography of the fabricated unimorph piezocantilever and the experimental setup. The whole experimental setup is composed of:

- The fabricated unimorph piezocantilever.
- An optical sensor (from *Keyence* company) with a resolution of 10 nm and used to measure the deflection of the piezocantilever.
- A high-voltage (HV) amplifier.
- A dSPACE acquisition board and a computer to generate the input voltage and to acquire the measurements. The sampling time of the whole acquisition system is set to 0.2 ms. The Matlab-Simulink software is used to manage the input and output signals.

In this application,the experimental verifications are done in the frequency domain. More precisely, we plot in the same graph the experimental frequency response of the designed piezocantilever and the frequency response of the reference model $[G_d]$. If the magnitude of the experimental response is enclosed in that of $[G_d]$, our objective is reached. For the experiment, we apply a sine input voltage with an amplitude of $U = 20$ V and a frequency ranging between 1 Hz ($6.28[\frac{\text{rad}}{s}]$) and more than 1,500 Hz ($9,500[\frac{\text{rad}}{s}]$) to the designed piezocantilever. The resulting experimental magnitude is shown in Fig. 3.6. In the same figure, the envelope magnitude that corresponds to the magnitude of the interval reference model is also plotted. According to Fig. 3.6, the experimental magnitude obtained with the

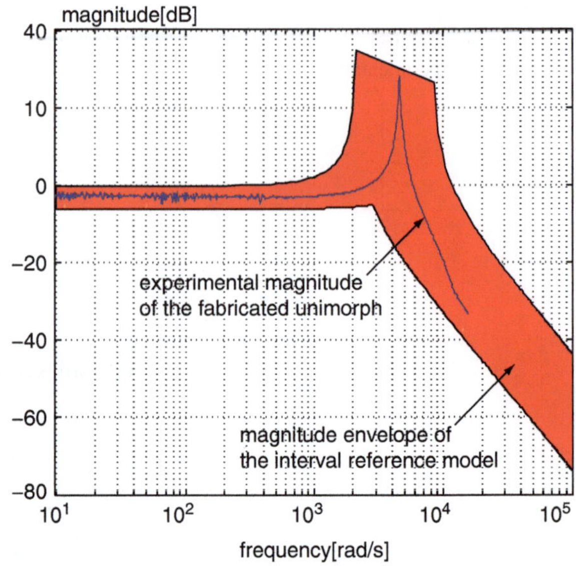

Fig. 3.6 Experimental magnitude of the unimorph and the desired magnitude

fabricated piezoelectric unimorph is enclosed in the desired magnitude of the interval reference model. Consequently, the method used to design a piezoelectric actuator by using the PIT efficiently provided the expected results and confirmed the theoretical results. Indeed, the performances obtained with the designed actuator lie within the performances imposed *a priori* as specifications.

3.5 Conclusion

In this chapter, the design of piezoelectric actuators based on the performance inclusion theorem has been presented. It has been shown that static and dynamic models of these piezoelectric actuators strongly depend on their geometrical sizes and physical properties. Then, our challenge was to design a (unimorph) piezoelectric actuator that would satisfy some imposed performances. Based on the inclusion performances theorem, the design problem has been formulated as a set-inversion problem. This later has been solved using interval techniques to find the geometrical sizes of the piezoelectric actuator. The main advantage of the proposed approach is that guaranteed solution or non-solution of the design problem can be obtained. The designed actuator was afterwards fabricated and characterized. The experimental results on the fabricated actuator validated the proposed method.

Acknowledgements This work is supported by the national ANR-JCJC C-MUMS-project (National young investigator project ANR-12-JS03007.01: Control of Multivariable Piezoelectric Microsystems with Minimization of Sensors).

References

1. A. Bergander, W. Driesen, T. Varidel, M. Meizoso, J.M. Breguet, Mobile *cm*3-microrobots with tools for nanoscale imaging and micromanipulation, in *Mechatronics & Robotics* (MechRob 2004), Aachen, Germany, 13–15 Sept 2004, pp. 1041–1047
2. S. Fatikow, T. Wich, H. Hulsen, T. Sievers, M. Jahnisch, Microrobot system for automatic nanohandling inside a scanning electron microscope. IEEE/ASME Trans. Mechatron. **12**(3), 244–252 (2007)
3. M. Rakotondrabe, Y. Haddab, P. Lutz, Development, modeling, and control of a micro/ nanopositioning 2-DOF stick-slip device. IEEE/ASME Trans. Mechatron. **14**(6), 733–745 (2009)
4. Physikinstrumente, Selection guide: linear flexure stages and actuators/nanopositioning systems, http://www.physikinstrumente.com/en/products/nanopositioning/nanopositioning_linear-stage_selection.php (2012)
5. K. Uchino, Piezoelectric ultrasonic motors: overview. Smart Mater. Struct. **7**(3), 273–285 (1998)
6. R. Lee, H.L. Li, Development and characterization of a rotary motor driven by anisotropic piezoelectric composite laminate. Smart Mater. Struct. **7**(3), 327–336 (1998)
7. S.H. Jun, S.M. Lee, S.H. Lee, H.E. Kim, K.W. Lee, Piezoelectric linear motor with unimorph structure by co-extrusion process. Sens. Act. A. **147**(1), 300–303 (2008)
8. G. Binnig, C.F. Quate, Ch. Gerber. Atomic force microscope. Phys. Rev. Lett. **56**(9), 930 (1986)
9. R. Perez, J. Agnus, C. Clevy, A. Hubert, N. Chaillet, Modeling, fabrication and validation of a high performance 2-DOF piezoactuator for micromanipulation. IEEE/ASME Trans. Mechatron. **10**(2), 161–171 (2005)
10. M. Rakotondrabe, A. Ivan, Development and dynamic modeling of a new hybrid thermo-piezoelectric micro-actuator. IEEE Trans. Robot. **26**(6), 1077–1085 (2010)
11. M. Grossard, C. Rotinat-Libersa, N. Chaillet, M. Boukallel, Mechanical and control-oriented design of a monolithic piezoelectric microgripper using a new topological optimization method. IEEE/ASME Trans. Mechatron. **1**(14), 32–45 (2009)
12. M. Rakotondrabe, Performances inclusion for stable interval systems, in *ACC (American Control Conference)*, San Francisco CA, USA, June–July 2011, pp. 4367–4372
13. R.E. Moore, *Interval Analysis* (Prentice-Hall, Englewood Cliffs, 1966)
14. L. Jaulin, M. Kieffer, O. Didrit, E. Walter, *Applied Interval Analysis* (Springer, London, 2001)
15. M. Rakotondrabe, *Piezoelectric Cantilevered Structures: Modeling, Control, and Measurement/Estimation Aspects* (Springer, Berlin 2013)
16. R.G. Ballas, *Piezoelectric Multilayer Beam Bending Actuators: Static and Dynamic Behavior and Aspects of Sensor Integration* (Springer, Berlin, 2007)
17. L. Jaulin, E. Walter, Set inversion via interval analysis for nonlinear bounded-error estimation. Automatica **29**(4), 1053–1064 (1993)

Chapter 4
Modeling and Robust H_∞ Control of a Nonlinear and Oscillating 2-dof Multimorph Cantilevered Piezoelectric Actuator

Micky Rakotondrabe

Abstract This chapter presents the characterization, modeling, and robust control of a nonlinear and oscillating 2-degrees of freedom (2-dof) piezoelectric cantilevered actuator. The actuator possesses a high resolution and a high bandwidth of the actuator, however, it is typified by a hysteresis and creep nonlinearities, a badly damped vibration and a strong coupling between the two axes. Based on the quadrilateral approach, a simple model which can account efficiently all these properties is proposed. Indeed, the model is linear followed by well-defined uncertainties and perturbations. In order to ensure certain performances, a robust standard H_∞ control technique is used to synthesize controllers for the 2-dof actuator. The experimental results confirm the efficiency of the proposed approach of modeling and control design.

4.1 Introduction

In several applications at the micro and nano scale, piezoelectric materials are well recognized and frequently used as core of many systems. Such recognition is due to their high bandwidth, their high resolution and their high force density. In addition to that, their control is easy (electrical supply) and they can also be used as sensors. One of the applications at the micro and nano scale that involves piezoelectric systems is microassembly and micromanipulation application [1]. In this, piezoelectric microgrippers are often used to pick-transport and to place a small object from a position to another one. Figure 4.1a presents a piezoelectric microgripper developed at the department of AS2M (FEMTO-ST Institute) [2]. It is composed of two piezoelectric cantilevers (actuators) that can be supplied and

M. Rakotondrabe (✉)
Automatic Control and Micro-Mechatronic Systems Department, AS2M,
FEMTO-ST Institute, 24 rue Alain Savary, Besançon 25000, France
e-mail: mrakoton@femto-st.fr

M. Rakotondrabe (ed.), *Smart Materials-Based Actuators at the Micro/Nano-Scale: Characterization, Control, and Applications*, DOI 10.1007/978-1-4614-6684-0_4,
© Springer Science+Business Media New York 2013

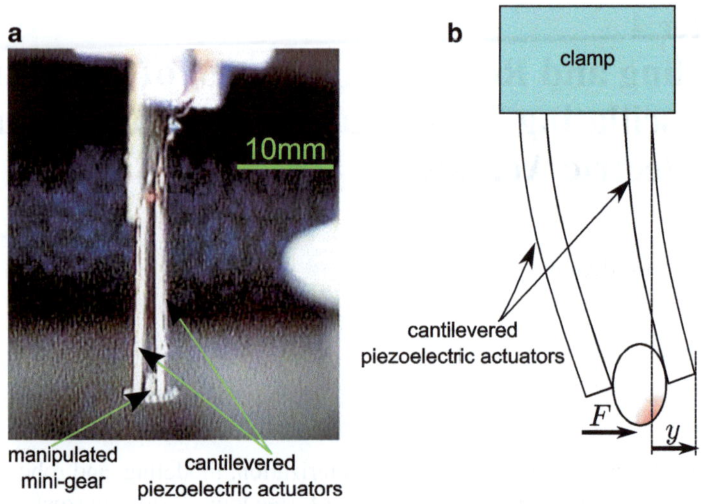

Fig. 4.1 (**a**): a photography of a piezoelectric microgripper. (**b**): principle of a piezoelectric microgripper manipulating a small object

controlled independently. Often, one actuator is controlled on force while the other is controlled on position (see Fig. 4.1b). This allows to manage the manipulation force (to avoid the destruction of the object or of the actuator, to estimate the mechanical characteristics of the object, etc.) by ensuring the precise positioning at the same time [2].

Although piezoelectric actuators can offer high resolution and high bandwidth, they are characterized by hysteresis and creep nonlinearities. Furthermore, piezoelectric actuators in microgrippers are typified by badly damped vibration which is due to their cantilevered structure. Consequently, they are subjected to a loss of their general performances: loss of the accuracy, increase in the settling time. The control of cantilevered piezoelectric actuators has therefore attracted the attention of researchers these last years. A survey on the different kinds of closed-loop control is devoted in [3]. While the modeling and closed-loop control of 1-degree of freedom (1-dof) cantilevered piezoelectric actuators are now well established, this is not yet the case for 2-dof actuators. In [4], the position control of a 2-dof bimorph piezoelectric actuator, i.e. composed of two piezoelectric layers, was presented. In this, the H_∞ robust control technique was efficiently used. In [5], an internal model control (IMC) technique was combined with a feedforward control of the hysteresis to control a 2-dof multimorph which is composed of several piezoelectric layers.

In this chapter, we present the modeling and position control of the 2-dof multimorph piezoelectric cantilever. For that, we propose to first decouple the bivariable system (two inputs two outputs) into two monovariable systems. Then, the hysteresis is approximated into a linear model with uncertain parameters. The main advantage relative to the work in [5] is that there is no need to model and compensate the hysteresis by feedforward, making the approach here more simple and direct. The standard H_∞ technique is afterwards used to synthesize a robust controller for each axis of the 2-dof system. The experimental results along the chapter demonstrate the efficiency of the proposed approach of modeling and control.

The chapter is organized as follows. Section 4.2 is devoted to the presentation of the 2-dof multimorph actuator while Sect. 4.3 to its characterization. In Sect. 4.4, we present the modeling and the identification procedure. Finally, Sect. 4.5 is devoted to the synthesis of the controllers.

4.2 Presentation of the 2-dof Cantilevered Piezoelectric Actuator

The 2-dof piezoelectric actuator is presented in Fig. 4.2a. With a rectangular section, it is made up of 36 piezoelectric layers and 2 passive layers (non-piezoelectric materials) glued themselves. An arrangement of the electrodes makes possible the obtention of bending of the actuator along y-axis and along z-axis. This arrangement of the electrodes is similar to the 2-dof bimorph piezoelectric actuators detailed in [6]. The actuator used in this chapter has a total active dimensions of: length $\times$ width $\times$ height $= 22\,\mathrm{mm} \times 1\,\mathrm{mm} \times 0.91\,\mathrm{mm}$ where each piezoelectric layer has a height of $20\,\mu\mathrm{m}$. The actuator can be seen as a bivariable system, i.e. having two inputs and two outputs, as in Fig. 4.2b. As indicated in the figure, we denote y and z the output bendings while U_y and U_z the input voltages.

In the sequel, the experimental setup to be used which is depicted in Fig. 4.3 is composed of the following elements:

- The 2-dof multimorph piezoelectric actuator.
- Two optical sensors (LK2420 from *Keyence*) which are used to measure the deflection (position) of y and of z axes. Each sensor is set to have a resolution of 50 nm.
- A computer and a dSPACE-board used to acquire the measured positions and to generate the control signals. The MATLAB-SIMULINK© software is employed to manage the different signals. The sampling time of the whole acquisition system is set to 0.2 ms which is well convenient for the system to be controlled. As the control signals U_y and U_z do not exceed 10 V, we do not use a voltage amplifier in this experimental setup.

4.3 Characterization

In this section, we characterize the 2-dof multimorph piezoelectric actuator. For that, the static property, the dynamics, and the creep are analyzed.

Fig. 4.2 (**a**): the 2-dof piezoelectric actuator. (**b**): the corresponding system block-scheme

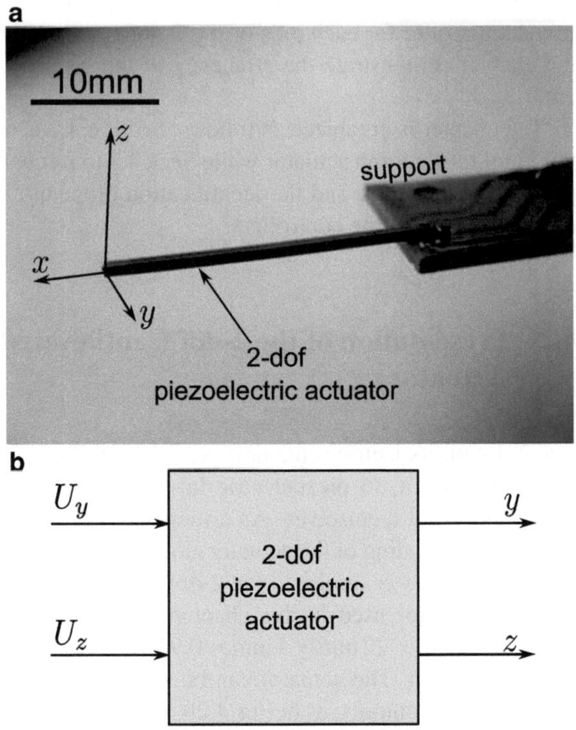

4.3.1 Static Characterization: Observation of the Hysteresis

The analysis of the static property of the actuator consists in observing the phase-plane (input–output plane). First, a sine input voltage U_y is applied while the voltage U_z is left to zero. The amplitude of U_y corresponds to the maximal range of use which is in this case equal to 10 V. The frequency should be chosen to be small in order to avoid the phase-lag effect, i.e. the effect of the dynamics to the static characteristic. However, it should not be too small enough in order to avoid the effect of the creep which occurs at very low frequencies [7,8]. Different *a priori* tests show that a frequency of 0.1 Hz is convenient for this actuator. The output deflection y resulting from the sine voltage U_y is reported and the *direct* characteristic (U_y, y) is plotted (see Fig. 4.4a). In the meantime, the effect of U_y to the z-axis, called coupling effect, is also plotted (see Fig. 4.4c). As we can see from these figures, the bending range of y is very interesting ($\pm 50\,\mu$m) for such low voltage (± 10 V). However, the bending versus voltage is typified by a hysteresis nonlinearity. We can also remark the presence of coupling (U_y, z) of up to $\pm 8\,\mu$m. Now, we set U_y to zero and apply a sine input voltage U_z with an amplitude of 10 V and a frequency of 0.1 Hz. The *direct* characteristic (U_z, z) is plotted in Fig. 4.4d while the coupling effect to the y-axis is plotted in Fig. 4.4b. In these, we deduce that the range for the z-axis is

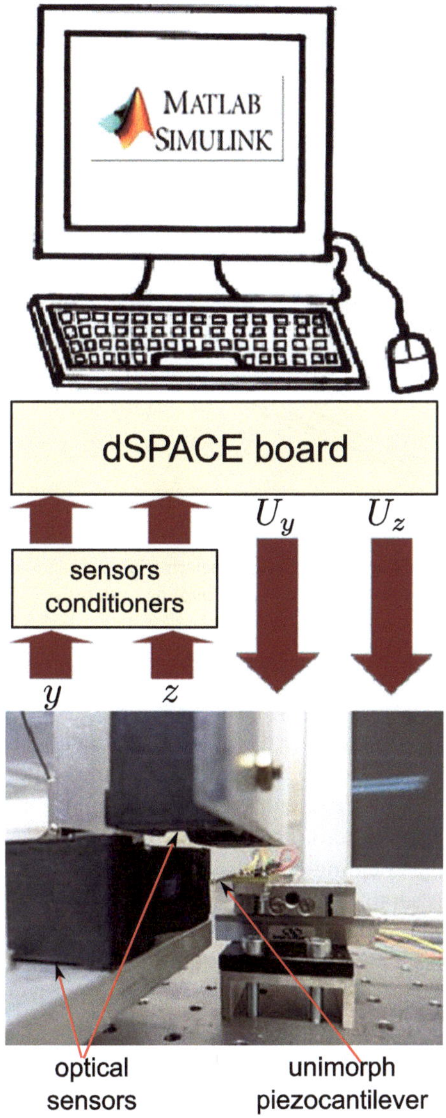

Fig. 4.3 The experimental setup

up to $\pm 35\,\mu$m. We also deduce the presence of hysteresis in this axis. Finally, the coupling effect of U_z to the y-axis, i.e. (U_z, y), is in excess of $\pm 16\,\mu$m. Notice that Fig. 4.4a–d also includes the internal loops of hysteresis which are obtained with sine input voltages of 5 V of amplitude.

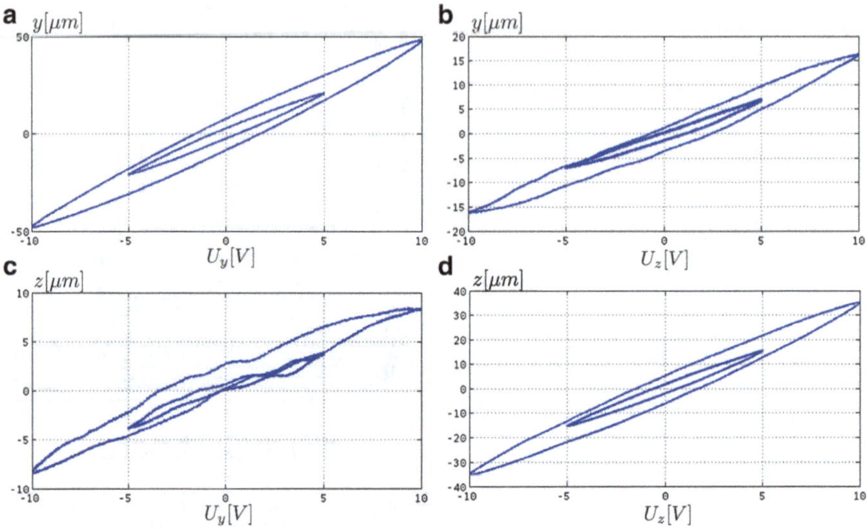

Fig. 4.4 Hysteresis characterization. (**a**) and (**d**): direct hysteresis for the *y* and for the *z*-axis respectively. (**b**) and (**c**): coupling hysteresis

4.3.2 Step Responses Characterization

To evaluate the dynamics of the actuator, we study its step responses. The aim is to analyze the transient parts of the response of the actuator: oscillation, overshoot, settling time, etc. First, a step input voltage $U_y = 10\,\text{V}$ is applied while U_z is left equal to zero. The direct step response, i.e. response $y(t)$, is plotted in Fig. 4.5a while the coupling response, i.e. response $z(t)$, is plotted in Fig. 4.5c. Then, the input U_y is set to zero and a step input voltage $U_z = 10\,\text{V}$ is applied. The response $z(t)$ (direct step response) is plotted in Fig. 4.5d while the response $y(t)$ which corresponds to the coupling is plotted in Fig. 4.5b. From these figures, we can deduce that the direct transfers ($U_y \to y$ and $U_z \to z$) are typified by strong oscillations. These oscillations drastically increase the settling times of the actuator even if this latter has a high bandwidth. Indeed, the frequency of oscillation is ≈625 Hz (which corresponds to a period of 1.6 ms) for the two axes; however, the settling time is more than 15 ms for the *y*-axis and more than 50 ms for the *z*-axis. Finally, we also remark the presence of oscillation on the couplings responses $U_y \to z$ and $U_z \to y$.

4.3.3 Creep Characterization

We now characterize the creep behavior of the 2-dof actuator. The creep is defined as the drift that appears after the end of the transient part when a step input voltage is applied [9]. This drift is observed for a long duration of period on the step response. It can also be observed at very low frequency if the input voltage is a periodic signal (sine or triangular).

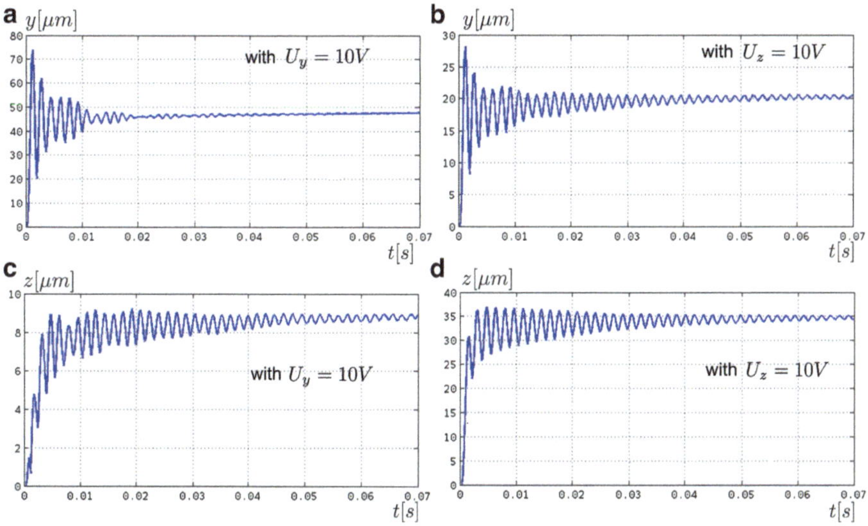

Fig. 4.5 Step responses characterization. (**a**) and (**d**): direct step responses for the y and for the z-axis respectively. (**b**) and (**c**): coupling step responses

Let us apply again a step input voltage $U_y = 10\,\text{V}$ to the actuator, set $U_z = 0$ and measure the output bendings $y(t)$ and $z(t)$ during 10 min. The creep of the direct transfer $U_y \rightarrow y$ is plotted in Fig. 4.6a while the creep of the coupling $U_y \rightarrow z$ is plotted in Fig. 4.6c. We remark that the "direct creep" is evaluated to be in excess of 15 μm in the studied duration and the "coupling creep" is negative. Now, we set U_y to zero and apply a step $U_z = 10\,\text{V}$. After reporting the output bendings y and z, they are plotted as in Fig. 4.6b and d, respectively. We see that the direct creep $U_z \rightarrow z$ is in excess of 12 μm and the coupling creep $U_z \rightarrow y$ is of about 3.5 μm. Figure 4.6a–d also present the creep evolution when the input voltages are 5 V. We deduce that the higher the voltage is, the higher the creep will be.

4.4 Modeling and Identification

In the previous section, we found that the 2-dof multimorph piezoelectric actuator was typified by hysteresis nonlinearity, a creep behavior, a strong oscillation, and couplings between the two axes. These characteristics directly affect the general behavior of the actuator. In particular, the hysteresis, the creep, and the coupling make it lose its accuracy while the oscillation may lead to unstability of the pick-transport-and place of objects during a precise positioning task. In this section, we provide a model of the 2-dof actuator in order to synthesize controllers in the next section for an enhancement of the actuator's performances. The behaviors of the actuator (hysteresis, creep, oscillation, and coupling) are accounted during

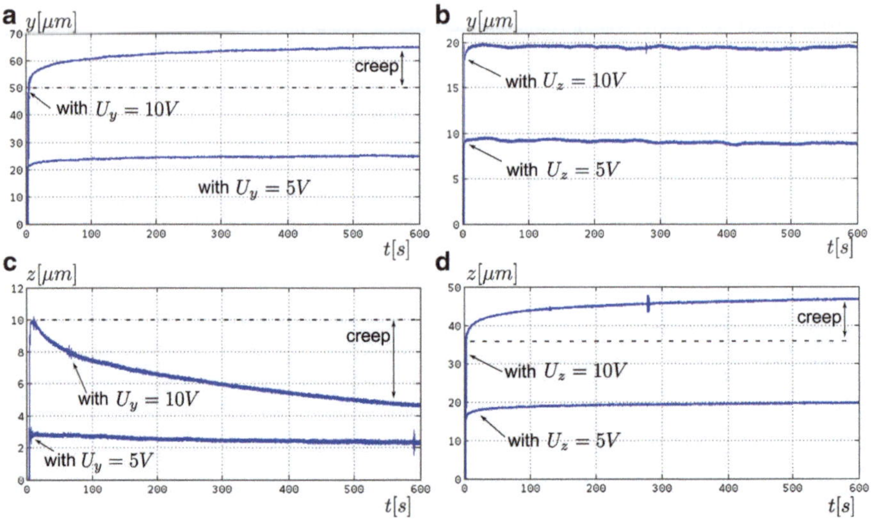

Fig. 4.6 Creep characterization. (**a**) and (**d**): direct creep for the y and for the z-axis respectively. (**b**) and (**c**): coupling creep

the modeling. In a spirit of derivation of simpler model however, we propose to approximate the hysteresis by the quadrilateral approximation method. This yields a linear approximation of the hysteresis followed by a well-modeled uncertainty. We also propose to model the coupling, the creep, and any eventual external disturbances such as manipulation force as a disturbance. At the end, we obtain two linear monovariable models (one for y-axis and one for z-axis) which are under uncertainties and under a disturbance. The proposed model has therefore the advantage to be simple (linear) which still takes into account the hysteresis, the creep, the coupling, and the oscillation.

4.4.1 General Formulation

Let Eq. (4.1) be a general nonlinear expression of the 2-dof actuator

$$\begin{pmatrix} y(s) \\ z(s) \end{pmatrix} = f\left(U_y(s), U_z(s), s\right) \Leftrightarrow \begin{cases} y(s) = f_y\left(U_y(s), U_z(s), s\right) \\ z(s) = f_z\left(U_y(s), U_z(s), s\right) \end{cases} \tag{4.1}$$

where s is the Laplace variable, $f_y\left(U_y(s), U_z(s), s\right)$, and $f_z\left(U_y(s), U_z(s), s\right)$ are nonlinear expressions dependent on U_z, U_y, and s for the output $y(s)$ and $z(s)$, respectively. The fact that the two functions can explicitly be dependent on s means that the models can be dynamic.

Assuming that each function $f_y(U_y(s), U_z(s), s)$, and $f_z(U_y(s), U_z(s), s)$ can be separated into two nonlinear functions exclusively or mainly dependent on $U_y(s)$ and on $U_z(s)$, Eq. (4.1) becomes:

$$\begin{cases} y(s) = f_{yy}(U_y(s), s) + f_{yz}(U_z(s), s) \\ z(s) = f_{zy}(U_y(s), s) + f_{zz}(U_z(s), s) \end{cases} \quad (4.2)$$

where $f_{yy}(U_y(s), s)$ and $f_{yz}(U_z(s), s)$ are the two nonlinear functions exclusively dependent on $U_y(s)$ and on $U_z(s)$, respectively, for the y-axis, and $f_{zy}(U_y(s), s)$ and $f_{zz}(U_z(s), s)$ are the two nonlinear functions exclusively dependent on $U_y(s)$ and on $U_z(s)$, respectively, for the z-axis. Such assumption has been validated for 2-dof bimorph actuator [4]. In other words, the above assumption means that the direct transfer and the coupling are additive.

When observing Fig. 4.5a (resp. Fig. 4.5d) which is the step response of the direct transfer in y-axis (resp. z-axis), we can say that this direct transfer is the sum of two signals: (1) the very quick transient part with a final value which is related to the hysteresis which we denote $f_{yy}^{\text{hyst}}(U_y(s), s)$ (resp. $f_{zz}^{\text{hyst}}(U_z(s), s)$), (2) and the creep signal which has a very low rate which we denote $f_{yy}^{\text{creep}}(U_y(s), s)$ (resp. $f_{zz}^{\text{creep}}(U_z(s), s)$). Hence, replacing the direct transfer $f_{yy}(U_y(s), s)$ (resp. $f_{zz}(U_z(s), s)$) in Eq. (4.2) by this fact, we obtain:

$$\begin{cases} y(s) = f_{yy}^{\text{hyst}}(U_y(s), s) + f_{yy}^{\text{creep}}(U_y(s), s) + f_{yz}(U_z(s), s) \\ z(s) = f_{zy}(U_y(s), s) + f_{zz}^{\text{hyst}}(U_z(s), s) + f_{zz}^{\text{creep}}(U_z(s), s) \end{cases} \quad (4.3)$$

Remind that $f_{yy}^{\text{hyst}}(U_y(s), s)$ and $f_{zz}^{\text{hyst}}(U_z(s), s)$ enclose the hysteresis and the dynamics of the actuator. They can be seen as dynamic hysteresis or rate-dependent hysteresis. However, it has been shown that the Hammerstein theorem can be applied to the dynamic hysteresis in piezoelectric actuators [7]. By this theorem, the dynamic hysteresis $f_{yy}^{\text{hyst}}(U_y(s), s)$ (resp. $f_{zz}^{\text{hyst}}(U_z(s), s)$) can be approximated by a static hysteresis $\Gamma_y(U_y(s))$ (resp. $\Gamma_z(U_z(s))$) followed by a linear dynamics $D_y(s)$ (resp. $D_z(s)$). Therefore, Eq. (4.3) becomes:

$$\begin{cases} y(s) = \Gamma_y(U_y(s)) D_y(s) + f_{yy}^{\text{creep}}(U_y(s), s) + f_{yz}(U_z(s), s) \\ z(s) = \Gamma_z(U_z(s)) D_z(s) + f_{zz}^{\text{creep}}(U_z(s), s) + f_{zy}(U_y(s), s) \end{cases} \quad (4.4)$$

Notice that the dynamics $D_y(s)$ and $D_z(s)$ are normalized, i.e. $D_y(0) = 1$ and $D_z(0) = 1$. Indeed, the statical gain is already described the nonlinearity Γ_y and Γ_z.

So far, we have not yet considered any external force applied to the actuator. An example of this external force is the manipulation force which is between the actuator and a manipulated object. An external force is a disturbance that can degrade the performances of the actuator: diminution of the precision, etc. Hence, it is essential to take it into account when calculating a controller. Considering the fact that mechanical effects are additive relative to electrical effects [9] even with a nonlinear behavior of the piezoelectric actuator, we add to Eq. (4.4) the mechanical terms and obtain:

$$\begin{cases} y(s) = \Gamma_y(U_y(s))D_y(s) + f_{yy}^{\text{creep}}(U_y(s),s) + f_{yz}(U_z(s),s) + s_{py}D_y(s)F_y(s) \\ z(s) = \Gamma_z(U_z(s))D_z(s) + f_{zz}^{\text{creep}}(U_z(s),s) + f_{zy}(U_y(s),s) + s_{pz}D_z(s)F_z(s) \end{cases} \tag{4.5}$$

where $F_y(s)$ (resp. $F_z(s)$) is the projection of the external force along y-axis (resp. z-axis) and s_{py} (resp. s_{pz}) is the elastic coefficient of the actuator along the same axis. As we can observe in Eq. (4.5), the mechanical term possesses a dynamics term $D_y(s)$ (resp. $D_z(s)$) similar to that of the electromechanical term $\Gamma_y(U_y(s))D_y(s)$ (resp. $\Gamma_y(U_z(s))D_z(s)$). In fact, as the dynamics is linked to the mechanical structure itself of the actuator, it is independent from the source of excitation. This fact has been verified experimentally, for instance, in [10].

For the sake of simplicity of reading, we remove all Laplace variable s of the signals in Eq. (4.5). Then:

$$\begin{cases} y = \Gamma_y(U_y)D_y(s) + f_{yy}^{\text{creep}}(U_y,s) + f_{yz}(U_z,s) + s_{py}D_y(s)F_y \\ z = \Gamma_z(U_z)D_z(s) + f_{zz}^{\text{creep}}(U_z,s) + f_{zy}(U_y,s) + s_{pz}D_z(s)F_z \end{cases} \tag{4.6}$$

Remind that $f_{yz}(U_z,s)$ (resp. $f_{zy}(U_y,s)$) is the coupling effect $U_z \to y$ (resp. $U_y \to z$). It is composed of the following three elements: 1) the coupling hysteresis that expresses the nonlinear static gain and that is characterized in Fig. 4.4b (resp. Fig. 4.4c), 2) the coupling transient part that is characterized in Fig. 4.5b (resp. Fig. 4.5c), 3) and the creep coupling that is characterized in Fig. 4.6b (resp. Fig. 4.6c). Similar to the direct transfer, the coupling transfer $f_{yz}(U_z,s)$ (resp. $f_{zy}(U_y,s)$) can be seen as the superposition of two signals: (1) the coupling dynamic hysteresis $f_{yz}^{\text{hyst}}(U_z,s)$ (resp. $f_{zy}^{\text{hyst}}(U_y,s)$) and the coupling creep $f_{yz}^{\text{creep}}(U_z,s)$ (resp. $f_{zy}^{\text{creep}}(U_y,s)$). Therefore, the model Eq. (4.6) can be rewritten as follows:

$$\begin{cases} y = \Gamma_y(U_y)D_y(s) + f_{yy}^{\text{creep}}(U_y,s) + f_{yz}^{\text{hyst}}(U_z,s) + f_{yz}^{\text{creep}}(U_z,s) + s_{py}D_y(s)F_y \\ z = \Gamma_z(U_z)D_z(s) + f_{zz}^{\text{creep}}(U_z,s) + f_{zy}^{\text{hyst}}(U_y,s) + f_{zy}^{\text{creep}}(U_y,s) + s_{pz}D_z(s)F_z \end{cases} \tag{4.7}$$

In the next two sub-sections, we give an approximate model of the static hysteresis term $\Gamma_y(U_y)$ and $\Gamma_z(U_z)$ and give the dynamics models $D_y(s)$ and $D_z(s)$. After having modeled these static and dynamics part, we will provide the final model of the 2-dof actuator.

4.4.2 Modeling and Identification of the Statical Part

First the statical part is modeled. This statical part is nonlinear and corresponds to the hysteresis $\Gamma_y(U_y)$ and $\Gamma_z(U_z)$, also called static hysteresis or rate-independent hysteresis. There are several approaches to model or to approximate a static

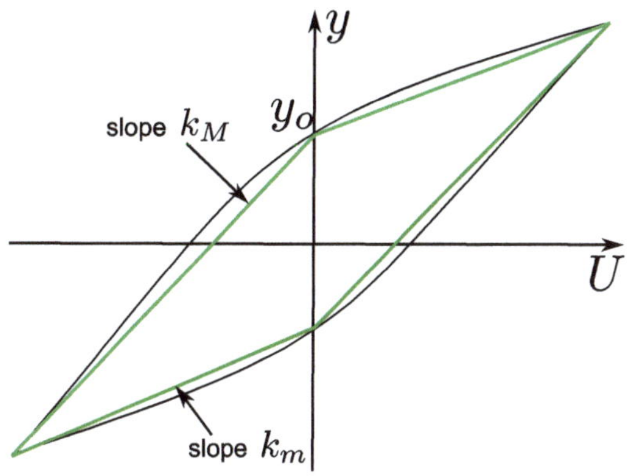

Fig. 4.7 The quadrilateral approximation of a hysteresis curve

hysteresis: the classical Preisach approach [11–13], the classical and the modified Prandtl–Ishlinskii approach [8, 14–16], the Bouc–Wen model [17], the quadrilateral approximation approach [7], etc. When using an open-loop (feedforward control scheme), a precise model such as that of the Preisach, the Prandtl–Ishlinskii, and the Bouc–Wen approaches is necessary. However, as we treat here a closed-loop control scheme, the objective is to employ a model as simple as possible which still accounts the hysteresis. The quadrilateral approximation is convenient for that.

4.4.2.1 The Quadrilateral Approximation

To explain the quadrilateral approach, consider a hysteretic system with input U and output y. The approach [7] consists to say that a static hysteresis is approximated by a quadrilateral as pictured in Fig. 4.7. In fact, this is a particular and the simplest approximation of the multilinear modeling of hysteresis [18].

From the quadrilateral, the maximal slope k_M and the minimal slope k_m among the four segments are identified. Then, an uncertain linear model is derived as follows:

$$y = \Gamma(U) \cong (k+\delta)U + y_o \tag{4.8}$$

where y_o is the offset (positive or negative) that varies between the minimal and the maximal offsets of the quadrilateral. In the case of closed-loop controller synthesis, y_o is directly taken equal to the maximal or the minimal, the one which has the highest norm. For instance, in Fig. 4.7, y_o corresponds to the positive offset. The statical gain k and the uncertainty δ are defined as follows:

$$\begin{cases} k = \frac{k_M+k_m}{2} \\ -\frac{(k_M-k_m)}{2} \leq \delta \leq \frac{(k_M-k_m)}{2} \end{cases} \tag{4.9}$$

Remind that Eq. (4.8) can be rewritten as follows:

$$y = \Gamma(U) \cong \left(1+\frac{\delta}{k}\right) kU + y_o \tag{4.10}$$

The latter equation is a representation of the uncertainty called "direct multiplicative uncertainty."

4.4.2.2 The Quadrilateral Approximation Applied to the 2-dof Piezoelectric Actuator

Now, we apply the quadrilateral approach in Eq. (4.10) to the hysteresis $\Gamma_y(U_y)$ in y-axis and the hysteresis $\Gamma_z(U_z)$ in z-axis defined in Eq. (4.6). We obtain:

$$\begin{cases} y=\left(1+\frac{\delta_y}{k_y}\right)k_yD_y(s)U_y+y_o+f_{yy}^{\text{creep}}(U_y,s)+f_{yz}^{\text{hyst}}(U_z,s)+f_{yz}^{\text{creep}}(U_z,s)+s_{py}D_y(s)F_y \\ z=\left(1+\frac{\delta_z}{k_z}\right)k_zD_z(s)U_z+z_o+f_{zz}^{\text{creep}}(U_z,s)+f_{zy}^{\text{hyst}}(U_y,s)+f_{zy}^{\text{creep}}(U_y,s)+s_{pz}D_z(s)F_z \end{cases} \tag{4.11}$$

where

$$\begin{cases} k_y=\frac{k_{yM}+k_{ym}}{2} \\ \frac{-(k_{yM}-k_{ym})}{2}\le\delta_y\le\frac{(k_{yM}-k_{ym})}{2} \end{cases} \text{ and } \begin{cases} k_z=\frac{k_{zM}+k_{zm}}{2} \\ \frac{-(k_{zM}-k_{zm})}{2}\le\delta_z\le\frac{(k_{zM}-k_{zm})}{2} \end{cases} \tag{4.12}$$

with:

- y_o is the offset, k_{yM} is the maximal slope, and k_{ym} is the minimal slope identified from the external loop of Fig. 4.4a. We find: $y_o = 8.3\,\mu\text{m}$, $k_{yM} = 5.63\,\mu\text{m}$, and $k_{ym} = 3.97\,\mu\text{m}$.
- z_o is the offset, k_{zM} is the maximal slope, and k_{zm} is the minimal slope identified from the external loop of Fig. 4.4d. We find: $z_o = -6\,\mu\text{m}$, $k_{zM} = 4.1\,\mu\text{m}$, and $k_{zm} = 2.9\,\mu\text{m}$.

which yields:

$$\begin{cases} k_y=4.8[\mu\text{m}] \\ -0.83[\mu\text{m}]\le\delta_y\le0.83[\mu\text{m}] \end{cases} \text{ and } \begin{cases} k_z=3.5[\mu\text{m}] \\ -0.6[\mu\text{m}]\le\delta_z\le0.6[\mu\text{m}] \end{cases} \tag{4.13}$$

4.4.3 Modeling and Identification of the Dynamics Part

The dynamics $D_y(s)$ and the dynamics $D_z(s)$ are identified from the step response in Fig. 4.5a and d, respectively. Remind that these dynamics should be normalized, i.e. $D_y(s=0)=1$ and $D_z(s=0)=1$.

The ARMAX-method (Auto Regressive Moving Average with eXternal inputs) of the Matlab software [19] is applied to the experimental data of Fig. 4.5a, d to perform the identification procedure. Different orders of models were tested and it is shown that from an order of 4, the precision of the identified dynamics does not increase substantially. Therefore, we use the identified 4th order models:

$$\begin{cases} D_y^{\text{init}}(s) = \dfrac{-0.563(s-6863)(s-30)\left(s^2+1.19s+4.96\times10^7\right)}{(s+2913)(s+27.8)\left(s^2+457s+1.49\times10^7\right)} \\ D_z^{\text{init}}(s) = \dfrac{0.347(s+6836)(s+39)\left(s^2-3304s+2.17\times10^7\right)}{(s+1063)(s+35)\left(s^2+91.8s+1.52\times10^7\right)} \end{cases} \quad (4.14)$$

The identified dynamics in Eq. (4.14) are not normalized. Indeed, we have: $D_y^{\text{init}}(s=0) \neq 1$ and $D_z^{\text{init}}(s=0) \neq 1$. To normalize them, we perform as follows:

$$\begin{aligned} D_y(s) &= \frac{D_y^{\text{init}}(s)}{D_y^{\text{init}}(0)} \\ D_z(s) &= \frac{D_z^{\text{init}}(s)}{D_z^{\text{init}}(0)} \end{aligned} \quad (4.15)$$

which finally gives the normalized dynamics:

$$\begin{cases} D_y(s) = \dfrac{-0.117(s-6863)(s-30)\left(s^2+1.19s+4.96\times10^7\right)}{(s+2913)(s+27.8)\left(s^2+457s+1.49\times10^7\right)} \\ D_z(s) = \dfrac{0.0989(s+6836)(s+39)\left(s^2-3304s+2.17\times10^7\right)}{(s+1063)(s+35)\left(s^2+91.8s+1.52\times10^7\right)} \end{cases} \quad (4.16)$$

4.4.4 The Final Model

Let us rewrite the model in Eq. (4.11) as follows:

$$\begin{cases} y = \left(1+\Delta_y W_{y\Delta}\right) k_y D_y(s) U_y + b_y \\ z = \left(1+\Delta_z W_{z\Delta}\right) k_z D_z(s) U_z + b_z \end{cases} \quad (4.17)$$

where b_y is an output disturbance for the y-axis and b_z is an output disturbance for the z-axis which are defined as follows:

$$\begin{cases} b_y = y_0 + f_{yy}^{\text{creep}}(U_y,s) + f_{yz}^{\text{hyst}}(U_z,s) + f_{yz}^{\text{creep}}(U_z,s) + s_{py} D_y(s) F_y \\ b_z = z_0 + f_{zz}^{\text{creep}}(U_z,s) + f_{zy}^{\text{hyst}}(U_y,s) + f_{zy}^{\text{creep}}(U_y,s) + s_{pz} D_z(s) F_z \end{cases} \quad (4.18)$$

where Δ_y and Δ_z are called normalized uncertainties and W_y and W_z are called weightings which are given by:

$$\begin{cases} -1 \le \Delta_y \le 1 \\ -1 \le \Delta_z \le 1 \end{cases} \text{and} \begin{cases} W_{y\Delta} = \dfrac{\left(k_{yM}-k_{ym}\right)}{2k_y} \\ W_{z\Delta} = \dfrac{\left(k_{zM}-k_{zm}\right)}{2k_z} \end{cases} \quad (4.19)$$

Denoting $G_y(s) = k_y D_y(s)$ and $G_z(s) = k_z D_z(s)$ which are called nominal models, we rewrite the "real models" in Eq. (4.20) as follows:

$$\begin{cases} y = \left(1+\Delta_y W_{y\Delta}\right) G_y(s) U_y + b_y \\ z = \left(1+\Delta_z W_{z\Delta}\right) G_z(s) U_z + b_z \end{cases} \quad (4.20)$$

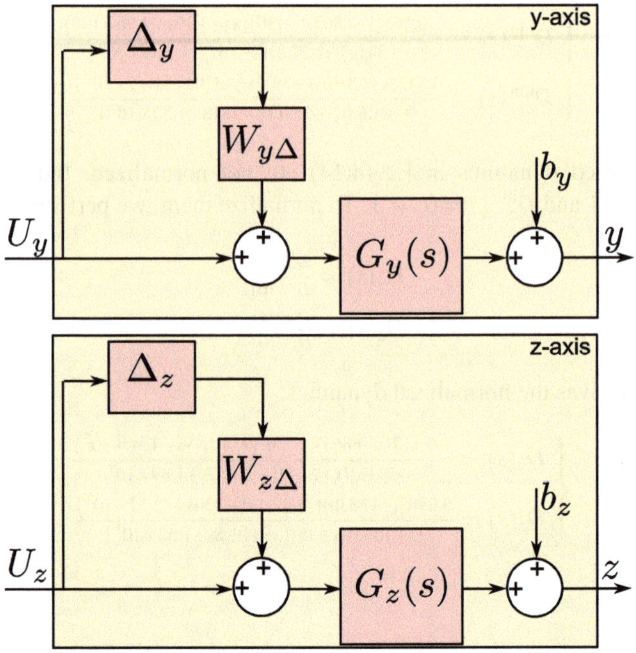

Fig. 4.8 The system to be controlled

Equation (4.20) is the final model to be controlled. It corresponds to two monovariable linear systems, each one with an input direct multiplicative uncertainty and an output disturbance. The block diagram corresponding to this final model is shown in Fig. 4.8.

In the sequel, we use the model in Eq. (4.20) to find a controller that will enhance the performances of the 2-dof piezoelectric actuator. The principal objective will be to track some input references, to reject the output disturbances, and to maintain all these performances although the presence of uncertainties in the model used to synthesize the controller. For that, we will use the robust H_∞ controller synthesis. In fact, this technique is well suited for the design of linear controllers when using uncertain models.

4.5 Robust Standard H_∞ Control Technique

Remind that the model to be used for the controller synthesis is given by Eq. (4.20). It corresponds to two monovariable systems: one system for the *y*-axis and one system for the *z*-axis. As the two monovariable systems are similar in structure, we will only present the controller synthesis for the *y*-axis.

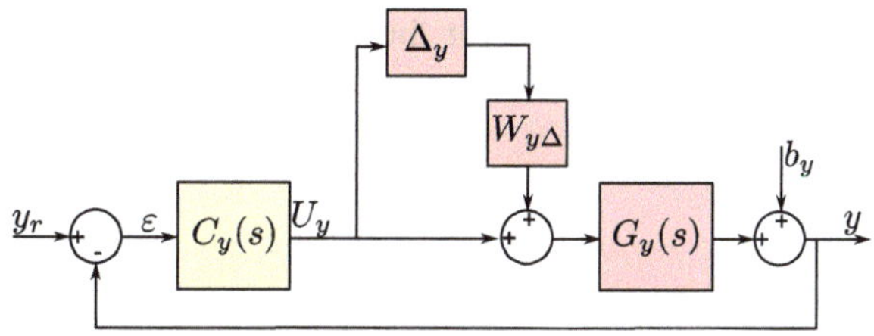

Fig. 4.9 The closed-loop system

4.5.1 Principle Scheme and Specifications

Reconsider the model of y-axis in Eq. (4.20) and its block diagram in Fig. 4.8 and introduce the controller $C_y(s)$ to be synthesized. We obtain the closed-loop scheme as shown in Fig. 4.9.

4.5.1.1 Specifications for the y-Axis

For this closed-loop we give the following specifications, i.e. specifications for the y-axis:

Tracking performances:

- The settling time should be less than or equal to 5 ms.
- The statical error should be less than or equal to 1 %.
- No overshoot should appear.

Command moderation:

In order to limit the voltage U_y applied to the actuator and therefore to avoid all possible material's destruction, we impose a maximal voltage of 10 V for any frequency and for any amplitude of reference within $y_r = 50\,\mu\text{m}$.

Rejection of the effect of the disturbance b_y:

- The settling time of the disturbance rejection should be less than or equal to 10 ms.
- When the disturbance b_y is maximal (worst case), its effect to the output y should not exceed 1 µm.
- No overshoot should appear during the disturbance rejection.

Robustness in presence of the uncertainty $\Delta_y W_{y_\Delta}$:

The above specifications should be ensured for any uncertainty contained in $\Delta_y W_{y\Delta}$.

4.5.1.2 Specifications for the z-Axis

The above specifications are for the y-axis. Concerning the specifications for the z-axis, we give:

Tracking performances:

- The settling time should be less than or equal to 15 ms.
- The statical error should be less than or equal to 1 %.
- No overshoot should appear.

Command moderation:

In order to limit the voltage U_y applied to the actuator and therefore to avoid all possible material's destruction, we impose a maximal voltage of 10 V for any frequency and for any amplitude of reference within $z_r = 35\,\mu\text{m}$.

Rejection of the effect of the disturbance b_z:

- The settling time of the disturbance rejection should be less than or equal to 10 ms.
- When the disturbance b_z is maximal (worst case), its effect to the output z should not exceed 1 µm.
- No overshoot should appear during the disturbance rejection.

Robustness in presence of the uncertainty $\Delta_z W_{z_\Delta}$:

The above specifications should be ensured although the presence of the uncertainty $\Delta_z W_{z_\Delta}$.

4.5.2 Standard Form and the Standard H_∞ Problem

In the standard H_∞ technique, during the controller synthesis, the specifications are accounted by introducing weighting for each specified point. For the y-axis, the tracking performances is systematically accounted by introducing a weighting $W_{y1}(s)$ to weight the error ε. The command moderation specification is accounted by using a weighting $W_{y2}(s)$ to weight the control signal U_y. Finally, for the disturbance rejection specification, $W_{yb}(s)$ is introduced at the perturbation signal b_y. We therefore obtain the weighted closed-loop scheme as in Fig. 4.10. In the scheme, the new outputs o_{y1} and o_{y2} are called weighted outputs while i_y indicates the new input.

The uncertainty presented in Fig. 4.10 is still in its initial structure (direct multiplicative structure). Yet, having an input direct multiplicative structure of uncertainty $\left\|\Delta_y\right\|_\infty \leq 1$ weighted by $W_{y\Delta}$, the condition of stability of a closed-loop is as follows (see for instance [20]):

$$\left\|S_y G_y C_y W_{y\Delta}\right\|_\infty < 1 \tag{4.21}$$

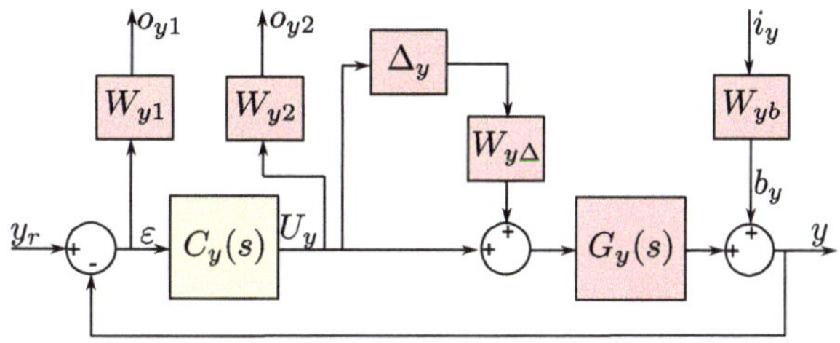

Fig. 4.10 The weighted closed-loop system

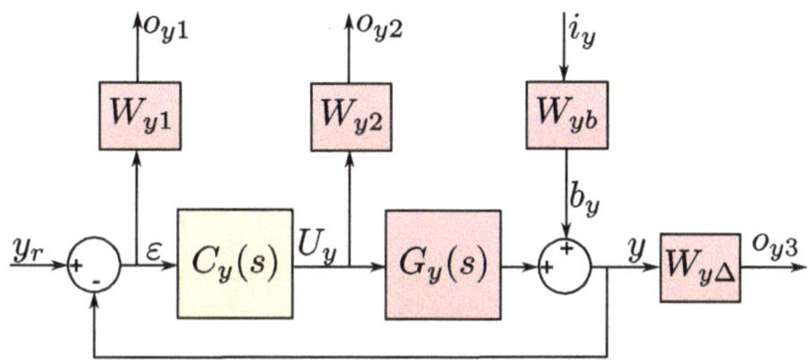

Fig. 4.11 Equivalent weighted closed-loop system when considering the stability condition with an input direct multiplicative structured uncertainty

where $S(s)$ is the sensitivity function and is given by:

$$S_y(s) = \frac{\varepsilon(s)}{y_r(s)} = \frac{1}{1 + C_y G_y} \tag{4.22}$$

where $S_y C_y G_y = H_y(s)$ is called complementary sensitivity function which links the output $y(s)$ and the input reference $y_r(s)$. It is given by:

$$H_y(s) = 1 - S_y = S_y C_y G_y = \frac{y(s)}{y_r(s)} = \frac{C_y G_y}{1 + C_y G_y} \tag{4.23}$$

However, Eq. (4.21) can be also interpreted as if we have a new complementary sensitivity function $\left(S_y C_y G_y W_{y\Delta}\right)$ that would link another output $o_{y3}(s)$ and the input reference $y_r(s)$. The new output $o_{y3}(s)$ is the result of weighting the output $y(s)$ by $W_{y\Delta}$. Therefore, Fig. 4.10 is equivalent to Fig. 4.11 when considering the stability condition of the uncertainty to which we are face.

In the standard H_∞ problem, a standard form is used. This standard form links the interconnection of an augmented system $\mathscr{P}_y(s)$ with a controller as pictured in Fig. 4.12a. The augmented system $P_y(s)$ is made up of the nominal system $G_y(s)$ augmented with the weighting functions. There are two types of inputs for $\mathscr{P}_y(s)$: (1) the first type is composed of all exogeneous inputs which are here the reference y_r and the disturbance i_y and (2) the second type is the input control signal U_y from the controller $C_y(s)$. There are also two types of outputs for $\mathscr{P}_y(s)$: (1) the first type is composed of the output signals from all weightings which are here o_{y1}, o_{y3}, and o_{y3} and (2) the second type is the output signal ε_y that will go to the controller. Figure 4.12b depicts the details of the standard scheme from the weighted closed-loop in Fig. 4.11.

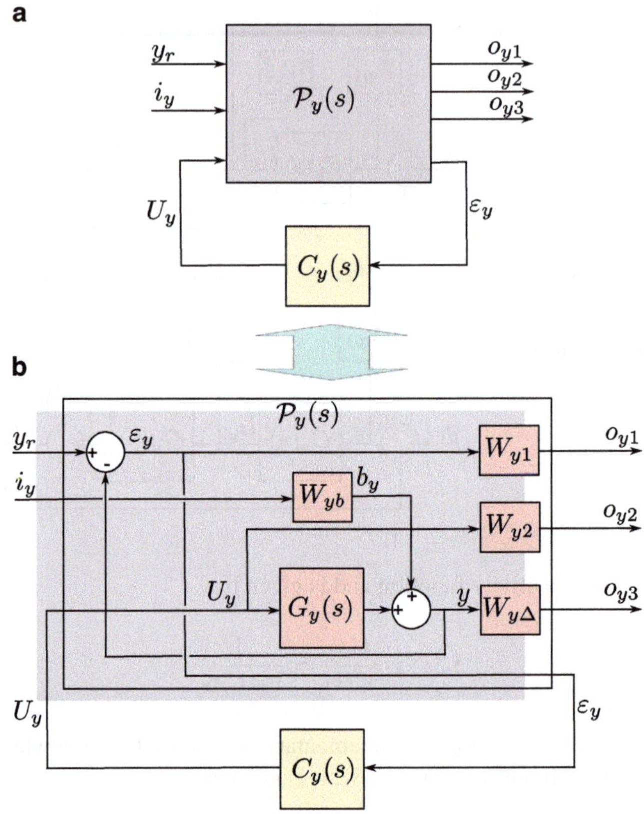

Fig. 4.12 (**a**): the standard form. (**b**): detailed block-diagram of the standard form

Having the standard scheme as in Fig. 4.12, the standard H_∞ problem can be expressed:

Problem 5.1 (The standard H_∞ problem [20]). Find the controller $C_y(s)$ such that:

- The interconnection in Fig. 4.12 is stable
- $\left\| F_{yl}\left(\mathscr{P}_y(s), C_y(s)\right) \right\|_\infty < \gamma_y$

where $\gamma_y > 0$ is called performances gain, and $F_{yl}\left(\mathscr{P}_y(s), C_y(s)\right)$ is called lower linear fractional transformation which is a matricial transfer function defined as follows:

$$\begin{pmatrix} o_{y1} \\ o_{y2} \\ o_{y3} \end{pmatrix} = F_{yl}\left(\mathscr{P}_y(s), C_y(s)\right) \begin{pmatrix} y_r \\ i_y \end{pmatrix} \tag{4.24}$$

However, from Fig. 4.11, we have the following equations:

$$\begin{cases} o_{y1} = W_{y1}S_y y_r - W_{y1}S_y W_{yb} i_y \\ o_{y2} = W_{y2}C_y S_y y_r - W_{y2}C_y S_y W_{yb} i_y \\ o_{y3} = W_{y\Delta} G_y C_y S_y y_r + W_{y\Delta} S_y W_{yb} i_y \end{cases} \tag{4.25}$$

Using Eqs. (4.24) and (4.25), we yield:

$$F_{yl}\left(\mathscr{P}_y(s), C_y(s)\right) = \begin{pmatrix} W_{y1}S_y & -W_{y1}S_y W_{yb} \\ W_{y2}C_y S_y & -W_{y2}C_y S_y W_{yb} \\ W_{y\Delta} G_y C_y S_y & W_{y\Delta} S_y W_{yb} \end{pmatrix} \tag{4.26}$$

By using Eq. (4.26), the second condition of Problem 5.1 can be rewritten as follows:

$$\left\| \begin{pmatrix} W_{y1}S_y & -W_{y1}S_y W_{yb} \\ W_{y2}C_y S_y & -W_{y2}C_y S_y W_{yb} \\ W_{y\Delta} G_y C_y S_y & W_{y\Delta} S_y W_{yb} \end{pmatrix} \right\|_\infty < \gamma_y \tag{4.27}$$

which is equivalent to:

$$\begin{matrix} \left\|W_{y1}S_y\right\|_\infty < \gamma_y & \left\|-W_{y1}S_y W_{yb}\right\|_\infty < \gamma_y \\ \left\|W_{y2}C_y S_y\right\|_\infty < \gamma_y & \left\|-W_{y2}C_y S_y W_{yb}\right\|_\infty < \gamma_y \\ \left\|W_{y\Delta} G_y C_y S_y\right\|_\infty < \gamma_y & \left\|W_{y\Delta} S_y W_{yb}\right\|_\infty < \gamma_y \end{matrix} \tag{4.28}$$

According to the Cauchy–Schwartz inequality, if we have the following conditions, conditions Eq. (4.28) can be satisfied:

$$\begin{matrix} \left\|W_{y1}\right\|_\infty \left\|S_y\right\|_\infty < \gamma_y & \left\|W_{y1}W_{yb}\right\|_\infty \left\|-S_y\right\|_\infty < \gamma_y \\ \left\|W_{y2}\right\|_\infty \left\|C_y S_y\right\|_\infty < \gamma_y & \left\|W_{y2}W_{yb}\right\|_\infty \left\|-C_y S_y\right\|_\infty < \gamma_y \\ \left\|W_{y\Delta}\right\|_\infty \left\|G_y C_y S_y\right\|_\infty < \gamma_y & \left\|W_{y\Delta}W_{yb}\right\|_\infty \left\|S_y\right\|_\infty < \gamma_y \end{matrix} \tag{4.29}$$

which are equivalent to:

$$\begin{matrix} \left\|S_y\right\|_\infty < \gamma_y \frac{1}{\left\|W_{y1}\right\|_\infty} & \left\|-S_y\right\|_\infty < \gamma_y \frac{1}{\left\|W_{y1}W_{yb}\right\|_\infty} \\ \left\|C_y S_y\right\|_\infty < \gamma_y \frac{1}{\left\|W_{y2}\right\|_\infty} & \left\|-C_y S_y\right\|_\infty < \gamma_y \frac{1}{\left\|W_{y2}W_{yb}\right\|_\infty} \\ \left\|G_y C_y S_y\right\|_\infty < \gamma_y \frac{1}{\left\|W_{y\Delta}\right\|_\infty} & \left\|S_y\right\|_\infty < \gamma_y \frac{1}{\left\|W_{y\Delta}W_{yb}\right\|_\infty} \end{matrix} \tag{4.30}$$

Conditions Eq. (4.30) can also be satisfied if we have the following conditions:

$$\begin{matrix} \left|S_y\right| < \gamma_y \left|\frac{1}{W_{y1}}\right| & \left|-S_y\right| < \gamma_y \left|\frac{1}{W_{y1}W_{yb}}\right| \\ \left|C_y S_y\right| < \gamma_y \left|\frac{1}{W_{y2}}\right| & \left|-C_y S_y\right| < \gamma_y \left|\frac{1}{W_{y2}W_{yb}}\right| \\ \left|G_y C_y S_y\right| < \gamma_y \left|\frac{1}{W_{y\Delta}}\right| & \left|S_y\right| < \gamma_y \left|\frac{1}{W_{y\Delta}W_{yb}}\right| \end{matrix} \tag{4.31}$$

Our objective now consists in finding the controller $C_y(s)$ that will satisfy the conditions in Eq. (4.31). In this, $1/W_{y1}(s)$, $1/W_{y2}(s)$, $1/W_{y\Delta}(s)$, $1/W_{y1}(s)W_{yb}(s)$, $1/W_{y2}(s)W_{yb}(s)$, and $1/W_{y\Delta}(s)W_{yb}(s)$ are called gabarits. In particular, $1/W_{y1}(s)$, $1/W_{y2}(s)$, and $1/W_{y1}(s)W_{yb}(s)$ are gabarits for the tracking performances, for the command moderation, and for the disturbance rejection, respectively. They are defined according to the specifications in Sect. 4.5.1. The objective is to find the controller such that the performances gain $\gamma_y > 0$ is as small as possible. If $\gamma_y > 1$, some of the specified performances will not be ensured by the controller. One reason to make this case, i.e. $\gamma_y > 1$, happens is when the specifications are too severe. In general, a magnitude plot (singular values plot) of the different transfer functions and of the gabarits defined in Eq. (4.31) permits to see which magnitude is not satisfied and how it is not satisfied. This allows to see which specification to be relaxed before recalculating the controller. If $\gamma \leq 1$, the specified performances will be ensured by the calculated controller.

Notice that the H_∞ problem for the z-axis is exactly similar to that of y-axis in structure. The same conditions than in Eq. (4.31) are therefore obtained for the z-axis, one should only replace subscript y into subscript z in all the inequalities.

4.5.3 Gabarits and Weighting Functions

To calculate the principal gabarits $1/W_{y1}(s)$, $1/W_{y2}(s)$ and $1/W_{y1}(s)W_{yb}(s)$, the specifications defined in Sect. 4.5.1 are used.

Derivation of the gabarit $1/W_{y1}(s)$: This gabarit is calculated from the specified tracking performances. Indeed, this gabarit is initially from the transfer function that relates the output o_{y1} and the input reference y_r (see Eq. (4.25)). A possible structure of gabarit that satisfies a tracking performances with a settling time of t_{ry}, a statical error of ε_{sy} and without overshoot is:

$$\frac{1}{W_{y1}(s)} = \frac{k_{ovy}s + (3\varepsilon_{sy}/t_{ry})}{s + (3/t_{ry})} \tag{4.32}$$

where $k_{ovy} = 1 + \theta_{ovy}$ and θ_{ovy} is the overshoot. If no overshoot is wanted, we have: $\theta_{ovy} = 0$.

Choosing the following numerical values: $t_{ry} = 5\,\text{ms}$ and $\varepsilon_{sy} = 0.01$ (see Sect. 4.5.1), we have:

$$\frac{1}{W_{y1}(s)} = \frac{s+6}{s+600} \tag{4.33}$$

From Eq. (4.33), the weighting $W_{y1}(s)$ is yielded:

$$W_{y1}(s) = \frac{s+600}{s+6} \tag{4.34}$$

Derivation of the gabarit $1/W_{y2}(s)$*:* This gabarit is calculated from the specified command moderation which mentions that a maximal voltage of 10 V is allowed for any frequency and any amplitude of reference input up to 50 μm. This yields:

$$\frac{1}{W_{y2}(s)} = \frac{U_{y\max}}{y_{r\max}} = \frac{10[\mathrm{V}]}{50[\mu\mathrm{m}]} = 0.2\left[\frac{\mathrm{V}}{\mu\mathrm{m}}\right] \tag{4.35}$$

which yields the weighting $W_{y2}(s)$:

$$W_{y2}(s) = 5\left[\frac{\mu\mathrm{m}}{\mathrm{V}}\right] \tag{4.36}$$

Derivation of the gabarit $1/W_{y1}(s)W_{yb}(s)$: This gabarit is calculated from the specified disturbance rejection which mentions that when the disturbance b_y is maximal (worst case), its effect to the output should not exceed 1 μm, i.e. maximal error is 1 μm. Remind from Eq. (4.18) that $b_y = y_o + f_{yy}^{creep}(U_y,s) + f_{yz}^{hyst}(U_z,s) + f_{yz}^{creep}(U_z,s) + s_{py}D_y(s)F_y$. To define the worst case b_{ywc} of the disturbance, we choose: $b_{ywc} = |y_o| + \left|f_{yy}^{creep}(U_y,s)\right| + \left|f_{yz}^{hyst}(U_z,s)\right| + \left|f_{yz}^{creep}(U_z,s)\right| + \left|s_{py}F_{y\max}\right|$, where y_o, $f_{yy}^{creep}(U_y,s)$ are obtained with the maximal or minimal voltage $U_y = \pm 10$ V while and $f_{yz}^{hyst}(U_z,s)$ and $f_{yz}^{creep}(U_z,s)$ are obtained with the maximal or minimal voltage $U_z = \pm 10$ V. We have: $|y_o| = |8.3\,\mu\mathrm{m}|$ (see Fig. 4.4a), $\left|f_{yy}^{creep}(U_y,s)\right| = |15\,\mu\mathrm{m}|$ (see Fig. 4.6a), $\left|f_{yz}^{hyst}(U_z,s)\right| = |16\,\mu\mathrm{m}|$ (see Fig. 4.4b), and $\left|f_{yz}^{creep}(U_z,s)\right| = |3.5\,\mu\mathrm{m}|$ (see Fig. 4.6b). In this application, we choose $F_y = 0$ N. Hence, the wanted maximal error ε_{bywc} due to the disturbance is derived as follows:

$$\varepsilon_{bywc} = \frac{|y_r - y|}{b_{ywc}} = \frac{1}{8.3+15+16+3.5+0} = 0.0234 \tag{4.37}$$

However as demanded in Sect. 4.5.1, the specifications require a settling time of $t_{rby} \leq 10$ ms and a zero overshoot for the disturbance rejection. Using the same structure than in Eq. (4.32) and using the statical error due to disturbance as given by Eq. (4.37), we propose (using $t_{rby} = 10$ ms):

$$\frac{1}{W_{y1}(s)W_{yb}(s)} = \frac{k_{bovy}s + (3\varepsilon_{bywc}/t_{rby})}{s + (3/t_{rby})} = \frac{s+7.009}{s+300} \tag{4.38}$$

where $k_{bovy} = 1 + \theta_{bovy}$ and θ_{bovy} is the overshoot. For a zero overshoot, we have $\theta_{bovy} = 0$.

By using Eqs. (4.34) and (4.38), we yield the weighting $W_{yb}(s)$

$$W_{yb}(s) = \frac{(s+6)(s+300)}{(s+7.009)(s+600)} \tag{4.39}$$

Derivation of the weighting $W_{y\Delta}$: The weighting $W_{y\Delta}$ is also required in Eq. (4.31) for the calculation of the controller. This weighting is given by Eq. (4.19). We have:

$$W_{y\Delta} = 0.1729 \tag{4.40}$$

The previously calculated gabarits and weightings are for the y-axis. The gabarits $1/W_{z1}(s)$, $1/Ww_{z2}(s)$ and $1/W_{z1}(s)W_{zb}(s)$ for the z-axis are calculated from the specified tracking performances, the command moderation, and the disturbance rejection, respectively, for the z-axis which are also given in Sect. 4.5.1. We have the following results.

Derivation of the gabarit $1/W_{z1}(s)$: This gabarit is derived from the specified tracking performances. Using the same structure than in Eq. (4.32) and considering the numerical value (maximal settling time of 15 ms, maximal statical error of 1 %, no overshoot) as in Sect. 4.5.1, we propose:

$$\frac{1}{W_{z1}(s)} = \frac{s+2}{s+200} \tag{4.41}$$

from which we derive the weighting $W_{z1}(s)$

$$W_{z1}(s) = \frac{s+200}{s+2} \tag{4.42}$$

Derivation of the gabarit $1/W_{z2}(s)$: This is calculated from the specified command moderation which states a maximal voltage of 10 V for any frequency and any amplitude of reference within $z_r = 35\,\mu\text{m}$ (see Sect. 4.5.1). We propose:

$$\frac{1}{W_{z2}(s)} = \frac{U_{z\max}}{z_{r\max}} = \frac{10[\text{V}]}{35[\mu\text{m}]} = 0.2857\left[\frac{\text{V}}{\mu\text{m}}\right] \tag{4.43}$$

from which we yield the weighting $W_{z2}(s)$:

$$W_{z2}(s) = 3.5\left[\frac{\mu\text{m}}{\text{V}}\right] \tag{4.44}$$

Derivation of the gabarit $1/W_{z1}(s)W_{zb}(s)$: This gabarit is defined from the specified disturbance rejection. The approach to constitute this gabarit is similar to that of the gabarit $1/W_{y1}(s)W_{yb}(s)$ for the y-axis. The disturbance being given by $b_z = z_0 + f_{zz}^{\text{creep}}(U_z,s) + f_{zy}^{\text{hyst}}(U_y,s) + f_{zy}^{\text{creep}}(U_y,s) + s_{pz}D_z(s)F_z$ (see Eq. (4.18)), we provide the worst case disturbance as equal to: $b_{zwc} = |z_0| + \left|f_{zz}^{\text{creep}}(U_z,s)\right| + \left|f_{zy}^{\text{hyst}}(U_y,s)\right| + \left|f_{zy}^{\text{creep}}(U_y,s)\right| + \left|s_{pz}F_{z\max}\right|$. The numerical values are: $|z_0| = |-6\,\mu\text{m}|$ (see Fig. 4.4d), $\left|f_{zz}^{\text{creep}}(U_z,s)\right| = |12\,\mu\text{m}|$ (see Fig. 4.6d), $\left|f_{zy}^{\text{hyst}}(U_y,s)\right| = |8\,\mu\text{m}|$ (see Fig. 4.4c) and $\left|f_{zy}^{\text{creep}}(U_y,s)\right| = |0\,\mu\text{m}|$ (see Fig. 4.6c). In this application, we choose $F_{z\max} = 0\,\text{N}$. Consequently, the specified maximal error due to this worst case of disturbance should be no more than:

$$\varepsilon_{bzwc} = \frac{|z_r - z|}{b_{zwc}} = \frac{1}{6+12+8+0+0} = 0.0385 \tag{4.45}$$

By combining the specified settling time $t_{rbz} \leq 10\,\text{ms}$ and the zero overshoot ($k_{bovz} = 1$) for the disturbance rejection (see Sect. 4.5.1) with the maximal error Eq. (4.45) and by using the same structure than that of y-axis as given in Eq. (4.38), we have the gabarit for the z-axis:

$$\frac{1}{W_{z1}(s)W_{zb}(s)} = \frac{k_{bovz}s + (3\varepsilon_{bzwc}/t_{rbz})}{s + (3/t_{rbz})} = \frac{s+11.54}{s+300} \tag{4.46}$$

Using Eqs. (4.42) and (4.46), we obtain the weighting $W_{zb}(s)$:

$$W_{zb}(s) = \frac{(s+2)\,(s+300)}{(s+11.54)\,(s+200)} \tag{4.47}$$

Derivation of the weighting $W_{z\Delta}$: The weighting $W_{z\Delta}$ is also required to compute the controller. This is given by Eq. (4.19). We have:

$$W_{y\Delta} = 0.1714 \tag{4.48}$$

4.5.4 Calculation of the Controllers

To calculate the controller $C_y(s)$ that satisfies the condition in Eq. (4.31), the controller $C_z(s)$ and optimal values of the performances gains γ_y and γ_z, the DGKF [21, 22] algorithm is used. The results provide controllers of order 7:

$$\begin{cases} C_y(s) = \dfrac{20088(s+2913)(s+600)(s+27)(s+6.8)\left(s^2+457s+1.5\times10^7\right)}{\left(s+2\times10^5\right)(s+682)(s+33)(s+7)(s+6)\left(s^2+7406s+2.5\times10^7\right)} \\ \gamma_y = 1.64 \end{cases} \tag{4.49}$$

$$\begin{cases} C_z(s) = \dfrac{5380(s+1063)(s+200)(s+35)(s+6.8)\left(s^2+91.8s+1.5\times10^7\right)}{\left(s+1.86\times10^4\right)(s+21.8)(s+2)\left(s^2+172s+1.15\times10^4\right)\left(s^2+2276s+2.37\times10^7\right)} \\ \gamma_z = 1.855 \end{cases} \tag{4.50}$$

The optimal performances gains γ_y and γ_z are strictly higher than one but the magnitudes (see Figs. 4.13 and 4.14) show that the overshoots of $|S_i|$ and $|C_iS_i|$ relative to the gabarits magnitudes $\left|\frac{1}{W_{i1}}\right|$, $\left|\frac{1}{W_{i2}}\right|$ and $\left|\frac{1}{W_{i1}W_{ib}}\right|$, where $i = \{y, z\}$ are negligible. The calculated controllers are therefore acceptable for our applications.

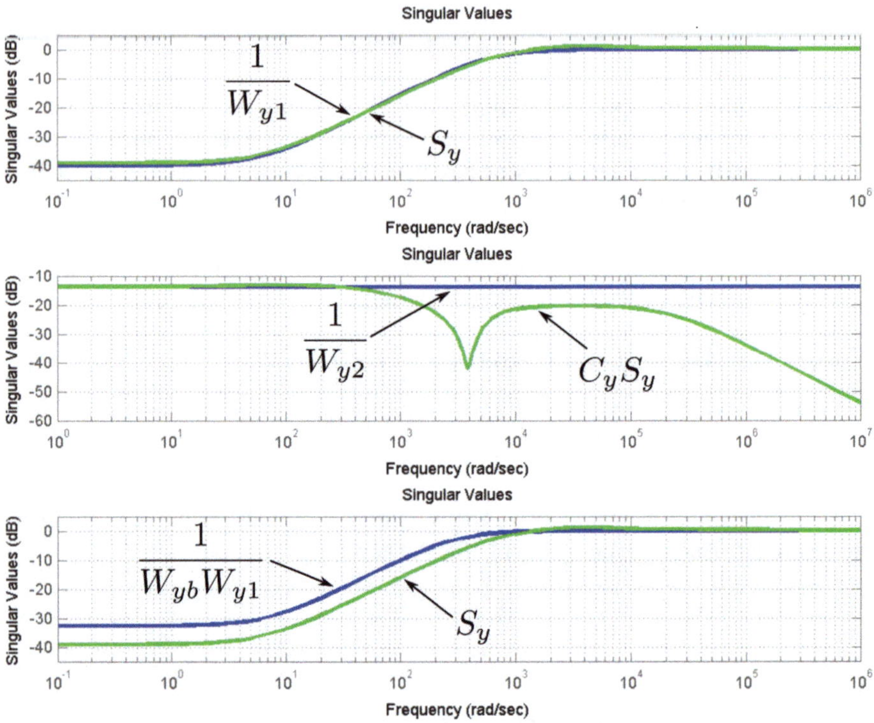

Fig. 4.13 Singular values plots for the y axis system

4.5.5 Controllers Implementation

The calculated controllers $C_y(s)$ and $C_z(s)$ were implemented in the Simulink-software as described in the setup-scheme presented in Fig. 4.3. The block diagram of the closed-loop composed of the two controllers and the 2-dof actuator to be controlled is pictured in Fig. 4.15. The experiments consist in applying a series of references steps y_r and z_r to the closed-loop. First a step reference $y_r = 50\,\mu\text{m}$ is applied at about $t = 1.2\,\text{s}$ while z_r is left equal to zero. As we can see, the output y tracks this input reference without vibration (Fig. 4.16a). Furthermore, we can observe the coupling effect $y_r \rightarrow z$ which is quickly rejected (Fig. 4.16b). Then, we apply a reference input $z_r = 50\,\mu\text{m}$ at time $t = 1.52\,\text{s}$. From Fig. 4.16b, we can see the output z reaches the reference z_r also without overshoot. The effect $z_r \rightarrow y$ can be seen in Fig. 4.16a where we observe that the coupling is quickly rejected.

In order to evaluate more precisely the performances, we give in Fig. 4.17 the zoom of the step responses. We yield that the settling time is 5.5 ms in the y-axis (Fig. 4.17a) while less than 4 ms in the z-axis (Fig. 4.17d), which will satisfy the specified settling times. We also observe from Fig. 4.17b, c that the disturbance rejection (coupling rejection) has a settling time much quicker than the specifications. These results clearly show the efficiency of the proposed controller.

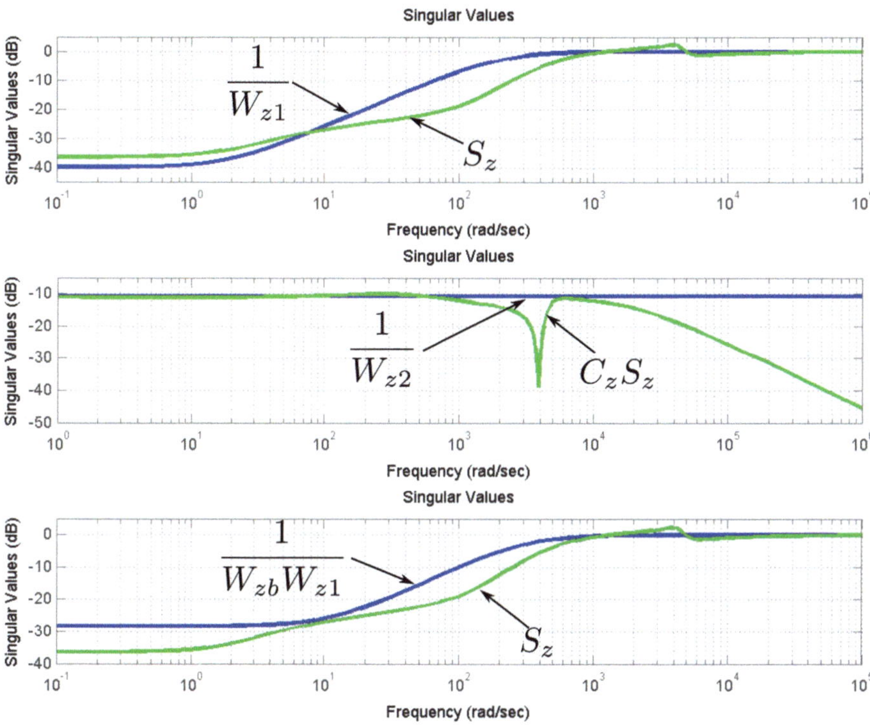

Fig. 4.14 Singular values plots for the z axis system

Fig. 4.15 Implementation of the controllers

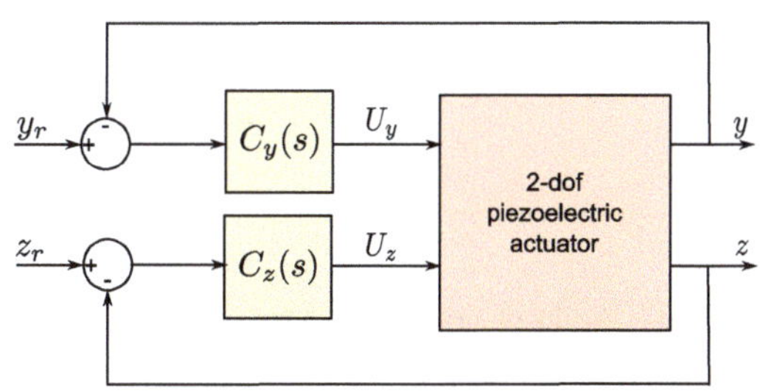

4.6 Conclusion

This chapter presented the characterization, modeling, and robust control of a 2-dof piezoelectric actuator. The actuator is a cantilevered piezoelectric structure that is able to bend along y- and along z-axes. As ceramic material, the actuator is typified by a strong hysteresis nonlinearity and a creep behavior. In addition to that, its structure makes the behavior of the actuator be with badly damped oscillation. Finally, the 2-dof piezoelectric actuator has a strong coupling between the two axes. All these properties makes the actuator lose the general performances (accuracy,

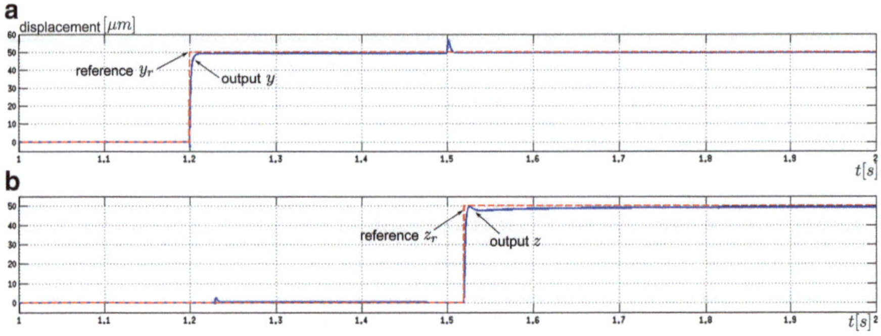

Fig. 4.16 Step responses of the controlled 2-dof piezoelectric actuator. (**a**): responses along y-axis. (**b**): responses along z-axis

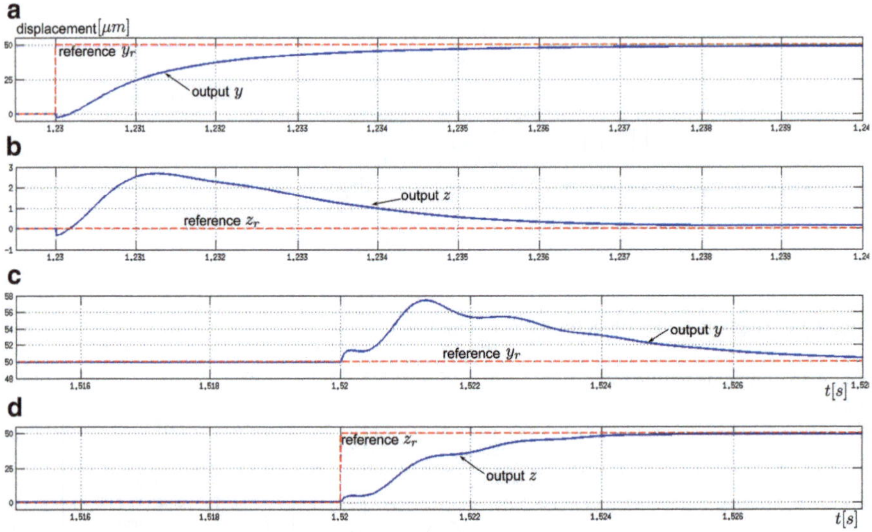

Fig. 4.17 Zoom of the step responses of the controlled 2-dof piezoelectric actuator. (**a**): step response along y-axis. (**b**): coupling effect on the z-axis. (**c**): coupling effect on the y-axis. (**d**): step response along z-axis

settling time, and damping) in spite of its high resolution and the high bandwidth. This chapter proposed therefore first a modeling that is able to track all these properties by remaining simple. Two monovariable linear models with well-defined uncertainties (related to the hysteresis) and well-defined disturbances (related to the coupling, the creep and eventual external force) were derived to track the 2-dof system. Afterwards, a robust standard H_∞ controller was synthesized for each axis. The calculated controllers permitted to ensure the specified performances even in presence of model uncertainties. Experimental results have shown the efficiency of the proposed method.

Acknowledgements This work is supported by the national ANR-Emergence MYMESYS-project (ANR-11-EMMA-006: High Performances Embedded Measurement Systems for multi-Degrees of Freedom Microsystems).

References

1. J. Agnus, N. Chaillet, C. Clévy, S. Dembélé, M. Gauthier, Y. Haddab, G. Laurent, P. Lutz, N. Piat, K. Rabenorosoa, M. Rakotondrabe, B. Tamadazte, Robotic microassembly and micromanipulation at FEMTO-ST. J. Micro. Bio. Robot. (JMBR), **8**(2), 91–106 (2013)
2. M. Rakotondrabe, C. Clévy, P. Lutz, Modelling and robust position/force control of a piezoelectric microgripper, in *IEEE—International Conference on Automation Science and Engineering (CASE)*, Scottsdale, AZ, USA, 2007, pp. 39–44
3. S. Devasia, E.E. Eleftheriou, R. Moheimani, A survey of control issues in nanopositioning. IEEE Trans. Contr. Syst. Technol. **15**(5), 802–823 (2007)
4. M. Rakotondrabe, K. Rabenorosoa, J. Agnus, N. Chaillet, Robust feedforward-feedback control of a nonlinear and oscillating 2-dof piezocantilever. IEEE Trans. Autom. Sci. Eng. (T-ASE) **8**(3), 506–519 (2011)
5. M. Rakotondrabe, J. Agnus, P. Lutz, Feedforward and IMC-feedback control of a nonlinear 2-DOF piezoactuator dedicated to automated micropositioning tasks, in *IEEE—CASE (International Conference on Automation Science and Engineering)*, Trieste, Italy, August 2011, pp. 393–398
6. R. Pérez, J. Agnus, C. Clévy, A. Hubert, N. Chaillet, Modeling, fabrication and validation of a high performance 2-DOF piezoactuator for micromanipulation. IEEE/ASME Trans. Mechatron. (T-mech), **10**(2), 161–171 (2005)
7. M. Rakotondrabe, Y. Haddab, P. Lutz, Quadrilateral modelling and robust control of a nonlinear piezoelectric cantilever. IEEE Trans. Contr. Syst. Technol. (T-CST) **17**(3), 528–539 (2009)
8. M. Rakotondrabe, C. Clévy, P. Lutz, Complete open loop control of hysteretic, creeped and oscillating piezoelectric cantilever. IEEE Trans. Autom. Sci. Eng. (TASE) **7**(3), 440–450 (2010)
9. M. Rakotondrabe, *Piezoelectric Cantilevered Structures: Modeling, Control and Measurement/Estimation Aspects* (Springer, Berlin, 2013)
10. M. Rakotondrabe, Y. Haddab, P. Lutz, Modelling and H-inf force control of a nonlinear piezoelectric cantilever, in *IEEE/RSJ—IROS (International Conference on Intelligent Robots and Systems)*, San Diego, CA, USA, October–November 2007, pp. 3131–3136
11. S. Devasia, E.E. Eleftheriou, R. Moheimani, A survey of control issues in nanopositioning. IEEE Trans. Contr. Syst. Technol. **15**(5), 802–823 (2007)
12. D. Croft, G. Shed, S. Devasia, Creep, hysteresis and vibration compensation for piezoactuators: atomic force microscopy application. ASME J. Dyn. Syst. Meas. Contr. **123**(1), 35–43 (2001)
13. A. Dubra, J. Massa, C.l. Paterson, Preisach classical and nonlinear modeling of hysteresis in piezoceramic deformable mirrors. Opt. Express **13**(22), 9062–9070 (2005)
14. W.T. Ang, P.K. Kholsa, C.N. Riviere, Feedforward controller with inverse rate-dependent model for piezoelectric actuators in trajectory-tracking applications. IEEE/ASME Trans. Mechatron. **12**(2), 134–142 (2007)
15. B. Mokaberi, A.A.G. Requicha, Compensation of scanner creep and hysteresis for AFM nanomanipulation. IEEE Trans. ASE **5**(2), 197–208 (2008)
16. M. Rakotondrabe, Classical Prandtl–Ishlinskii modeling and inverse multiplicative structure to compensate hysteresis in piezoactuators, in *ACC (American Control Conference)*, Montréal, Canada, June 2012, pp. 1646–1651
17. M. Rakotondrabe, Bouc–Wen modeling and inverse multiplicative structure to compensate hysteresis nonlinearity in piezoelectric actuators. IEEE Trans. ASE **8**(2), pp. 428–431 (2011)

18. M. Rakotondrabe, Y. Haddab, P. Lutz, Plurilinear modeling and discrete μ-synthesis control of a hysteretic and creeped unimorph piezoelectric cantilever, in *IEEE—ICARCV (International Conference on Automation, Robotics, Control and Vision)*, Singapore, December 2006, pp. 57–64
19. L. Ljung, System identification toolbox, for use with Matlab. The Matworks (1995)
20. K. Zhou, J.C. Doyle, K. Glover, *Robust and Optimal Control* (Prentice Hall, Englewood Cliffs, 1995)
21. K. Glover, J.C. Doyle, State-space formulae for all stabilizing controllers that satisfy an H_∞-norm bound and relations to risk sensitivity. Syst. Contr. Lett. **11**, 167–172 (1988)
22. J.C. Doyle, K. Glover, P.K. Khargonekar, B.A. Francis, State-space solutions to standard H_2 and H_∞ control problems. IEEE Trans. Automat. Contr. AC **34**(8), 831–846 (1989)

Chapter 5
A Hybrid Control Approach to Nanopositioning

Tomas Tuma, Abu Sebastian, John Lygeros, and Angeliki Pantazi

Abstract Precise position control on the nanometer and subnanometer scale, referred to as nanopositioning, is a key enabler for nanoscale science and engineering. In nanopositioning, feedback control is essential to meet the stringent requirements on accuracy, stability, and repeatability in the presence of model uncertainties and environmental disturbances. In this chapter, we review a new hybrid control approach to nanopositioning which is based on the combination of a continuous-time control law with impulsive modifications of the controller states. By using impulsive control, the limitations of conventional linear controllers can be overcome, such as the inherent trade-off between closed-loop bandwidth and resolution. We review the related literature, present an in-depth analysis of the stability and performance characteristics of impulsive control, and verify the theoretical conclusions experimentally using a custom-built atomic force microscope.

5.1 Introduction

The burgeoning field of nanoscale science and nanotechnology was arguably started in the 1980s with the invention of the scanning tunneling microscope [1] and the atomic force microscope (AFM) [2], two instruments which form the foundation of scanning probe microscopy (SPM), allowing us to interact with matter on scales down to the size of a single atom. Scanning probe techniques are currently

T. Tuma • A. Sebastian (✉) • A. Pantazi
IBM Research - Zurich, 8803 Rüschlikon, Switzerland
e-mail: uma@zurich.ibm.com; ase@zurich.ibm.com; agp@zurich.ibm.com

J. Lygeros • T. Tuma
ETH Zurich, 8092 Zürich, Switzerland
e-mail: lygeros@control.ee.ethz.ch

M. Rakotondrabe (ed.), *Smart Materials-Based Actuators at the Micro/Nano-Scale: Characterization, Control, and Applications*, DOI 10.1007/978-1-4614-6684-0_5,
© Springer Science+Business Media New York 2013

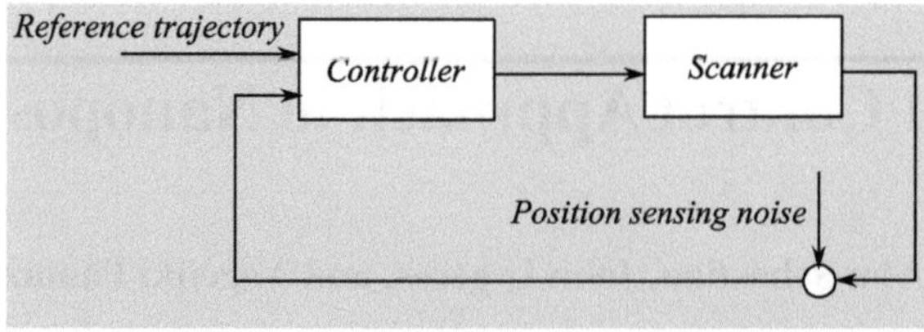

Fig. 5.1 A schematic of a feedback control loop as often encountered in nanopositioning. The controller forces the scanner to follow the desired reference trajectory by generating an actuation signal according to a control law based on the reference trajectory and a feedback measurement signal of the scanner position. The measurements are typically affected by sensing noise

used in the exploration of molecular structure [3], fabrication of nanometer-scale objects [4], observation of biological phenomena in real time [5], high-density data storage [6, 7], and characterization of semiconductor devices [8].

In SPM, precise control of motion with a subnanometer accuracy is of key importance. Nanometer-scale objects and their properties, such as topographical, electrical, and magnetic properties, are examined by means of a microfabricated cantilever with an atomically sharp tip at its end. Typically, the tip of the cantilever is positioned with subnanometer accuracy relative to the sample by means of a nanopositioner, and the sample properties are inferred from the forces acting on the tip. For example, to resolve the topography of a surface, the cantilever tip can be moved in a raster-based pattern across a rectangular scan area, and a three-dimensional estimate of the topography is obtained from analyzing the deflection of the tip. The accuracy of the nanopositioner, or scanner, is vital for the overall performance of the microscope.

In achieving the stringent accuracy requirements in nanopositioning, the concept of feedback control plays a pivotal role. Feedback controllers are essential in compensating for exogenous disturbances, such as building vibrations, temperature fluctuations, and manufacturing imperfections. Moreover, actuation techniques with ultra-high resolution, such as those using piezoelectric actuators, often exhibit significant nonlinearities (e.g., hysteresis, creep, and drift) which need to be compensated for by means of control. If high-speed operation is required, the complex dynamical behavior of the scanner also needs to be taken into account to prevent unwanted scanner-induced vibrations.

Figure 5.1 shows a schematic of a feedback loop as often encountered in nanopositioning. Here, the scanner is controlled by means of a feedback and/or feedforward controller whose inputs are the reference trajectory and the measurement signal. Based on the reference and the measurement signals, the controller executes a control law and makes the scanner follow the desired reference trajectory. The measurement signal is affected by additive measurement noise.

This chapter reviews a novel hybrid control approach to nanopositioning in which the controller combines a continuous-time control law with a discrete control law based on impulsive changes of the controller state. By using this impulsive state multiplication (ISM), some principal limitations of linear control can be overcome, such as the trade-off between the closed-loop bandwidth and the positioning resolution. Moreover, impulsive control is an important theoretical concept which links hybrid control with some of the recent nonlinear control techniques for nanopositioning and enables new and elegant ways to analyze and improve them. Despite its nonlinear character, impulsive control is easy to implement and can significantly improve the performance of existing nanopositioning systems. The exposition in this chapter is based on the theoretical and experimental results published in [9–12].

First, we briefly review the landscape of feedback control for nanopositioning in Sect. 5.2. Section 5.3 introduces the concept of impulsive control and its particular type, ISM, and analyzes the stability of linear systems with impulses. Section 5.4 presents control architectures based on impulsive feedback control, including feedback controllers for tracking piecewise constant and piecewise affine reference signals, and analyzes their performance. Section 5.5 discusses an inherent connection between impulsive control and the recently published signal transformation approach (STA) to nanopositioning, including techniques to significantly improve the transient performance of STA. Section 5.6 contains extensive experimental results, which demonstrate the properties of impulsive control and STA in a custom-built high-speed AFM. Section 5.7 concludes the chapter.

5.2 Feedback Control for Nanopositioning

In the design of feedback controllers for nanopositioning, specific control challenges must be taken into account, such as the high amount of measurement noise and the complex dynamical behavior of nanopositioners. In what follows, we review some of the recent linear and nonlinear control approaches in nanopositioning. The review includes an extensive account on the existing linear and nonlinear feedback control techniques in nanopositioning, which provides the necessary context for introducing the concept of impulsive control in Sect. 5.3.

5.2.1 Linear Feedback Control

Linear feedback controllers have been widely used in nanopositioning because of their conceptual and implementational simplicity [13–15]. Conventional linear feedback controllers are often based on the proportional, integral, and derivative (PID) control components. However, PID control often cannot meet the increasing requirements on robustness, accuracy, and speed. For instance, PI controllers inherently impose a trade-off between robustness and performance which becomes prohibitive in complex high-speed nanopositioning devices, especially when high bandwidth is of critical importance. Hence, a significant research effort has been devoted to developing advanced linear feedback architectures for nanopositioning and design methodologies.

Resonant control [16–18] is another technique which is used in nanopositioning as an alternative to conventional PI controllers. In integral resonant control, the controller consists of a constant feed-through term and a first-order integral feedback controller. Resonant controllers are well suited for damping of vibrations in systems with collocated actuators and sensors and are robust and simple to implement.

Advanced linear control techniques can improve the performance and design trade-offs of feedback control for nanopositioning, but they are subject to inherent constraints [19,20]. For example, there is a fixed mathematical relationship between the closed-loop sensitivity and complementary sensitivity transfer functions which dictates a trade-off between the tracking bandwidth and the positioning resolution [21]. At the same time, constraints such as Bode's integral law limit the shape of the closed-loop sensitivity transfer function. Consequently, bandwidth, positioning resolution, and robustness requirements often pose a significant challenge in linear feedback control.

In two-degree-of-freedom (2DOF) control [22], the control laws which act on the reference and measurement signals are specified independently. This is in contrast to the conventional single degree of freedom controller, which typically operates on the difference between the reference and the measurement signal. By using 2DOF controllers, the transfer functions among the reference signal, measurement noise, and output can be designed independently, which allows better design trade-offs.

5.2.2 Feedback Control of Repetitive Reference Signals

In many nanopositioning control problems, the reference signals are determined by the physics of the underlying application and can be exploited in the control design. In particular, the reference signals in SPM are often repetitive. For example, in raster scanning SPM, the tip is positioned relative to the sample along consecutive scan lines, which typically requires repetitive reference signals such as triangular or saw waveforms.

Repetitive reference signals can be tracked with tailored feedback controllers. In adaptive control [23], the control law itself is adapted online to improve the tracking performance. In repetitive control [24, 25], the feedback loop is augmented with a signal generator which recreates the *a priori* known reference signal, typically by means of a pure delay element. By doing so, the repetitive tracking error can be reduced after a certain number of iterations at the price of increased settling time, implementation complexity, and robustness issues. Iterative learning control (ILC) [26] is a similar learning-based concept in which, in contrast to repetitive control, the initial conditions are reset at every iteration and hence, discontinuous operation is possible. ILC-based controllers can achieve nearly perfect tracking but might result in a significant computation complexity which imposes requirements on the digital signal-processing hardware.

5.2.3 *Feedforward Control and Reference Signal Shaping*

Feedback control is often combined with feedforward control in which the dynamics and nonlinear characteristics of the actuator, such as creep and hysteresis, are anticipated, modeled, and inverted [27]. Feedforward control can significantly improve the closed-loop performance. However, it depends on the availability of accurate models and can be severely affected by changes in the plant dynamics. In some cases, nonlinear characteristics can be suppressed or inverted by using special hardware, such as in the case of charge-driven piezoelectric actuators. This simplifies the control design but increases the complexity and cost of the hardware.

A special case of feedforward control is command pre-shaping, i.e. shaping or modifying of the reference signal. For instance, vibrations induced by the reference signal can be reduced by using impulse input sequences [28]. In raster scanning SPM, shaping of the turnaround points of triangular waveforms has been extensively used and matured into a useful technique [29]. Most recently, alternative SPM scan trajectories which result in dramatically different reference signals have been proposed, such as spiral [30–32], cycloid [33], and Lissajous [34] scan trajectories. These trajectories can also be obtained by mathematical optimization [35].

5.2.4 *Hybrid Feedback Control*

Hybrid systems are dynamical systems that combine continuous and discrete dynamics [36]. They arise naturally in sampled digital control architectures and in applications where dynamical systems are combined with discrete logic. For example, in the framework of switched systems [37], multiple linear feedback controllers are switched depending on various algebraic and logical conditions, such as the operating points of the plant. Hybrid control systems are a particularly promising technique for advanced control of mechanical systems: on the one hand, like nonlinear systems, they have the potential of overcoming the fundamental limitations of linear feedback control; on the other hand, they allow the use of well-established linear control techniques in combination with nonlinear control laws.

One of the earliest developments of this kind in the field of applied hybrid control is the Clegg integrator [38]. Clegg observed that in a feedback loop with a single integrator, the phase lag due to the integral component can be significantly decreased if the state of the integrator is reset to zero whenever its input reaches zero. Denoting the integrator state $x(t)$ and its input $e(t)$, the dynamics of a Clegg integrator in state-space are

$$\dot{x}(t) = e(t) \text{ when } e(t) \neq 0 \quad (5.1)$$

$$x(t) := 0 \text{ for } e(t) = 0 \quad (5.2)$$

For instance, if $e(t)$ is the tracking error, resetting the integrator state when the tracking error is zero reduces the unnecessary overshoot caused by $x(t) \neq 0$ when the error is already removed. The effect of the reset control law is quite significant: the magnitude response of the integral controller remains unchanged, the phase lag decreases from 90° to only about 50°.

More generally, the Clegg integrator belongs to the class of reset control systems which have been further generalized and studied both theoretically and experimentally [39–41]. The concept of reset control is conceptually close to that of impulsive control, which will be studied in the remainder of this chapter.

5.3 Impulsive Control

Dynamical systems which evolve continuously but are subject to sudden impulses or impacts arise naturally and are one of the basic hybrid phenomena. Examples include colliding particles, bouncing balls, and systems with mechanical impacts. In a general treatment, impulsive systems have been studied thoroughly [42]. This chapter reviews a control concept recently introduced in nanopositioning and based on impulsive changes of the states of a linear feedback controller. We shall introduce a particular type of impulsive control namely, ISM, and analyze the stability and performance of the resulting hybrid impulsive system.

5.3.1 Impulsive State Multiplication

In ISM, the state of a dynamical system is multiplied by given factors at discrete time instants.

Let $\{t_i\}_{i=1}^{\infty}$ denote a sequence of time instants such that $0 = t_1 < t_2 < \cdots < t_i < \cdots$. Let $\{Q_i\}_{i=1}^{\infty}$ be a sequence of real square multiplication matrices $Q_i \in \mathbb{R}^{n \times n}$ for $i = 1, 2, \ldots$. The multiplication matrices will be referred to as state multiplication matrices (SMM), and in what follows, they will be assumed to be diagonal

$$Q_i = \begin{bmatrix} q_{i1} & 0 & \ldots & 0 \\ 0 & q_{i2} & \ldots & 0 \\ \ldots & & & \\ 0 & \ldots & & q_{in} \end{bmatrix} \tag{5.3}$$

for $i = 1, 2, \ldots$; n is the number of states in the state vector.

Definition 1. Let K be a linear, time-invariant system with state space matrices (A, B, C, D). *K with ISM* is a dynamical system that evolves according to the following equations:

$$\dot{x}(t) = Ax(t) + Bu(t) \text{ when } t \neq t_i, i = 1, 2, 3, \ldots \tag{5.4}$$

$$x(t_i) := Q_i x(t_i^-) \text{ for } i = 1, 2, 3, \ldots \tag{5.5}$$

$$y(t) = Cx(t) + Du(t) \tag{5.6}$$

where $t \in \mathbb{R}_+$, $x_K(t) \in \mathbb{R}^n$, $u(t) \in \mathbb{R}^m$, $y(t) \in \mathbb{R}^p$, $A \in \mathbb{R}^{n\times n}$, $B \in \mathbb{R}^{n\times m}$, $C \in \mathbb{R}^{p\times n}$, $D \in \mathbb{R}^{1\times m}$ and will be denoted as

$$\mathrm{ISM}(K, \{t_i\}_{i=1}^{\infty}, \{Q_i\}_{i=1}^{\infty}). \tag{5.7}$$

The assumption on the diagonality of SMM is not critical; however, it simplifies the analysis because it excludes any state coupling at the time of the multiplication.

5.3.2 *Stability*

Linear systems with ISM are subject to instantaneous state changes at predefined time instants. Because a stable linear system can be destabilized with ISM, the stability of the resulting hybrid system must be studied carefully with respect to the impulse magnitudes and timing. In the following, we present a bounded-input-bounded-output stability theorem for ISM-based systems.

Theorem 1. *Consider a system with ISM*

$$\mathrm{ISM}(K, \{t_i\}_{i=1}^{\infty}, \{Q_i\}_{i=1}^{\infty})$$

Assume that

$$0 < t_{i+1} - t_i < \Theta$$

and that the input signal is bounded, i.e.

$$\|u(t)\|_\infty < \infty$$

If there exists $q < 1$ such that

$$\left\| e^{A(t_{i+1}-t_i)} Q_i \right\| < q$$

for $i = 1, 2, 3 \ldots$ then

$$\|x(t)\|_\infty < \infty$$

The proof of Theorem 1 is relatively straightforward and can be found in [9]. The theorem parallels the results known from the theory of impulsive systems as elaborated, e.g., in [42]. While the theorem guarantees bounded-input-bounded-state stability when only an upper bound on the time between impulses is assumed, stronger results can be obtained, e.g., when the average time between the impulses is

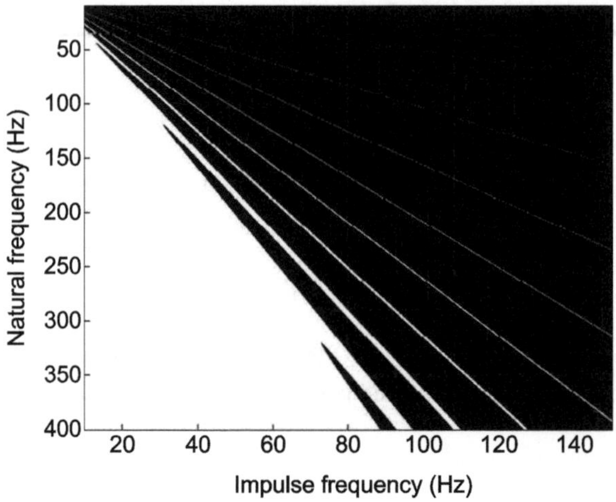

Fig. 5.2 Sufficient condition for stability of a second-order system under regular impulses. The *plots* show the quantity $\|e^{AT}Q_i\|$ as a function of the frequency of the impulses applied and the natural frequency and the damping coefficient of the plant, respectively. *White* corresponds to values smaller than one, *black* corresponds to values greater than or equal to one. The system is guaranteed to be stable in the *white* parameter region. Figure ©IEEE 2012, reprinted from [9] with permission

considered. This approach is well known in switched systems [37]. Theorem 1 sheds light onto the relation between the frequency and the magnitude of the impulses and the dynamics of the linear system required for stability. For example, if the time between the impulses is fixed, $T := t_{i+1} - t_i$ for $i \in \mathbb{N}$, and the magnitude of the impulses is upper bounded by Q such that $\|Q_i\| \leq Q$ for $i \in \mathbb{N}$, the stability condition can be rewritten as

$$\|e^{AT}\| < 1/Q. \tag{5.8}$$

Hence, if the stability condition applies, the possible locations of the system poles are limited. It can be shown that an upper bound on the real part of the system poles is

$$\Re(\lambda) \leq \frac{1}{T}\log\|e^{AT}\| < -\frac{\log Q}{T}. \tag{5.9}$$

For $Q > 1$, i.e., when the impulses magnify some of the states, $\log Q > 0$ and the real parts of the poles are enforced to be less than a negative number i.e. proportional to the frequency of the impulses. When $Q \leq 1$, the upper bound shifts to nonnegative numbers, suggesting that poles with a nonnegative real part may be present in a stable impulsive system. Indeed, such systems can be found, for instance, by constantly counteracting a state growth with an appropriate impulsive control law.

In the following, we demonstrate the stability condition for a second-order system with natural frequency ω_n and damping coefficient $\xi = 0.3$. Figure 5.2

shows the stability criterion as a function of the impulse frequency $1/T$ and the plant parameter ω_n. The impulse magnitude is uniform and fixed as $Q_i = \begin{bmatrix} 2 & 0 \\ 0 & 2 \end{bmatrix}$. The white region depicts the parameter space in which stability is guaranteed, the dark region depicts the parameter space in which $\|e^{AT}Q_i\| > 1$ and stability is not guaranteed. It is apparent that a high damping coefficient and/or a high natural frequency are vital for stability; stability is not guaranteed if fast impulses are applied to plants which are relatively "slow" or insufficiently damped. It also becomes clear that the frequency of the impulses may interfere with the natural frequency of the plant. For instance, the line-shaped stability regions in Fig. 5.2 arise when the impulses align with the natural frequency ω_n, e.g., when $T = c/\omega_n$ for some $c \in \mathbb{N}$. In such a case, the stability may be guaranteed even for very high impulse frequencies.

5.4 Impulsive Control for Feedback Systems

In this section, the concept of ISM is applied to feedback control. A novel control architecture is presented in which a linear feedback controller is extended into a hybrid system with ISM.

The control architecture is based on a one degree of freedom, single-input-single-output feedback loop which combines linear and impulsive control as shown in Fig. 5.3. The aim of the controller, K, is to force the plant, P, to track the reference signal, r. The controller consists of a linear feedback and a linear feedforward component, K_{FB} and K_{FF}, respectively. The feedback component is subject to impulsive state changes by the ISM block which can be driven either by the reference or by the measurement signals. The control and measurement signals are affected by the input disturbance signal, d_{i}, the output disturbance signal, d_{o}, and the measurement noise, n.

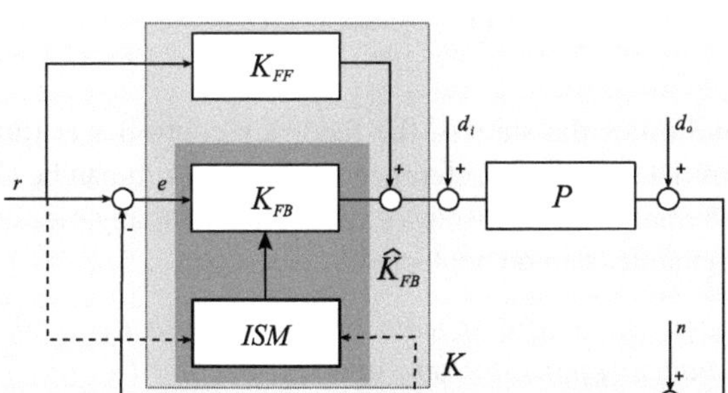

Fig. 5.3 Control architecture combining linear and impulsive control. Plant P is controlled by controller K comprising a feedforward component, K_{FF}, and a feedback component, K_{FB}. The *ISM* block impulsively modifies the states of K_{FB} based on the reference or measurement signals. Figure ©IEEE 2012, reprinted from [9] with permission

In what follows, we present a control design for tracking piecewise constant and piecewise affine reference signals that is based on the control architecture shown in Fig. 5.3. Piecewise constant and piecewise affine reference signals are widely used in nanopositioning and specially in SPM. The control architectures we present can significantly improve the tracking bandwidth for these reference signals without increasing the sensitivity to measurement noise. Thereby, they overcome one of the fundamental limitations of linear feedback systems. Finally, a general methodology is presented for the design of more complex control architectures, such as those for feedback control with multiple control objectives.

5.4.1 Tracking of Piecewise Constant Signals

In the following, a linear feedback control loop with a single integrator for tracking of piecewise constant signal is extended with ISM-based control laws.

Definition 2. A piecewise constant signal $r(t)$, $t \geq 0$ is determined by a series of time instants $\{t_i\}_{i=1}^{\infty}$ and values $\{v_i\}_{i=0}^{\infty}$ such that $r(t) = v_i$ when $t \in [t_{i-1}, t_i)$ for $i \in \mathbb{N}$.

First, consider the control scheme of Fig. 5.3 with $ISM = 0$, $K_{\text{FF}} = 0$, and P with no poles at the origin and assume that the controller involves a single integrator. In such a control scheme, a step signal can be tracked with a zero steady-state error. Because a piecewise constant signal can be viewed as a series of steps, it can also be tracked if the control loop has a sufficient bandwidth. The tracking will result in a short transient effect at the points of step changes and near-zero tracking error elsewhere.

By using impulsive control, the tracking performance of the single integral controller $K_{\text{FB}} = k/s$ can be improved without increasing the controller bandwidth. This can be done by using the state multiplication matrix (SMM).

$$Q_i = \left[\frac{v_i}{v_{i-1}}\right] \tag{5.10}$$

for $i \in \mathbb{N}$, which multiplies the state of the feedback controller at time instants t_i, $i \in \mathbb{N}$, by the ratio of the successive reference signal values. It can be shown that by using this SMM, the tracking error decays exponentially for any piecewise constant signal if the plant dynamics can be neglected:

Theorem 2. *Consider the control scheme in Fig. 5.3 with* $K_{\text{FF}} = 0$, $K_{\text{FB}} = \frac{k}{s}$. *If* $P(s) = 1$, *all disturbance signals are zero and* $r(t) = v_i$ *for* $t \in [t_{i-1}, t_i)$, $i \in \mathbb{N}$, *the controller*

$$\hat{K}_{\text{FB}} = \text{ISM}\left(\frac{k}{s}, \{t_i\}_{i=1}^{\infty}, \{\left[\frac{v_i}{v_{i-1}}\right]\}_{i=0}^{\infty}\right)$$

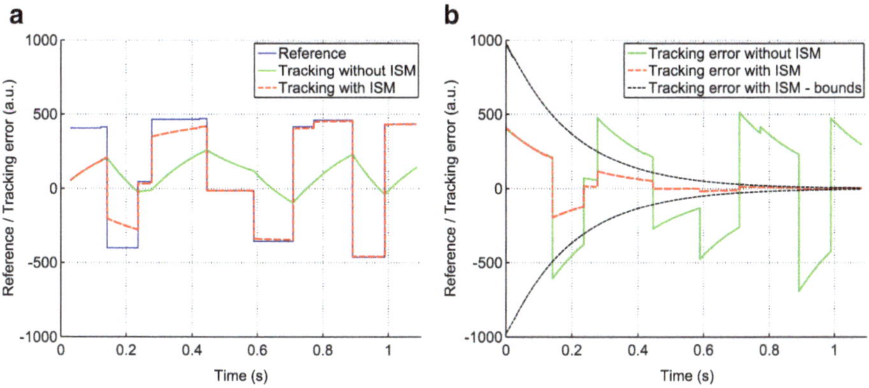

Fig. 5.4 Tracking of a piecewise constant reference signal with and without impulsive control (ISM). With impulsive control, the tracking error converges to zero despite the low bandwidth of the controller. The controller bandwidth governs the error bounds (shown in (**b**)), which decay exponentially to zero. (**a**) System output. (**b**) Tracking error. Figure ©IEEE 2012, reprinted from [9] with permission

tracks the reference signal with error

$$e(t) = v_i e^{-kt}.$$

Theorem 2 bounds the tracking error of an impulsive system with a single integral controller and a piecewise constant reference on an ideal plant. It says that the reference is tracked with an error that decays exponentially to zero.

We illustrate the results of Theorem 2 through simulations. A piecewise constant reference signal (shown in Fig. 5.4a as thin blue curve) was randomly generated and tracked on an ideal plant, $P(s) = 1$, with a single integral feedback controller $K(s) = 5/s$. The bandwidth of the controller is less than 1 Hz, which is very low compared with the bandwidth of the reference signal. Consequently, the reference signal is tracked poorly and with a large tracking error, as shown in Fig. 5.4a and Fig. 5.4b by the thick green curve. When K is equipped with ISM, the resulting feedback controller $\hat{K}$ is the nonlinear system

$$\hat{K} = \mathrm{ISM}\left(\frac{5}{s}, \{t_i\}_{i=1}^{\infty}, \left\{\left[\frac{v_i}{v_{i-1}}\right]\right\}_{i=0}^{\infty}\right). \tag{5.11}$$

In controller $\hat{K}$, the impulsive control law multiplies the state of the low bandwidth integrator by the ratio of the successive values of the reference signal. By doing so, a near-perfect tracking performance is achieved after a short transition period (shown in Fig. 5.4a, b as dashed red curve). The duration of the transition period and the magnitude of the tracking error can be derived from Theorem 2. The bounds on the tracking error (up to the scaling by v_i) are indicated in Fig. 5.4b by the solid black curve.

The assumption of $P(s) = 1$ is critical for the validity of Theorem 2 and translates into the requirement of a sufficient bandwidth and damping of the plant in the frequency region where impulses are applied. In many practical scenarios, this requirement is fulfilled: nanopositioners are often designed specifically for high bandwidth [43] and their dynamics may in addition be improved by active or passive damping [44,45]. Also, the linearity of the nanopositioner is important. For nanopositioners based on intrinsically nonlinear actuators, such as piezoelectric tube actuators or piezo-stack actuated flexure stages, a linearizing stage has to precede the application of impulsive control. Alternatively, in the case of piezoelectric actuation, using a charge amplifier instead of the conventional voltage amplifier to drive them can bring about sufficient linearity [46].

5.4.2 Tracking of Piecewise Affine Signals

Similar to piecewise constant signals, piecewise affine signals can be tracked by extending a linear feedback loop with an ISM-based control law.

Definition 3. A piecewise affine signal $r(t), t \geq 0$, is determined by a series of time instants $\{t_i\}_{i=1}^{\infty}$, offsets $\{v_i\}_{i=0}^{\infty}$ and slopes $\{w_i\}_{i=0}^{\infty}$ such that

$$r(t) = v_i + w_i(t - t_{i-1}) \tag{5.12}$$

when $t \in [t_{i-1}, t_i)$ for $i \in \mathbb{N}$.

Definition 3 defines a piecewise affine signal as a sequence of ramp-like signals with successive segments that do not necessarily share their endpoints. In addition, we will assume that the ramp always has a nonzero slope, i.e., $w_i \neq 0$ for $i \in \mathbb{N}$. The class of piecewise affine signals includes some of the signals widely used in nanopositioning, such as triangular waveforms with constant or varying speed and amplitude.

In the control framework of Fig. 5.3, piecewise affine reference signals can be tracked using a feedforward gain compensation and a single integral feedback controller. If the plant, P, has a nonzero, finite DC gain k_P which is known exactly, feedforward compensation with a constant term is applicable, i.e. $K_{\mathrm{FF}} = \frac{1}{k_P}$. With this feedforward compensation, it can be shown that a single integral feedback controller $K_{\mathrm{FB}}(s) = \frac{k}{s}$ tracks a ramp reference signal with a zero steady-state error. Consequently, given sufficient bandwidth of the system, also a piecewise affine reference can be tracked.

By employing impulsive control, fast piecewise affine signals can be tracked even with low bandwidth K_{FB}. The ISM control law that can do this originates from the control architecture for piecewise constant signals derived in Sect. 5.4.1; the problems are linked because tracking piecewise affine signals inherently involves tracking piecewise constant signals. This can be shown for various families of

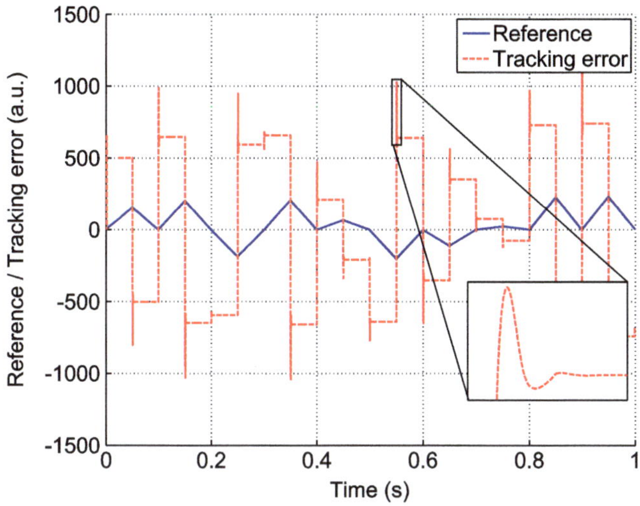

Fig. 5.5 Tracking of a piecewise affine signal on a second-order positioner (simulation). The tracking error is approximately piecewise constant and is determined by the slope of the reference signal. Figure ©IEEE 2012, reprinted from [9] with permission

plants, e.g., for linear, time-invariant plants with a nonzero, finite gain and proper transfer function [9]. For these plants, the steady-state tracking error under a ramp input is constant and proportional to the slope of the ramp. If the bandwidth of the plant is sufficiently high, such that the transients can be neglected, the tracking error can be approximated by a constant. Hence, because a piecewise affine signal is just a sequence of ramp signals, the tracking error for a piecewise affine reference signal is approximately a piecewise constant signal, provided the bandwidth of the plant is sufficiently high. Therefore, the tracking error can be reduced by applying the techniques presented in Sect. 5.4.1.

Figure 5.5 illustrates these observations in simulation for a second-order plant with natural frequency 1 kHz and damping ratio 0.5. A random triangular waveform with different amplitudes and slopes was applied to the plant and the output was subtracted from the input to obtain the tracking error. The tracking error is approximately piecewise constant, with short transients at the points where the reference signal changes the slope. The character of the transients depends on the bandwidth and damping of the plant.

Having realized that the tracking error can be approximated by a piecewise constant signal whose values are proportional to the slope of the ramp, w_i, and applying the results presented in Sect. 5.4.1, the SMM for tracking piecewise affine signals becomes

$$Q_i = \left[\frac{w_i}{w_{i-1}}\right] \tag{5.13}$$

for $i \in \mathbb{N}$. This SMM multiplies the state of the feedback controller by the ratio of the successive slopes of the reference signal. By doing so, the tracking error for piecewise affine signals can be reduced under the same conditions as discussed in Sect. 5.4.1, namely, for sufficiently damped plants the tracking error diminishes according to the exponential bounds of Theorem 2.

5.4.3 Feedback Control with Multiple Control Objectives

In the control architecture of Fig. 5.3, the ISM component is able to simultaneously execute multiple impulsive control laws which are based on different input signals and affect different states of the linear controller. Such a configuration is needed when more complex or multiple control objectives have to be met. A general way to design an ISM-based feedback control architecture in such a case is to partition the states of the linear controller and apply different ISM control laws to different parts of the controller.

Consider that the feedback controller, K_{FB}, is decomposed as

$$K_{\mathrm{FB}}(s) = \sum_{j=1}^{N} K_{\mathrm{FB}}^{j}(s), \tag{5.14}$$

where K_{FB}^{j}, $j = 1, \dots, N$ are the controller components. Assume that the realization of the i-th transfer function is $(A_{\mathrm{FB}}^{i}, B_{\mathrm{FB}}^{i}, C_{\mathrm{FB}}^{i})$ and consider the realization of the sum of transfer functions with

$$A_{\mathrm{FB}} = \begin{bmatrix} A_{\mathrm{FB}}^{1} & 0 & \dots & & & & \\ 0 & A_{\mathrm{FB}}^{2} & 0 & \dots & & & \\ \dots & & & & & & \\ 0 & \dots & 0 & A_{\mathrm{FB}}^{j} & 0 & \dots & \\ \dots & & & & & & \\ & & & & & & A_{\mathrm{FB}}^{N} \end{bmatrix}$$

and the state partitioned as

$$x_{\mathrm{FB}} = (x_{\mathrm{FB}}^{1} \dots x_{\mathrm{FB}}^{N})$$

where $x_{\mathrm{FB}}^{j} \in \mathbb{R}^{n(j)}$, $j = 1, \dots, N$, is the state of K_{FB}^{j} with the dimension $n(j) \in \mathbb{N}$. Accordingly, the input matrix

$$B_{\mathrm{FB}} = \left[B_{\mathrm{FB}}^{1}\ B_{\mathrm{FB}}^{2} \dots B_{\mathrm{FB}}^{N} \right]^{T} \tag{5.15}$$

where $B_{\mathrm{FB}}^{j} \in \mathbb{R}^{n(j)\times 1}$ is the input matrix of K_{FB}^{j}. The outputs of the controller components are added,

$$C_{\mathrm{FB}} = \left[C_{\mathrm{FB}}^{1}\; C_{\mathrm{FB}}^{2} \ldots C_{\mathrm{FB}}^{N} \right], \tag{5.16}$$

where $C_{\mathrm{FB}}^{j} \in \mathbb{R}^{1\times n(j)}$ is the output matrix of K_{FB}^{j}.

Assume that ISM is applied to the partitioned feedback controller,

$$\hat{K}_{\mathrm{FB}} := \mathrm{ISM}(K_{\mathrm{FB}}, \{t_i\}_{i=1}^{\infty}, \{Q_i\}_{i=1}^{\infty}). \tag{5.17}$$

The resulting impulsive controller, $\hat{K}_{\mathrm{FB}}$, has dynamics that evolve between the impulses, $t \neq t_i$, according to

$$\begin{aligned} \dot{\hat{x}}_K(t) &= A_{\mathrm{FB}}\hat{x}_K(t) + B_{\mathrm{FB}}e(t) \\ y(t) &= C_{\mathrm{FB}}\hat{x}_K(t), \end{aligned}$$

where $\hat{x}_K$ denotes the state of K and $e(t)$, $y(t)$ denote the system input and output, respectively. At the time of impulses, $t = t_i$, the state of K is multiplied

$$\hat{x}_K(t_i) := Q_i\hat{x}_K(t_i^-).$$

As the state $\hat{x}_K$ is partitioned according to the decomposition of the feedback controller, it is straightforward to partition the SMM:

$$Q_i = \begin{bmatrix} Q_i^1 & 0 & \ldots & & & & \\ 0 & Q_i^2 & 0 & \ldots & & & \\ \ldots & & & & & & \\ 0 & \ldots & 0 & Q_i^j & 0 & \ldots & \\ \ldots & & & & & & \\ & & & & & & Q_i^N \end{bmatrix}$$

In the partitioned SMM, $Q_i^j \in \mathbb{R}^{n(j)\times n(j)}$ is the SMM corresponding to the controller component K_{FB}^{j}. The advantage of the partitioning is that distinct impulsive control laws can be applied to the controller components. In the special case when Q_i^j is an identity matrix of appropriate dimensions, the corresponding states evolve linearly. When $Q_i^j = 0$, the corresponding states are reset to zero.

The method described above is general and can be applied to a wide variety of problems. The control architectures for tracking piecewise constant and piecewise affine signals presented above are one example thereof. In [9], an example in which multiple control objectives must be met simultaneously is elaborated. Experimental results using an ISM-based feedback loop to track a high-bandwidth reference signal in presence of significant input disturbance are presented in Sect. 5.6.

5.5 Relation Between Impulsive Control and Signal Transformation Approach

Recently, a new nonlinear control scheme for nanopositioning has been reported called STA [47]. STA is based on transformation of the reference, measurement, and control signals in a feedback loop. For example, triangular reference signals are tracked in STA by transforming the triangular waveforms into a ramp signal and vice versa. Interestingly, there is an inherent connection between STA and impulsive control which is analyzed in this section. The analysis has some important implications in the theoretical understanding of STA and leads to significant improvements in the implementation and performance of STA.

5.5.1 Signal Transformation Approach

STA is a control technique in which a linear feedback system is equipped with a pair of possibly time-invariant, nonlinear transformation functions. Consider the control scheme in Fig. 5.6. The transformation functions Φ, Φ^{-1} are mutually inverse and translate the signals between a low-bandwidth and high-bandwidth representation. Before the high speed reference signal enters the feedback loop, it is transformed by Φ into the low-bandwidth representation seen by the linear feedback controller K. The control effort that K generates is translated by Φ^{-1} before it is applied to the positioner to ensure that the desired reference signal is tracked. The measured output of the positioner is transformed by Φ back into the low-bandwidth representation.

The key idea of STA is that if the low-bandwidth representation of r is tracked accurately by K, the positioner follows the desired reference signal even if the bandwidth of K is much lower than what would be needed to track r in a conventional control architecture. The control framework of STA is general; specific transformation functions can be designed for the reference signals of interest.

One particular case is a triangular reference signal, which can be transformed into a ramp signal and vice versa by means of an affine, time-variant transformation [48, 49] as follows. Assume that the reference signal, $r(t)$, has frequency $f = 1/T$ and range $[0,A]$. For $t \geq 0$, divide the time into intervals corresponding to the half periods of the triangular waveform, $[(i-1)\frac{T}{2}, i\frac{T}{2})$ for $i \in \mathbb{N}$. In the odd

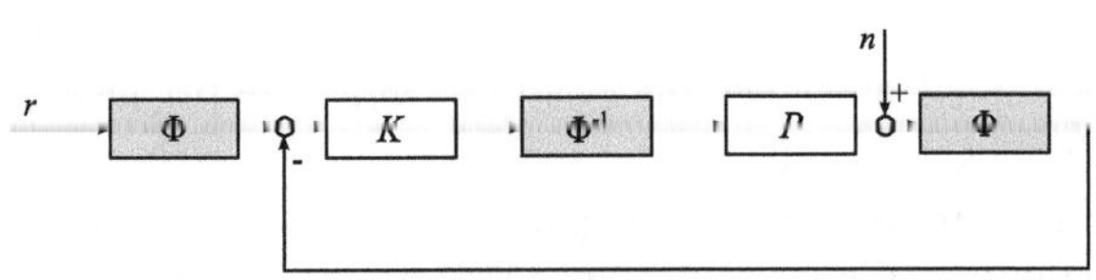

Fig. 5.6 Signal transformation approach to nanopositioning. A pair of mutually inverse transformation functions Φ, Φ^{-1} are used to transform the signals between a low-bandwidth and high-bandwidth representation. Figure ©Elsevier 2012, reprinted from [11] with permission

and even time intervals, $r(t)$ is strictly increasing and decreasing, respectively, and is governed by the equation

$$r(t) = (-1)^{(i-1)} \frac{2A}{T} \left(t - \lfloor \frac{i}{2} \rfloor T \right) \tag{5.18}$$

for $t \in [(i-1)\frac{T}{2}, i\frac{T}{2})$, $i \in \mathbb{N}$.

Definition 4. The forward and the inverse signal transformation operator, for conversion between a triangular waveform of frequency $1/T$ and range $[0,A]$ and a ramp signal with slope $2A/T$ are

$$\Phi(x(t),t) = (-1)^{(i-1)} x(t) + 2A \lfloor \frac{i}{2} \rfloor \tag{5.19}$$

for $t \in [(i-1)\frac{T}{2}, i\frac{T}{2})$, $i \in \mathbb{N}$.

$$\Phi^{-1}(x(t),t) = (-1)^{(i-1)} x(t) + (-1)^i 2A \lfloor \frac{i}{2} \rfloor \tag{5.20}$$

for $t \in [(i-1)\frac{T}{2}, i\frac{T}{2})$, $i \in \mathbb{N}$.

In the STA control scheme of Fig. 5.6, the above transformation functions can be used to facilitate the tracking of fast triangular waveforms by means of a low bandwidth double integral controller K. This is particularly beneficial in nanopositioning, wherein a low bandwidth K can be used for a control design with low sensitivity to the measurement noise, n.

5.5.2 *ISM and Multiplicative Signal Transformation*

The link between ISM and STA is based on the fact that multiplicative impulsive changes to the state of a linear system are equivalent to multiplicative transformation of the system input and output. This important fact is stated precisely in the following.

Consider the control scheme in Fig. 5.7. The input and output of a linear, time-invariant system K, are transformed by a pair of signal transformation operators J^{-1} and J, respectively. Assume that the time, $t \geq 0$, is divided into intervals $[t_{i-1}, t_i)$, $i \in \mathbb{N}$. The transformation operators are piecewise constant over the time intervals and multiply the signals as

$$\begin{aligned} J(u(t),t) &:= j_i u(t) \\ J^{-1}(u(t),t) &:= j_i^{-1} u(t) \end{aligned} \tag{5.21}$$

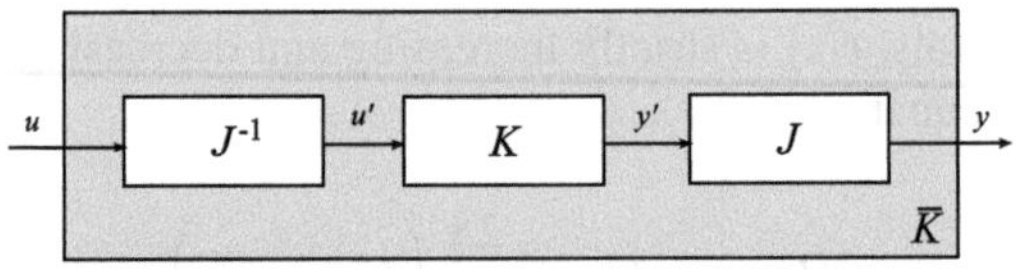

Fig. 5.7 Multiplicative signal transformation. The input and output of linear system K are transformed by a pair of mutually inverse multiplicative transformation operators. This scheme is equivalent to ISM applied to system K, thereby creating a link between ISM and STA. Figure ©Elsevier 2012, reprinted from [11] with permission

for $t \in [t_{i-1}, t_i)$, $i \in \mathbb{N}$. The multiplication factors j_i, $i \in \mathbb{N}$, are nonzero real numbers. At any given time point $t \in [t_{i-1}, t_i)$, the input signal, $u(t)$ is transformed into $u'(t) := j_i^{-1} u(t)$ and enters K. The output of K, $y'(t)$, is transformed as $y(t) = j_i y'(t)$. Hence, the input and output signals entering K are pre-multiplied and post-multiplied, respectively, by mutually inverse scalar factors.

The multiplicative signal transformation shown in Fig. 5.7 is tightly related to ISM. In fact, the following theorem holds:

Theorem 3. *For a given input $u(t)$, multiplication factors $\{j_i\}_{i=1}^{\infty}$ and times $\{t_i\}_{i=1}^{\infty}$, the output $y(t)$ of system $\bar{K}$ is the same as the output, $z(t)$, of the impulsive system*

$$\hat{K} := \mathrm{ISM}\left(K, \{t_i\}_{i=1}^{\infty}, \{\frac{j_{i+1}}{j_i} I\}_{i=1}^{\infty}\right) \tag{5.22}$$

for all times $t \geq 0$.

In other words, multiplicative signal transformation by factors j_i and $1/j_i$ is equivalent to ISM with SMM $Q_i = \frac{j_{i+1}}{j_i} I$. This fact can be shown by induction; the proof can be found in [11].

5.5.3 *Tracking of Triangular Waveforms*

Based on the relation between multiplicative impulsive changes to the state of a linear system and multiplicative signal transformation of its input and output presented in the preceding section, STA and ISM for tracking of triangular waveforms can be rigorously linked.

To reveal the connection between ISM and STA, the STA control scheme for tracking triangular waveforms (see Sect. 5.5.1) can be decomposed and simplified in a series of steps. This provides useful insights into the inherent properties of the scheme and reveals its relation to impulsive control. A detailed exposition is presented in [11]; in what follows, only the basic results are presented.

First, define the signal transformation operator

$$J(x(t), t) := (-1)^{i-1} x(t) \tag{5.23}$$

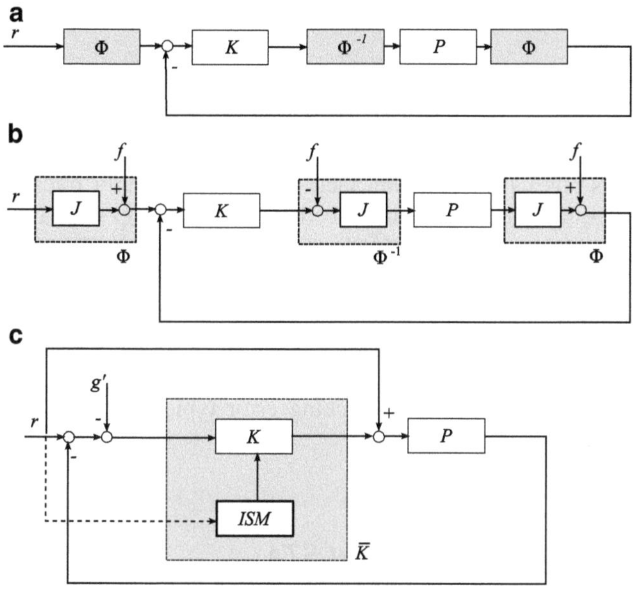

Fig. 5.8 Tracking of triangular waveforms: the relation between STA and ISM. (a) shows an STA control scheme with signal transformation blocks Φ and Φ^{-1}. (b) shows a control scheme equivalent to that shown in (a) with the signal transformation blocks decomposed. (c) shows an equivalent ISM-based control scheme

for $t \in [(i-1)\frac{T}{2}, i\frac{T}{2})$, $i \in \mathbb{N}$ where $1/T$ is the frequency of the triangular reference signal. The operator J inverts the sign of the input signal in the even half periods of the reference signal, i.e., when the triangular waveform is decreasing. It preserves the sign of the input signal in the odd half periods. Hence, J is the basic element of the transformation between a triangular waveform and a ramp signal.

Using the transformation operator J and defining an auxiliary stair-like signal $f(t)$, we realize that the signal transformation operators defined in Definition 4 can be expressed as

$$\Phi(x(t),t) = J(x(t),t) + f(t) \tag{5.24}$$

$$\Phi^{-1}(x(t),t) = J(x(t) - f(t),t) \tag{5.25}$$

Therefore, the control scheme of Fig. 5.8a is equivalent to the control scheme of Fig. 5.8b, with the transformation blocks decomposed.

By applying basic algebraic rules for signal flow diagrams and Theorem 3, the control scheme of Fig. 5.8b can be transformed into the control scheme shown in Fig. 5.8c, where the disturbance signal

$$g'(t) = (-1)^{i-1}\frac{2A}{k_1 T}e^{-\frac{k_2}{k_1}t} \tag{5.26}$$

for $t \in [(i-1)\frac{T}{2}, i\frac{T}{2})$, $i \in \mathbb{N}$, and the feedback controller is an ISM-based impulsive system

$$\mathrm{ISM}(K, \{i\frac{T}{2}\}_{i=1}^{\infty}, \{-I\}_{i=1}^{\infty}) \tag{5.27}$$

while yielding the same input–output relation.

The analysis reveals an important fact: the STA scheme for tracking triangular waveforms is equivalent to a negative feedback control loop with a constant-term feedforward connection and an impulsive feedback controller. The sign of the controller state is reversed at each turnaround point of the triangular reference signal. Furthermore, the analysis reveals the presence of a fixed bounded signal, $g'(t)$, which enters the feedback loop externally. The signal is implicit in the definition of the transformation operators and decays exponentially to zero; however, it is responsible for the large transient tracking error typically observed in the STA control loop.

5.5.4 Transient Performance of STA

The analysis presented in the preceding sections shows that the transient tracking error of STA for tracking triangular waveforms is strongly affected by the disturbance signal $g'(t)$ entering the feedback loop, see Fig. 5.8c. The disturbance signal is intrinsically generated by the signal transformation operators and depends on the reference signal, $r(t)$, and the double-integral feedback controller, K:

$$K(s) = \frac{k_1 s + k_2}{s^2} \tag{5.28}$$

Exact knowledge of the disturbance signal allows us to estimate the tracking error of STA analytically. It can be shown [11] that the tracking error induced by $g'(t)$ corresponds to a natural response of a second-order system with "damping coefficient" k_1 and "spring coefficient" k_2, with k_1 and k_2 being the integral gains of the feedback controller K. For example, if $k_2 > k_1^2/4$, the equation of the tracking error induced by the intrinsic disturbance signal is

$$|e_{g'}(t)| = \frac{2A/T}{\sqrt{k_2 - \frac{k_1^2}{4}}} e^{-\frac{k_1}{2}t} \sin t\sqrt{k_2 - \frac{k_1^2}{4}} \tag{5.29}$$

The equation says that the magnitude of the tracking error induced by $g'(t)$ is a sinusoidal waveform that exponentially decays to zero. It also shows that it is possible to design the feedback controller so that the transient response is "critically damped" by choosing gains k_1, k_2 that preserve the equation $k_2 - \frac{k_1^2}{4} = 0$.

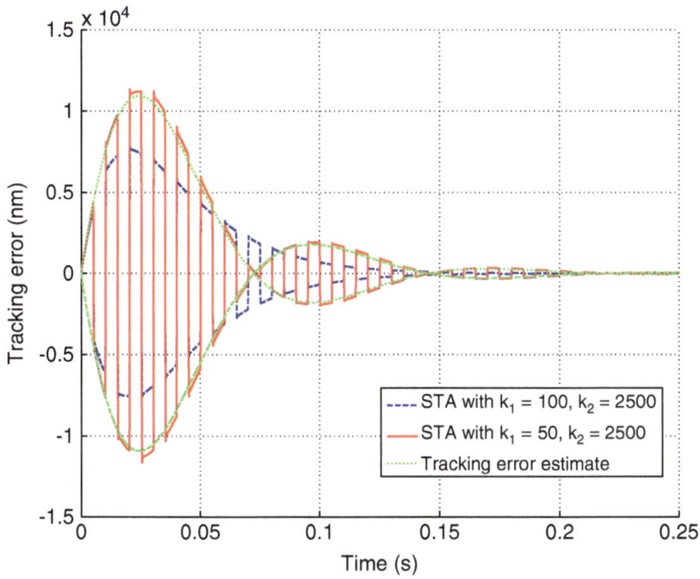

Fig. 5.9 The transient tracking error of STA as a function of the feedback controller gains. By choosing the controller gains carefully, a critically damped transient response can be achieved (*dashed blue line*). Oscillations may occur, however, for certain combinations of controller gains (*bold red line*). The transient tracking error can be quantified accurately (*dotted green line*). Figure ©Elsevier 2012, reprinted from [11] with permission

In the following example, the effect of k_1 and k_2 on the transient tracking error of STA is demonstrated. Assume that P is a second-order positioner with natural frequency 3,000 Hz and damping coefficient 0.8, and the reference signal is a triangular waveform of 100 Hz frequency and 5 µm amplitude. We will compare the tracking performance of the STA control architecture of Fig. 5.8a if two different feedback controllers are used. In the first controller, the integral gain coefficients are $k_1 = 100$ and $k_2 = 2{,}500$, so that $k_2 - \frac{k_1^2}{4} = 0$. In the second controller, $k_1 = 50$ and $k_2 = 2{,}500$. Figure 5.9 shows the tracking error for the first and second controller as dashed blue and solid red lines, respectively. As predicted by theory, the first controller exhibits a transient tracking error that converges exponentially to zero without oscillations. The tracking error of the second configuration also decays exponentially to zero, but, with undesirable transient oscillations. For the second configuration, the estimate of the transient tracking error based on (5.29) is shown as dashed green line. Equation (5.29) captures the tracking error exactly if $P(s) = 1$; for plants that are well damped and sufficiently fast, such as the one used in the simulation, (5.29) can be used as an accurate estimate of the tracking error.

The transient response of STA can be improved in several ways. In the most straightforward approach, STA can be simplified by removing the parasitic disturbance signal $g'(t)$ completely: as the preceding analysis shows, STA can be implemented by merely inverting the signs of the input and the output of the feedback controller and using a constant feedforward term. Also, an equivalent performance can be achieved with the ISM-based control scheme of Fig. 5.8c. Other ways to improve the performance of STA are discussed in [11].

5.6 Experiments

In this section, the concept of impulsive control and its properties are demonstrated in an experimental setup. The experimental results are obtained on a custom-built AFM, a type of a scanning probe microscope.

A schematic of a common AFM setup is shown in Fig. 5.10. The sample under investigation is mounted on a nanopositioner, or scanner, which can move the sample in three degrees of freedom. To enable high-speed, high-resolution motion in the presence of external disturbances and modeling uncertainties, the scanner is typically equipped with highly sensitive noncontact sensors and operated in a feedback loop. During operation, the sample is brought into contact with the tip of a microfabricated cantilever and positioned relative to the tip along a two-dimensional scan trajectory, e.g., along a raster-based, spiral or Lissajous pattern. Properties of the sample, such as its topographical, electrical, or magnetic properties, are inferred from the nanoscale forces acting on the cantilever tip as the tip is moved across

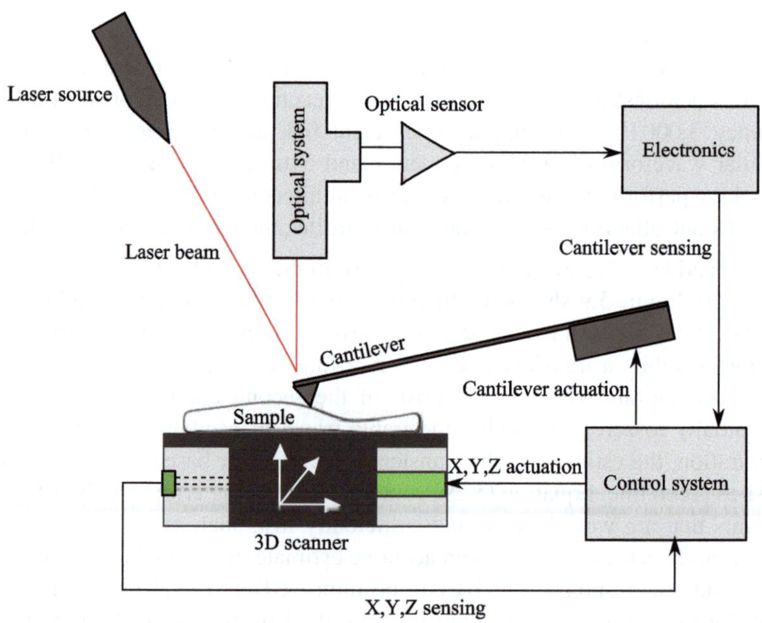

Fig. 5.10 Schematic of an atomic force microscope

the sample. In the schematic shown in Fig. 5.10, the deflection of the cantilever is estimated by means of an optical read-out system in which a laser beam is reflected from the cantilever tip, passes through an optical system and detected by means of a special-purpose optical sensor. The deflection of the cantilever provides an accurate estimate of the topography of the sample.

The feedback loop which controls the scanner motion is of key importance for the accuracy and speed of the AFM instrument. In the experiments described below, the feedback loop is equipped with an ISM-based hybrid controller or a linear feedback controller for comparison and used for high-speed AFM imaging. In the imaging experiments, a raster-based scan trajectory is followed by actuating the scanner in two orthogonal directions. In the fast, x-direction, a triangular reference signal is applied to scan the sample in consecutive scan lines; in the slow, y-direction, the position of the scanner is increased in discrete steps at the end of every scan line. In the x-scan direction, the nanopositioner uses a voice-coil actuator which is particularly suitable for this task because of its clean dynamics and linearity. In the y- and z-scan directions, a piezo-electric actuator is used. In the imaging experiments, thermomechanical cantilevers were used [50,51]. The sample consists of nanolitograpically patterned nanostructures on silicon surface.

For the stability and performance of impulsive control and STA, the flat frequency response of the nanopositioner is of key importance. To that end, the mechanical resonances of the nanopositioner were damped by inverting its dynamics. The resulting transfer function could be approximated as $P(s) \approx 1$ in the frequency region below approx. 250 Hz. Owing to the voice-coil actuation principle, the dynamics of the scanner are highly linear, which is important for impulsive control. For positioners with significant nonlinearities, such as piezo-actuated positioners, additional compensation techniques might have to be employed, such as feedforward inversion [27] and charge amplifiers [46]. A magnetoresistive sensor [52] with a bandwidth exceeding 10 kHz was used to sense the position of the nanopositioner.

5.6.1 *Impulsive Control for Tracking Piecewise Affine Signals*

In this experiment, the performance of a linear feedback loop with and without ISM is compared and its effect on the quality of AFM imaging is demonstrated.

The control architecture used was the one shown in Fig. 5.3, with P denoting the shaped positioner dynamics along the fast scan axis. The feedback controller, K_{FB} was chosen as a simple integral controller such that the closed loop system had a very low bandwidth below 1 Hz. A benefit of such a scheme is that it minimizes the impact of measurement noise. The feedforward term K_{FF} was unity as the DC gain of the shaped positioner is 1.

The imaging was performed over an area of $3.5\,\mu\mathrm{m} \times 1.5\,\mu\mathrm{m}$ in a raster pattern. In the first case, the controller was only the linear feedback system as described above. In the second case, the linear feedback system was equipped with an ISM-based

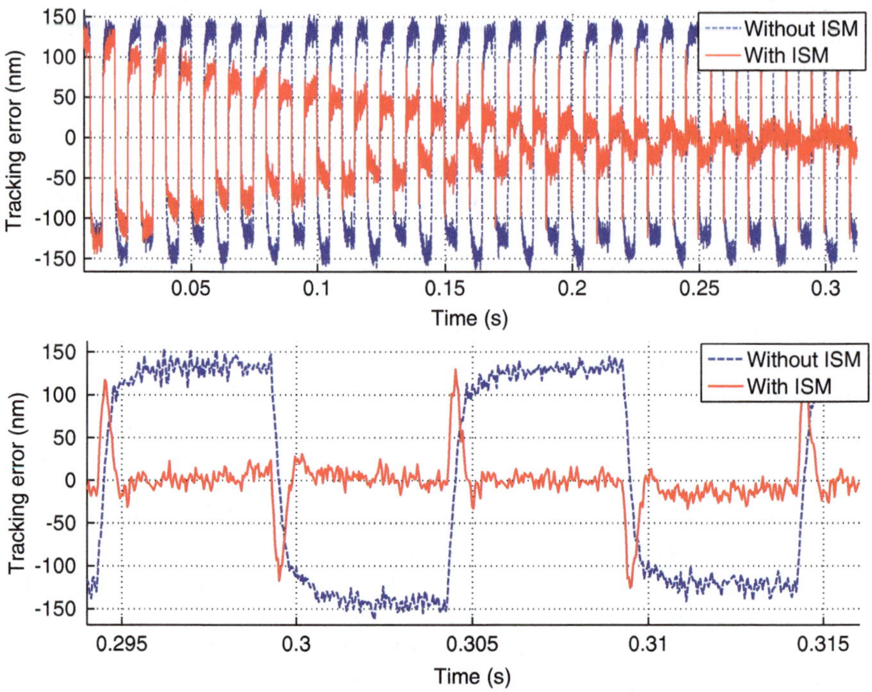

Fig. 5.11 Comparison of tracking error for closed loop without and with ISM. The *upper panel* shows the settling period of ISM, the *lower panel* shows the steady-state performance. The frequency of the reference signal was 100 Hz. Figure ©IOP Publishing 2011, reprinted from [10] with permission

impulsive control law as described in Sect. 5.4.2. Because the reference signal is a constant-amplitude, constant-frequency triangular waveform, the ISM control law simply inverted the sign of the accumulator of K at every turnaround point of r.

The tracking performance of the closed loop system with and without ISM is compared in Fig. 5.11. Using the ISM-based feedback controller, after a short settling time (top panel) near-zero steady-state error (bottom panel) was achieved; the convergence time corresponds to the rise time of the feedback loop $K/(1+K)$ under a step input which is consistent with the estimates presented in Sect. 5.5. The reduction of the positioning error is truly remarkable as it is enabled by the impulsive control law; the bandwidth of the linear control remains very low. This has a significant effect on the noise sensitivity of the control loop, an important aspect in nanopositioning as we shall see in the next section.

Figure 5.12 demonstrates how the tracking performance affects AFM imaging by comparing the images obtained using open loop control (on the left), feedback control without ISM (in the middle), and feedback control with ISM (on the right). The closed-loop image without ISM is similar to the open-loop image because of the very low bandwidth of the controller. The tracking lag resulted in an uncorrected,

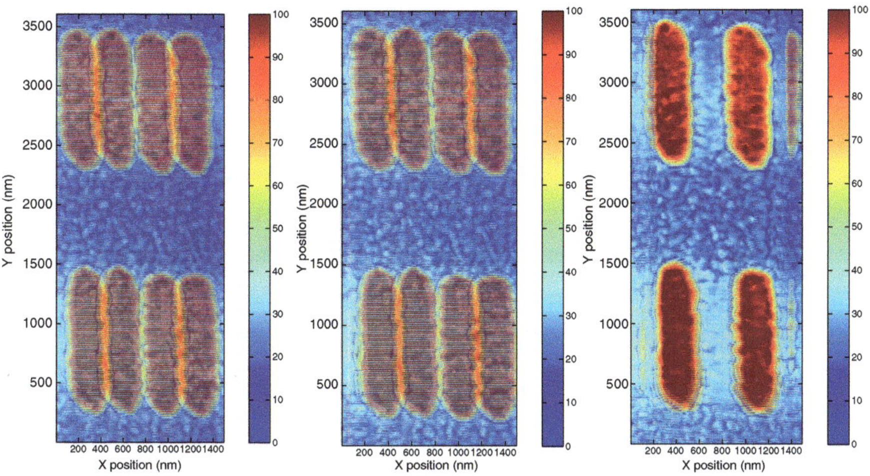

Fig. 5.12 The images of nanopatterns obtained with three different control schemes. The image on the *left* was acquired in open loop, using only the shaping of the positioner dynamics. The image in the *middle* was obtained using feedback control without ISM. The image on the *right* was acquired using feedback control with ISM. The height is in nanometers. The images were obtained in approx. 2 s. Figure ©IOP Publishing 2011, reprinted from [10] with permission

unknown shift in the image which was different for forward and backward scans. On the other hand, ISM removed the tracking lag almost perfectly without increasing the bandwidth of K. This not only resulted in detecting the real positions of the nanopatterns but also yielded consistent data in both scan directions, speeding up the imaging by a factor of two.

5.6.2 Sensitivity to Measurement Noise

In high-speed nanopositioning, measurement noise can significantly deteriorate the positioning accuracy. For example, in the control architecture of Fig. 5.3, additive measurement noise affects the measurement signal, enters the feedback loop, and is projected into the motion of the nanopositioner. An estimate of the nanopositioner motion induced by the measurement noise is an important figure in assessing the feedback control scheme and its suitability for nanopositioning. In the following, the nanopositioner motion induced by measurement noise is estimated for a high-bandwidth linear controller and an ISM-based controller of equivalent tracking performance.

The character and frequency spectrum of the measurement noise are determined by the position sensor. In what follows, a relatively noisy magnetoresistive position sensor is used to demonstrate the effect of ISM. The spectral characteristics of the measurement noise are presented in Fig. 5.13 and are largely dominated by *1/f* noise

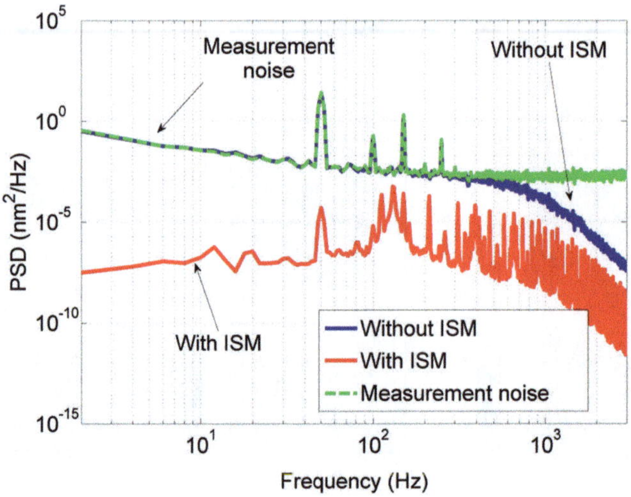

Fig. 5.13 Frequency spectra of the positioner motion induced by measurement noise. The measurement noise (*dashed green curve*) significantly affects the positioner motion if a high-bandwidth linear controller is used (*solid blue curve*). By using impulsive control, the positioning error introduced is significantly smaller (*red curve*) without compromising the tracking performance. Figure ©IEEE 2012, reprinted from [9] with permission

as is common for this type of sensor; spurious peaks occur at frequencies which are multiples of 50 Hz and are due to the ambient electrical noise. The standard deviation is approx. 9 nm over the frequency range from 0 Hz to 3 kHz.

For the purposes of a fair comparison, the ISM-based controller used in Sect. 5.6.1 was compared with a linear high gain integrator $K_{\mathrm{FB}}(s) = 1{,}530/s$. By increasing the gain of the integrator, the steady-state performance of both control schemes was made equivalent. The ISM-based and the linear controller were used to track a reference signal of frequency 130 Hz.

For the linear controller, the simulated spectral characteristics of the estimated positioning error are shown in Fig. 5.13 in blue. Because of the high bandwidth of the controller, the measurement noise at low frequencies significantly affects the motion of the positioner and induces a positioning error with a standard deviation of more than 8 nm over the frequency range examined.

For the ISM-based control scheme, the standard deviation of the positioning error is less than 1 nm over the frequency range from 0 Hz to 3 kHz. The spectral characteristics are shown in Fig. 5.13 in red. This improvement in resolution is achieved even though the tracking performance is not compromised; in fact, the transient tracking performance of the ISM-based controller is even better than that of the linear controller. However, there are additional peaks in the frequency spectrum which are induced by the impulsive changes of the controller state. In particular, the spectral component at 130 Hz corresponds to the frequency of the signal tracked. Further spectral components at frequencies which are multiples of 130 Hz occur because of the discontinuous evolution of the state. However, these spectral components have only minimal influence on the overall tracking performance. Moreover, they are not strongly affected by the closed-loop dynamics because the impulsive control law is of feedforward type and is fully determined by the reference signal.

5.6.3 Multiobjective Impulsive Control: Tracking and Disturbance Rejection

Impulsive control can be applied to feedback controllers with a complex structure, such as those designed for achieving multiple control objectives simultaneously. To demonstrate this, a feedback controller for the control architecture of Fig. 5.3 is presented which can track a fast triangular waveform in the presence of a sinusoidal input disturbance.

The feedback controller consists of a tracking component, a disturbance rejection component and an ISM-based component. Using the notation introduced in Sect. 5.4.3, the controller takes the form

$$K_{\mathrm{FB}}(s) = K_{\mathrm{FB}}^{R}(s) + K_{\mathrm{FB}}^{Di}(s), \tag{5.30}$$

where $K_{\mathrm{FB}}^{R}(s) = k/s$ is the tracking component, here a single integrator with gain k, and K_{FB}^{Di} rejects the sinusoidal input disturbance. For example, K_{FB}^{Di} can be a peak filter or a model-based controller. To track the fast triangular waveform with low gain k, K_{FB}^{R} is extended with an impulsive control law as in the previous experiments. On the other hand, given the type of the disturbance, the states of K_{FB}^{Di} are not affected by the impulsive control law and evolve linearly.

This feedback controller was used for high-speed AFM imaging in the presence of significant input disturbance that would cause scanner motion of more than 500 nm in the absence of control. Figure 5.14 compares the images taken in open-loop mode and with the feedback controller. In open-loop mode, the image is heavily distorted by the input disturbance, as is evident in Fig. 5.14a. By using the proposed controller, the effect of the input disturbance is minimized and at the same time, excellent tracking performance is achieved (Fig. 5.14b).

5.6.4 Transient Performance of STA and Impulsive Control

Section 5.5 provided a detailed analysis of the relation between impulsive control and the recently published STA. It was shown that for triangular waveform tracking, both schemes are inherently linked and essentially equivalent. Importantly, one of the conclusions in the theoretical analysis was that impulsive control has

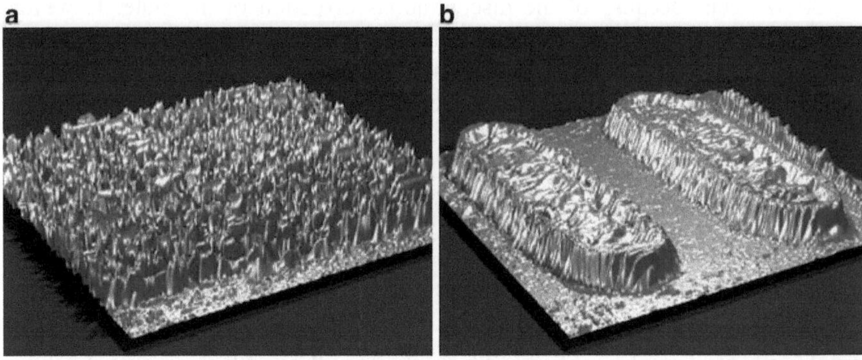

Fig. 5.14 High-speed AFM images of approx. $1{,}000 \times 250 \times 100$ nm titanium nitride nanopatterns taken in the presence of a significant input disturbance. The image in (**a**) was acquired in open-loop; the image in (**b**) was acquired using a feedback controller with combined control objectives for tracking and disturbance rejection. It demonstrates the combination of impulsive and linear control: the good tracking performance was obtained with only a low-gain integrator with ISM. Moreover, at the same time, the effect of the input disturbance was minimized. Figure ©IEEE 2012, reprinted from [9] with permission

significantly less transient tracking error than STA while having the same steady-state performance. In the following, experiments are presented which compare impulsive control with STA.

The STA control architecture was implemented as shown in Fig. 5.8a. A double integral controller

$$K(s) = \frac{50}{s} + \frac{1{,}000}{s^2} \tag{5.31}$$

was used together with the pair of signal transformation operators Φ, Φ^{-1} as defined in Sect. 5.5.1. By doing so, the triangular reference signal was transformed into a ramp signal which was tracked by the feedback controller K.

In the control architecture based on impulsive control, the control scheme of Fig. 5.3 was used with the feedforward term $K_{\mathrm{FF}}(s) = 1$ and the feedback controller

$$K_{\mathrm{FB}}(s) = \frac{50}{s} \tag{5.32}$$

As the DC gain of the positioner was known exactly, a single integrator was sufficient to provide a near-zero tracking error in the steady state. The state of the feedback controller was multiplied by -1 at every turnaround point of the triangular reference signal. The factor of -1 is the ratio of the successive slopes of the single frequency triangular signal.

Figure 5.15 compares the tracking performance of STA and impulsive control. As predicted by theory, the transient phase of STA is dominated by a large overshoot which exponentially decays to zero. Because of the overshoot, the amplitude of the reference signal had to be limited so that the scanner stroke was not exceeded

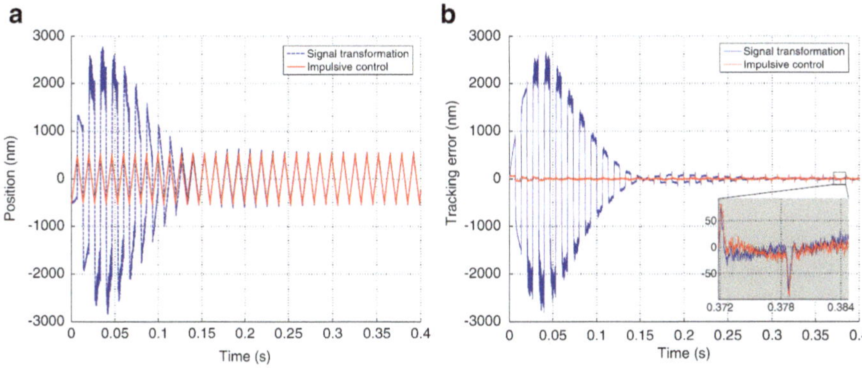

Fig. 5.15 Tracking of a fixed frequency triangular waveform with a control architecture based on impulsive control in *blue* and signal transformation approach in *red*. The frequency of the reference signal was 75 Hz and the amplitude was 1 µm. (**a**) Tracking output (position). (**b**) Tracking error. Figure ©Elsevier 2012, reprinted from [11] with permission

in the transient phase. The transient effects also induced unwanted mechanical vibrations. The transient phase of the impulsive control scheme did not suffer from the overshoot present in STA and quickly achieved the convergence as discussed in Sect. 5.5.4. The steady state performance of both schemes was comparable.

5.7 Conclusion

We have reviewed a novel hybrid control approach to nanopositioning that is based on the concept of impulsive control. In impulsive control, the states of a feedback controller are changed abruptly at discrete instances in time. We have analyzed the stability of impulsive control systems and introduced feedback control architectures based on impulsive control, including control architectures for tracking piecewise constant and piecewise affine signals and multiobjective hybrid feedback control. We discussed the inherent connection between impulsive control and the recently published STA to nanopositioning. The theoretical findings are supported by experiments in which impulsive control was used in a high-speed AFM. The experiments show that by using impulsive control, the tracking error in AFM can be minimized without increasing the bandwidth of the linear controller, thereby speeding up the AFM imaging process by a factor of two without increasing the sensitivity of the feedback loop to measurement noise. We have also demonstrated impulsive feedback control in the presence of a significant input disturbance and experimentally compared the performance of impulsive control and the signal transformation approach. The theoretical and experimental results indicate that impulsive control for nanopositioning has a significant potential for practical applications and at the same time fosters new research directions in the theoretical understanding of hybrid feedback control.

Acknowledgements We thank Urs Egger and Walter Häberle for their support with the mechanical and electronic hardware used in the experiments. Special thanks go to Haris Pozidis and Evangelos Eleftheriou for their support of this work.

References

1. G. Binnig, H. Rohrer, C. Gerber, E. Weibel, Surface studies by scanning tunneling microscopy. Phys. Rev. Lett. **49**(1), 57–61 (1982)
2. G. Binnig, C. Quate, C. Gerber, Atomic force microscope. Phys. Rev. Lett. **56**(9), 930–933 (1986)
3. L. Gross, F. Mohn, N. Moll, B. Schuler, A. Criado, E. Guitián, D. Peña, A. Gourdon, G. Meyer, Bond-order discrimination by atomic force microscopy. Science **337**(6100), 1326–1329 (2012)
4. D. Pires, J.L. Hedrick, A.D. Silva, J. Frommer, B. Gotsmann, H. Wolf, M. Despont, U. Duerig, A.W. Knoll, Nanoscale three-dimensional patterning of molecular resists by scanning probes. Science **328**, 732–735 (2010)
5. T. Ando, High-speed atomic force microscopy coming of age. Nanotechnology **23**, 062001 (2012)
6. E. Eleftheriou, T. Antonakopoulos, G. Binnig, G. Cherubini, M. Despont, A. Dholakia, U. Durig, M. Lantz, H. Pozidis, H. Rothuizen, P. Vettiger, Millipede - a MEMS-based scanning-probe data-storage system. IEEE Trans. Magn. **39**(2) 938–945 (2003)
7. A. Pantazi, A. Sebastian, T.A. Antonakopoulos, P. Baechtold, A.R. Bonaccio, J. Bonan, G. Cherubini, M. Despont, R.A. DiPietro, U. Drechsler, U. Duerig, B. Gotsmann, W. Haeberle, C. Hagleitner, J.L. Hedrick, D. Jubin, A. Knoll, M.A. Lantz, J. Pentarakis, H. Pozidis, R.C. Pratt, H. Rothuizen, R. Stutz, M. Varsamou, D. Wiesmann, E. Eleftheriou, Probe-based ultrahigh-density storage technology. IBM J. Res. Develop. **52**(4.5), 493–511 (2008)
8. R.A. Oliver, Advances in AFM for the electrical characterization of semiconductors. Rep. Progr. Phys. **71**(7), 076501 (2008)
9. T. Tuma, A. Pantazi, J. Lygeros, A. Sebastian, Nanopositioning with impulsive state multiplication: a hybrid control approach. IEEE Trans. Contr. Syst. Technol. (2012, to appear)
10. T. Tuma, A. Sebastian, W. Häberle, J. Lygeros, A. Pantazi, Impulsive control for fast nanopositioning. Nanotechnology **22**, 135501 (2011)
11. T. Tuma, A. Pantazi, J. Lygeros, A. Sebastian, Comparison of two non-linear control approaches to fast nanopositioning: impulsive control and signal transformation. Mechatronics **22**, 302–309 (2012)
12. T. Tuma, A. Pantazi, J. Lygeros, A. Sebastian, Impulsive control for nanopositioning: stability and performance, in *Proceedings of the 14th International Conference on Hybrid Systems: Computation and Control*, ACM, pp. 173–180 (2011)
13. S.M. Salapaka, M.V. Salapaka, Scanning probe microscopy. IEEE Contr. Syst. Mag. **28**(2), 65–83 (2008)
14. D. Abramovitch, S. Andersson, L. Pao, G. Schitter, A tutorial on the mechanisms, dynamics, and control of atomic force microscopes, in *Proceedings of the American Control Conference*, IEEE, pp. 3488–3502 (2007)
15. S. Devasia, E. Eleftheriou, S.O.R. Moheimani, A survey of control issues in nanopositioning. IEEE Trans. Contr. Syst. Technol. **15**(5), 802–823 (2007)
16. S. Aphale, A. Fleming, S. Reza Moheimani, Integral resonant control of collocated smart structures. Smart Mater. Struct. **16**, 439 (2007)
17. A. Fleming, S. Aphale, S. Moheimani, A new method for robust damping and tracking control of scanning probe microscope positioning stages. IEEE Trans. Nanotechnol. **9**(4), 438–448 (2010)

18. A. Sebastian, A. Pantazi, S.O.R. Moheimani, H. Pozidis, E. Eleftheriou, Achieving sub-nanometer precision in a MEMS-based storage device during self-servo write process. IEEE Trans. Nanotechnol. **7**(5), 586–595 (2008)
19. G. Schitter, R. Stark, A. Stemmer, Fast contact-mode atomic force microscopy on biological specimen by model-based control. Ultramicroscopy **100**(3), 253–257 (2004)
20. S. Salapaka, A. Sebastian, J.P. Cleveland, M.V. Salapaka, High bandwidth nano-positioner: a robust control approach. Rev. Sci. Instrum. **73**(9), 3232–3241 (2002)
21. A. Sebastian, S. Salapaka, Design methodologies for robust nano-positioning. IEEE Trans. Contr. Syst. Technol. **13**(6), 868–876 (2005)
22. C. Lee, S.M. Salapaka, Robust broadband nanopositioning: fundamental trade-offs, analysis, and design in a two-degree-of-freedom control framework. Nanotechnology **20**(3), 035501 (2009)
23. S. Bashash, N. Jalili, Robust adaptive control of coupled parallel piezo-flexural nanopositioning stages. IEEE/ASME Trans. Mechatron. **14**(1), 11–20 (2009)
24. S. Hara, Y. Yamamoto, T. Omata, M. Nakano, Repetitive control system: a new type servo system for periodic exogenous signals. IEEE Trans. Automat. Contr. **33**(7), 659–668 (1988)
25. Y. Shan, K. Leang, Repetitive control with Prandtl-Ishlinskii hysteresis inverse for piezo-based nanopositioning, in *Proceedings of the American Control Conference*, IEEE, pp. 301–306 (2009)
26. D. Bristow, M. Tharayil, A. Alleyne, A survey of iterative learning control. IEEE Contr. Syst. Mag. **26**(3), 96–114 (2006)
27. K. Leang, Q. Zou, S. Devasia, Feedforward control of piezoactuators in atomic force microscope systems. IEEE Contr. Syst. Mag. **29**, 70–82 (2009)
28. N.C. Singer, W.P. Seering, Preshaping command inputs to reduce system vibration. J. Dyn. Syst. Meas. Contr. **112**(1), 76–82 (1990)
29. A. Fleming, A. Wills, Optimal periodic trajectories for band-limited systems. IEEE Trans. Contr. Syst. Technol. **17**(3), 552–562 (2009)
30. I. Mahmood, S. Reza Moheimani, Fast spiral-scan atomic force microscopy. Nanotechnology **20**, 365503 (2009)
31. A. Kotsopoulos, T. Antonakopoulos, Nanopositioning using the spiral of archimedes: the probe-based storage case. Mechatronics **20**(2), 273–280 (2010)
32. A. Kotsopoulos, A. Pantazi, A. Sebastian, T. Antonakopoulos, High-speed spiral nanopositioning, in *Proceedings of IFAC world congress*, IFAC, pp. 2018–2023 (2011)
33. Y. Yong, S. Moheimani, I. Petersen, High-speed cycloid-scan atomic force microscopy. Nanotechnology **21**, 365503 (2010)
34. T. Tuma, J. Lygeros, V. Kartik, A. Sebastian, A. Pantazi, High-speed multiresolution scanning probe microscopy based on Lissajous scan trajectories. Nanotechnology **23**, 185501 (2012)
35. T. Tuma, J. Lygeros, A. Sebastian, A. Pantazi, Optimal scan trajectories for high speed scanning probe microscopy, in *Proceedings of the 2012 American Control Conference*, IEEE, pp. 3791–3796 (2012)
36. R. Goebel, R. Sanfelice, A. Teel, Hybrid dynamical systems. IEEE Contr. Syst. Mag. **29**(2), 28–93 (2009)
37. D. Liberzon, *Switching in Systems and Control*. Ser. Systems & Control: Foundations & Applications (Birkhäuser, Boston, 2003)
38. J.C. Clegg, A nonlinear integrator for servomechanisms. Trans. AIEE, Part II. Appl. Ind. **77**(2), 41–42 (1958)
39. I. Horowitz, P. Rosenbaum, Non-linear design for cost of feedback reduction in systems with large parameter uncertainty. Int. J. Contr. **21**, 977–1001 (1975)
40. D. Nesic, L. Zaccarian, A.R. Teel, Stability properties of reset systems. Automatica **44**, 2019–2026 (2008)
41. D. Wu, G. Guo, Y. Wang, Reset integral-derivative control for HDD servo systems. IEEE Trans. Contr. Syst. Technol. **15**(1), 161–167 (2007)

42. D.D. Bainov, P.S. Simeonov, *Systems with Impulse Effect: Stability, Theory and Applications*. Ser. Ellis Horwood Series: Mathematics and Its Applications. Chichester, UK (Ellis Horwood, 1989)
43. G. Schitter, K.J. Astrom, B.E. DeMartini, P.J. Thurner, K.L. Turner, P.K. Hansma, Design and modeling of a high-speed AFM-Scanner. IEEE Trans. Contr. Syst. Technol. **15**(5), 906–915 (2007)
44. S.O.R. Moheimani, B.J.G. Vautier, Resonant control of structural vibration using charge-driven piezoelectric actuators. IEEE Trans. Contr. Syst. Technol. **13**(6), 1021–1035 (2005)
45. A.J. Fleming, S.O.R. Moheimani, Sensorless vibration suppression and scan compensation for piezoelectric tube nanopositioners. IEEE Trans. Contr. Syst. Technol. **14**(1), 33–44 (2006)
46. A. Fleming, S. Moheimani, A grounded-load charge amplifier for reducing hysteresis in piezoelectric tube scanners. Rev. Sci. Instrum. **76**(7), 073707 (2005)
47. A. Sebastian, S.O.R. Moheimani, Signal transformation approach to fast nanopositioning. Rev. Sci. Instrum. **80**(7), 076101-1–076101-3 (2009)
48. A. Bazaei, S.O.R. Moheimani, A. Sebastian, An analysis of signal transformation approach to triangular waveform tracking. Automatica **47**(4), 838–847 (2011)
49. A. Bazaei, Y. Yong, S. Moheimani, A. Sebastian, Tracking of triangular references using signal transformation for control of a novel AFM scanner stage. IEEE Trans. Contr. Syst. Technol. **20**(2), 453–464 (2012)
50. H. Rothuizen, M. Despont, U. Drechsler, C. Hagleitner, A. Sebastian, D. Wiesmann, Design of power-optimized thermal cantilevers for scanning probe topography sensing, in *Proceedings of IEEE 22nd International Conference on Micro Electro Mechanical Systems*, IEEE, pp. 603–606 (2009)
51. A. Sebastian, D. Wiesmann, Modeling and experimental identification of silicon microheater dynamics: a systems approach. IEEE/ASME J. Microelectromech. Syst. **17**(4), 911–920 (2008)
52. V. Kartik, A. Sebastian, T. Tuma, A. Pantazi, H. Pozidis, D. Sahoo, High-bandwidth nanopositioner with magnetoresistance based position sensing. Mechatronics **22**, 295–301 (2012)

Chapter 6
Interval Modeling and Robust Feedback Control of Piezoelectric-Based Microactuators

Sofiane Khadraoui, Micky Rakotondrabe, and Philippe Lutz

Abstract This chapter presents the modeling and the control of piezoelectric-based microactuators. Typified by uncertainties of models, we propose to use intervals to bound the uncertain parameters. These uncertainties are particularly due to the difficulties to perform precise identification and to the high sensivity of the systems at the micro/nanoscale. In order to account the models uncertainties, we propose therefore to combine interval tools and classical control theory to derive robust controllers. Experimental results confirm the predicted theory and demonstrate the efficiency of the proposed method.

6.1 Introduction

This chapter presents the control of piezoelectric actuators used in microgrippers generally dedicated to micromanipulation or to microassembly. Piezoelectric actuators are well recognized for their high resolution (submicrometric), their high bandwidth (up to several tens of kiloHertz), their high force density, and for their ease of *control* (control signal is electrical). However, like other microactuators (thermal, electrostatic, etc.), piezoelectric microactuators suffer from the high sensivity face to the environment due to their small sizes. For instance, small mechanical vibrations or small thermal noises surrounding the microactuators would generate nonnegligible unwanted movement of them. All these make the used models have uncertain or varying parameters and consequently may lead to the loss of performances or even the loss of stability during the utilization of the actuators.

S. Khadraoui • M. Rakotondrabe (✉) • P. Lutz
Automatic Control and Micro-Mechatronic Systems Department, AS2M,
FEMTO-ST Institute, 24 rue Alain Savary, Besançon 25000, France
e-mail: sofiane.khadraoui@femto-st.fr; mrakoton@femto-st.fr; plutz@femto-st.fr

M. Rakotondrabe (ed.), *Smart Materials-Based Actuators at the Micro/Nano-Scale: Characterization, Control, and Applications*, DOI 10.1007/978-1-4614-6684-0_6,
© Springer Science+Business Media New York 2013

In order to achieve the required performances in micromanipulation and microassembly tasks, linear modeling with Δ-matrix uncertainties has been used and classical robust control laws (H_2, H_∞, and μ-synthesis) were applied for each piezocantilever [4, 14–16]. The efficiency of these advanced methods was proved in several applications (SISO and MIMO microsystems). However their major disadvantage is the derivation of high-order controllers which are time consuming and which limit their embedding possibilities, as required for real packaged microsystems.

An alternative possibility to classical robust control laws is the use of interval analysis which is a way to model the parametric uncertainties. The principle of the controller design is therefore based on the combination of the interval arithmetic with a linear control theory. In addition to its principle simplicity to model the uncertain parameters, the main advantage is the derivation of low-order controllers.

In this chapter, interval tools are used to design robust controllers for piezoelectric microactuators and to check *a posteriori* their performances. Two methods are proposed for the control design, a method based on the *Performances Inclusion Theorem* [13] and a method based on the combination of the H_∞ and interval tools. Experimental results demonstrate the efficiency of the proposed approaches and show their real interest for uncertain systems such as piezoelectric microactuators.

The chapter is organized as follows. We give first some preliminaries on interval tools in Sect. 6.2. Section 6.3 is devoted to the design of robust controller using the *Performances Inclusion Theorem* while the method based on the combination of H_∞ tool and interval tools is presented in Sect. 6.4. In Sect. 6.5, we present the *a posteriori* performances analysis still by using H_∞ tool and interval tools. Finally, the experimental results are presented in Sect. 6.6.

6.2 Preliminaries on Intervals

6.2.1 Definitions

We remind here some basics on intervals that will be used in the rest of the chapter. The readers who are interested to see more in details the techniques of intervals are suggested to read the references [6, 12].

A real interval $[x]$ is a closed interval such that

$$[x] = [x^-, x^+] \tag{6.1}$$

where x^- and x^+ are called lower bound and upper bound, respectively. We have, $x^- \leq x^+$. Having $x^- = x^+$ means that the interval $[x]$ is degenerate. By convention, a degenerate interval $[a] = [a, a]$ is identified by the real number a. The designation *point number* is similar to the designation *degenerate interval number*. While the set of real point numbers is $\mathbb{R}$, the set of real intervals (or real interval numbers) is $\mathbb{IR}$.

Instead of using the notation in (6.1), one can also identify a real interval number by its midpoint mid $([x])$ and its radius rad $([x])$ such that

$$\begin{cases} \text{mid}\,([x]) = \frac{(x^+ + x^-)}{2} \\ \text{rad}\,([x]) = \frac{(x^+ - x^-)}{2} = \frac{w([x])}{2} \end{cases} \tag{6.2}$$

where $w([x])$ is the width of the interval.

6.2.2 Operations on Intervals

In the arithmetics of intervals, the basic operations are extended to interval numbers. Consider two intervals $[x] = [x^-, x^+]$ and $[y = y^-, y^+]$. So we have

$$[x] + [y] = [x^- + y^-, x^+ + y^+] \tag{6.3}$$

and

$$[x] - [y] = [x^- - y^+, x^+ - y^-] \tag{6.4}$$

Consequently, we have, $[x] - [x] \neq 0$, except for $x^- = x^+$.

The multiplication and division are defined as follows

$$\begin{aligned} [x]\,.\,[y] = [&\min(x^-y^-, x^-y^+, x^+y^-, x^+y^+), \\ &\max(x^-y^-, x^-y^+, x^+y^-, x^+y^+)] \end{aligned} \tag{6.5}$$

and

$$[x]\,/\,[y] = [x]\,.\,[1/y^+, 1/y^-],\ 0 \notin [y] \tag{6.6}$$

We say that an interval $[x]$ is included in an interval $[y]$, i.e. $[x] \subset [y]$, if and if only $[x] \cap [y] = [x]$. We have $[x] > [y]$ if $x^- > y^+$. The real interval $[x]$ is said to be positive if $x^- > 0$. The distributive law does not hold in general for interval. However, the following relation, called subdistributivity, holds, $[x]([y] + [z]) \subseteq [x][y] + [x][z]$. In addition, if $[x] + [y] = [x] + [z]$, the cancellation law for addition holds, and $[y] = [z]$. The same property holds for multiplication, if $[x][y] = [x][z]$ and $0 \notin [x]$, thus $[y] = [z]$.

If f is a function $f : \mathbb{R} \to \mathbb{R}$, then its interval counterpart $[f]$ satisfies

$$[f]([x]) = [\{f(x) : x \in [x]\}] \tag{6.7}$$

The interval function $[f]$ is called inclusion function because $f([x]) \subseteq [f]([x])$, for all $[x] \in \mathbb{IR}$. An inclusion function $[f]$ is thin if for any degenerate interval $[x] = x$, $[f](x) = f(x)$. It is minimal if for any $[x]$, $[f]([x])$ is the smallest interval that contains $f([x])$. The minimal inclusion function for f is unique and is denoted by $[f]^*([x])$.

An easy way to compute an inclusion function for f is to replace each variable x in the expression of f by $[x]$ and all operations on points by their interval counterpart. Thus, one obtains the natural inclusion function.

6.2.3 Interval Systems

An interval system is a transfer function representation, a state space representation or a differential representation where the parameters are intervals. For an interval transfer function, which is the interest of this chapter, the representation is as follows

$$[G](s) = \frac{[b_m]s^m + \cdots + [b_1]s^1 + [b_0]}{[a_n]s^n + \cdots + [a_1]s^1 + [a_0]} = \frac{\sum_{l=0}^{m}[b_l]s^l}{\sum_{k=0}^{n}[a_k]s^k} \tag{6.8}$$

where s is the Laplace variable and where $m \leq n$, n being the order of the interval system $[G](s)$. The parameters $[a_k]$ and $[b_l]$ are considered to be constant real intervals in order to assume linear time invariant (LTI) systems. The notation $[G](s)$ shall be used if the intervals $[a_k]$ and $[b_l]$ are known. Instead, the notation $[G]([a_k],[b_l],s)$ is used when they are unknown and to be sought for.

The notion of inclusion of systems should also be defined. Consider two interval systems having the same polynomials degrees m and n, i.e. having the same structure

$$[G_1](s) = \frac{\sum_{l=0}^{m}[b_{1l}]\cdot s^l}{\sum_{k=0}^{n}[a_{1k}]\cdot s^k}, \quad [G_2](s) = \frac{\sum_{l=0}^{m}[b_{2l}]\cdot s^l}{\sum_{k=0}^{n}[a_{2k}]\cdot s^k} \tag{6.9}$$

$[G_1](s) \subseteq [G_2](s)$ is equivalent to saying that for any $s \in [0,\infty)$, we have $[G_1] \subseteq [G_2]$.

Lemma 2.1. *If $[b_{1l}] \subseteq [b_{2l}]$ and $[a_{1k}] \subseteq [a_{2k}]$, $\forall k,l$, then $[G_1](s) \subseteq [G_2](s)$.*

Proof. See [13].

6.2.4 The Performances Inclusion Theorem [13]

Consider two interval systems having the same polynomials degrees m and n

$$[G_1](s) = \frac{\sum_{l=0}^{m}[b_{1l}]\cdot s^l}{\sum_{k=0}^{n}[a_{1k}]\cdot s^k}, \quad [G_2](s) = \frac{\sum_{l=0}^{m}[b_{2l}]\cdot s^l}{\sum_{k=0}^{n}[a_{2k}]\cdot s^k} \tag{6.10}$$

The performances inclusion theorem (PIT) which will be used to further design a controller is composed of two results.

Theorem 2.1. *The performances inclusion in the frequency domain*

$$\text{if}\begin{cases}[a_{1k}]\subseteq[a_{2k}],\ \forall k=1,\ldots,n\\ \text{and}\\ [b_{1l}]\subseteq[b_{2l}],\ \forall l=1,\ldots,m\end{cases}$$

$$\Rightarrow\begin{cases}[\rho]\left([G_1](j\omega)\right)\subseteq\rho\left([G_2](j\omega)\right)\\ \text{and}\\ [\varphi]\left([G_1](j\omega)\right)\subseteq\varphi\left([G_2](j\omega)\right)\end{cases}$$

Theorem 2.2. *The performances inclusion in the time domain*

$$\text{if}\begin{cases}[a_{1k}]\subseteq[a_{2k}],\ \forall k=1,\ldots,n\\ \text{and}\\ [b_{1l}]\subseteq[b_{2l}],\ \forall l=1,\ldots,m\end{cases}$$

$$\Rightarrow [g_1](t)\subseteq[g_2](t)$$

where

- $[\rho]\left([G_i](j\omega)\right)$ is the modulus of the system $[G_i]$.
- $[\varphi]\left([G_i](j\omega)\right)$ is the argument.
- $[g_i](t)$ is the impulse response.

Proof. See [13].

6.3 PIT-Based Robust Control Design

Consider the feedback system shown in Fig. 6.1, where an uncertain system modeled by an interval transfer function $[G](s,[\mathbf{a}],[\mathbf{b}])$ is controlled by a controller $[C](s)$. $y_c(t)$ is the reference input, $y(t)$ is the output signal, and $u(t)$ is the input control signal.

Let us define the SISO interval system $[G](s,[\mathbf{a}],[\mathbf{b}])$ as follows

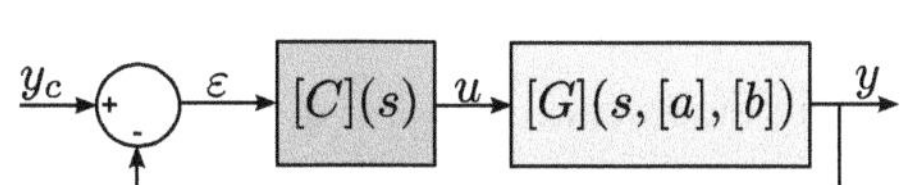

Fig. 6.1 A unity feedback interval control system

$$[G](s,[\mathbf{a}],[\mathbf{b}]) = \frac{[N](s,[\mathbf{b}])}{[D](s,[\mathbf{a}])}, \tag{6.11}$$

where $[N](s,[\mathbf{b}])$ and $[D](s,[\mathbf{a}])$ are known polynomial with interval coefficients

$$\begin{aligned} [D](s,[\mathbf{a}]) &= [a_0] + [a_1]s + [a_2]s^2 + \cdots + [a_n]s^n \\ [N](s,[\mathbf{b}]) &= 1 + [b_1]s + [b_2]s^2 + \cdots + [b_m]s^m \end{aligned} \tag{6.12}$$

with $m \leq n$ and the interval vectors $[\mathbf{a}]$ and $[\mathbf{b}]$ are defined by

$$[\mathbf{a}] = \begin{pmatrix} [a_0] \\ [a_1] \\ [a_2] \\ \vdots \\ [a_n] \end{pmatrix} \quad [\mathbf{b}] = \begin{pmatrix} 1 \\ [b_1] \\ [b_2] \\ \vdots \\ [b_m] \end{pmatrix}$$

The natural question in control design approaches for interval systems is: How can one derive a candidate controller for which the closed-loop system of Fig. 6.1 meets some performance requirements whatever the coefficients a_i and b_j ranging in their intervals $[a_i]$ and $[b_j]$ (for $i = 0,\ldots,n$ and $j = 1,\ldots,m$), respectively. This point will be presented next.

Let us define a controller $[C](s,[\theta])$ with a prior knowledge on its order $l \leq k$ as follows

$$[C](s,[\theta]) = \frac{[N_c](s)}{[D_c](s)} \tag{6.13}$$

where the interval polynomials $[D_c](s)$ and $[N_c](s)$ are given as follows

$$\begin{aligned} [D_c](s) &= [c_0] + [c_1]s + [c_2]s^2 + \cdots + [c_k]s^k \\ [N_c](s) &= [d_0] + [d_1]s + [d_2]s^2 + \cdots + [d_l]s^l \end{aligned} \tag{6.14}$$

with the interval parameters vector of the controller $[\theta] = ([c_0],\ldots,[c_k],[d_0],\ldots,[d_l])^T$ is assumed to be unknown.

Let us denote the closed-loop model of Fig. 6.1 by $[H_{\text{cl}}](s,[\mathbf{p}],[\mathbf{q}])$. This latter can be computed using the interval model (6.11) and the imposed controller (6.13) as follows

$$[H_{\text{cl}}](s,[\mathbf{p}],[\mathbf{q}]) = \frac{1}{\frac{1}{[C](s,[\theta])[G](s,[\mathbf{a}],[\mathbf{b}])} + 1} \tag{6.15}$$

where the interval vectors $[\mathbf{q}]$ and $[\mathbf{p}]$ are function of the intervals $[\mathbf{a}]$, $[\mathbf{b}]$, and $[\theta]$.

The closed-loop form given in (6.15) allows to avoid a multi-occurrence of the interval terms $[G](s,[\mathbf{a}],[\mathbf{b}])$ and $[C](s,[\theta])$ which can produce an overestimation during the closed-loop computation.

After replacing $[G](s,[\mathbf{a}],[\mathbf{b}])$ and $[C](s,[\theta])$ in (6.15), we get

$$[H_{\text{cl}}](s,[\mathbf{p}],[\mathbf{q}]) = \frac{[N](s,[\mathbf{b}])[N_c](s)}{[N](s,[\mathbf{b}])[N_c](s)+[D](s,[\mathbf{a}])[D_c](s)} \tag{6.16}$$

which can be written after developing as follows

$$[H_{\text{cl}}](s,[\mathbf{p}],[\mathbf{q}]) = \frac{[N_{\text{cl}}](s,[\mathbf{q}])}{[D_{\text{cl}}](s,[\mathbf{p}])} \tag{6.17}$$

with

$$\begin{aligned} [D_{\text{cl}}](s,[\mathbf{p}]) &= [p_0]+[p_1]s+[p_2]s^2+\cdots+[p_r]s^r \\ [N_{\text{cl}}](s,[\mathbf{q}]) &= 1+[q_1]s+[q_2]s^2+\cdots+[q_e]s^e \end{aligned} \tag{6.18}$$

where $e=m+l$, $r=n+k$, and

$$[\mathbf{p}] = \begin{pmatrix} [p_0] \\ [p_1] \\ [p_2] \\ \vdots \\ [p_r] \end{pmatrix} \quad [\mathbf{q}] = \begin{pmatrix} 1 \\ [q_1] \\ [q_2] \\ \vdots \\ [q_e] \end{pmatrix}$$

Consider a family of wanted closed-loop behaviors described by a known interval transfer function, called interval reference model. If the controller defined in (6.13) for a given θ ensures that the set of all possible closed-loop plants (6.17) is included in the set of all feasible reference models, then robust performances are achieved.

Let's denote by $[H](s,[\overline{\mathbf{p}}],[\overline{\mathbf{q}}])$ the interval reference model that describes the required performance measures. Also, let Θ be the set of admissible values of the controller parameters allowing to ensure required performances. Thus, the design problem to be addressed can be viewed as finding the set Θ for which the following inclusion holds [7, 8, 11], i.e., robust performances achieve.

$$\Theta = \{\theta \in \mathscr{D} \,|\, [H_{\text{cl}}](s,[\mathbf{p}],[\mathbf{q}]) \subseteq [H](s,[\overline{\mathbf{p}}],[\overline{\mathbf{q}}])\} \tag{6.19}$$

where $\mathscr{D}$ is the definition domain of θ.

Assume that an interval reference model is available and can be defined as follows

$$[H](s,[\overline{\mathbf{p}}],[\overline{\mathbf{q}}]) = \frac{[\overline{N}](s,[\overline{\mathbf{q}}])}{[\overline{D}](s,[\overline{\mathbf{p}}])} \tag{6.20}$$

where

$$\begin{aligned} [\overline{D}](s,[\overline{\mathbf{p}}]) &= [\overline{p}_0]+[\overline{p}_1]s+[\overline{p}_2]s^2+\cdots+[\overline{p}_{\overline{r}}]s^{\overline{r}} \\ [\overline{N}](s,[\overline{\mathbf{q}}]) &= 1+[\overline{q}_1]s+[\overline{q}_2]s^2+\cdots+[\overline{q}_{\overline{e}}]s^{\overline{e}} \end{aligned} \tag{6.21}$$

such as $\overline{e} \leq \overline{r}$ and

$$[\overline{\mathbf{p}}] = \begin{pmatrix} [\overline{p}_0] \\ [\overline{p}_1] \\ [\overline{p}_2] \\ \vdots \\ [\overline{p}_{\overline{r}}] \end{pmatrix} \quad [\overline{\mathbf{q}}] = \begin{pmatrix} 1 \\ [\overline{q}_1] \\ [\overline{q}_2] \\ \vdots \\ [\overline{q}_{\overline{e}}] \end{pmatrix}$$

In order to check the inclusion $[H_{\text{cl}}](s,[\mathbf{p}],[\mathbf{q}]) \subseteq [H](s,[\overline{\mathbf{p}}],[\overline{\mathbf{q}}])$ by applying the parameter by parameter inclusion as given in the PIT theorem in Sect. 6.2.4, the interval reference model $[H](s,[\overline{\mathbf{p}}],[\overline{\mathbf{q}}])$ must have the same structure than the closed-loop transfer $[H_{\text{cl}}](s,[\mathbf{p}],[\mathbf{q}])$ defined in (6.17). For that, let's assume that the interval polynomials $[\overline{D}](s,[\overline{\mathbf{p}}])$ and $[\overline{N}](s,[\overline{\mathbf{q}}])$ of the interval reference model have the same order as in the polynomials $[D_{\text{cl}}](s,[\mathbf{p}])$ and $[N_{\text{cl}}](s,[\mathbf{q}])$, respectively, as follows

$$\begin{aligned} [\overline{D}](s,[\overline{\mathbf{p}}]) &= [\overline{p}_0]+[\overline{p}_1]s+[\overline{p}_2]s^2+\cdots+[\overline{p}_r]s^r \\ [\overline{N}](s,[\overline{\mathbf{q}}]) &= 1+[\overline{q}_1]s+[\overline{q}_2]s^2+\cdots+[\overline{q}_e]s^e \end{aligned} \tag{6.22}$$

According to the PIT theorem in Sect. 6.2.4, if the following set of inclusions

$$\begin{cases} [q_j] \subseteq [\overline{q}_j], \text{ for } j=1,\ldots,e \\ [p_i] \subseteq [\overline{p}_i], \text{ for } i=0,\ldots,r \end{cases} \tag{6.23}$$

hold, then the set of all possible closed-loop plants $[H_{\text{cl}}](s,[\mathbf{p}],[\mathbf{q}])$ belong to the set of all admissible plants $[H](s,[\overline{\mathbf{p}}],[\overline{\mathbf{q}}])$, and therefore the performances defined by $[H_{\text{cl}}](s,[\mathbf{p}],[\mathbf{q}])$ are included in those of the wanted closed-loop $[H](s,[\overline{\mathbf{p}}],[\overline{\mathbf{q}}])$. As a result, the controller $[C](s,[\theta])$ that guarantees the above inclusions will effectively ensures the required performances for any system $G(s)$ in the interval model $[G](s,[\mathbf{a}],[\mathbf{b}])$.

Remark 1. The interval vectors $[\overline{\mathbf{p}}]$ and $[\overline{\mathbf{q}}]$ are known and they can be easily computed from the required specifications, while the interval parameters $[p_i]$ and $[q_j]$ (for $i=0,\ldots,r$ and $j=1,\ldots,e$) depend on the controller parameters which are unknown.

Finally, the design problem given in (6.19) can be reduced as finding the set-solution Θ of the admissible values of the controller parameters that ensure the following set of inclusions

Table 6.1 SIVIA algorithm for solving a set-inversion problem [5,6]

Step	SIVIA(in: $[\mathbf{p}]$, $[\mathbf{q}]$, $[\overline{\mathbf{p}}]$, $[\overline{\mathbf{q}}]$, $[\theta]$, ε; inout: $\underline{\Theta}$, $\overline{\Theta}$)
1	if $[[\mathbf{p}]([\theta]),[\mathbf{q}]([\theta])] \bigcap [[\overline{\mathbf{p}}],[\overline{\mathbf{q}}]] = \emptyset$ return;
2	if $[[\mathbf{p}]([\theta]),[\mathbf{q}]([\theta])] \subseteq [[\overline{\mathbf{p}}],[\overline{\mathbf{q}}]]$ then
	$\{\underline{\Theta} := \underline{\Theta} \bigcup [\theta]; \overline{\Theta} := \overline{\Theta} \bigcup [\theta]\}$ return;
4	if $width([\theta]) < \varepsilon$ then $\{\overline{\Theta} := \overline{\Theta} \bigcup [\theta]\}$; return;
5	bisect $[\theta]$ into $L([\theta])$ and $R([\theta])$;
6	SIVIA($[\mathbf{p}]$, $[\mathbf{q}]$, $[\overline{\mathbf{p}}]$, $[\overline{\mathbf{q}}]$, $L([\theta])$, ε; $\underline{\Theta}$, $\overline{\Theta}$);
	SIVIA($[\mathbf{p}]$, $[\mathbf{q}]$, $[\overline{\mathbf{p}}]$, $[\overline{\mathbf{q}}]$, $R([\theta])$, ε; $\underline{\Theta}$, $\overline{\Theta}$).

$$\Theta = \left\{ \theta \in \mathscr{D} \left| \begin{cases} [q_j]([\theta]) \subseteq [\overline{q}_j], \forall j = 1, \ldots, m+l \\ [p_i]([\theta]) \subseteq [\overline{p}_i], \forall i = 0, \ldots, n+k \end{cases} \right. \right\} \tag{6.24}$$

where $\mathscr{D}$ is the definition domain of θ.

The above problem described in (6.24) is known as a set-inversion problem which can be solved using interval techniques. The set inversion operation consists to compute the reciprocal image of a compact set called subpaving. The set-inversion algorithm SIVIA (more details are given in [5,6]) allows to solve the design problem given in (6.24) and provides an approximation with subpavings of the set solution Θ. This approximation is realized with an inner and outer subpavings, respectively, $\underline{\Theta}$ and $\overline{\Theta}$, such that $\underline{\Theta} \subseteq \Theta \subseteq \overline{\Theta}$. The subpaving Θ corresponds to the controller parameter vector for which the problem (6.24) holds. If $\Theta = \emptyset$, then it is guaranteed that no solution exists for (6.24).

We give in Table 6.1 the recursive SIVIA algorithm allowing to solve the control problem (6.24) with guaranteed solution. SIVIA algorithm requires a search box $[\theta_0]$ (possibly very large) also called initial box within which $\overline{\Theta}$ is guaranteed to belong. The inner and outer subpavings ($\underline{\Theta}$ and $\overline{\Theta}$) are initially empty. ε represents the wanted accuracy of computation.

Quite often we are interested to compute an inner approximation $\underline{\Theta}$ for which we are sure that $\underline{\Theta}$ is included in the set solution Θ, i.e., $\underline{\Theta} \subseteq \Theta$, but when no inner approximation exists i.e., $\underline{\Theta} = \emptyset$, it is possible to choose parameters inside the outer subpaving, i.e., choose $\theta \in \overline{\Theta}$.

Remark 2. The number of unknown parameters in (6.24) is $l+k+2$, while the number of inclusions is $r+e+1$. Since $e = m+l$ and $r = n+k$, we can write $r+e+1 \geq l+k+2$. Therefore, there are more inclusions than unknown variables. So, the set solution Θ can be obtained by the intersection of the set solution of each inclusion in (6.24) as follows

$$\Theta = \bigcap_{i=1}^{r+e+1} (set\_sol)_i$$

such as, $(set\_sol)_i$ is the set solution of the ith inclusion.

Remark 3. If the set-inversion problem is not feasible, i.e., $\Theta = \emptyset$, the initial box of the parameters must be changed and/or one must modify the controller structure and/or the required performance specifications.

6.4 Design of a Robust Controller by Combining Standard H_∞ and Interval Tools

In this part, another approach to design robust controllers for interval systems is proposed. The method is based on the standard H_∞ technique and interval tools. While the specifications and wanted performances are transcribed in terms of weighting transfers and the standard H_∞ is used to formulate the objective or problem, interval tools are used to compute the controllers.

Consider the closed-loop pictured in Fig. 6.1, where the controlled system $[G](s,[\mathbf{a}],[\mathbf{b}])$ is a general nth-order interval system defined by the following transfer function

$$[G](s,[\mathbf{a}],[\mathbf{b}]) = \frac{[b_0]+[b_1]s+[b_2]s^2+\cdots+[b_m]s^m}{[a_0]+[a_1]s+[a_2]s^2+\cdots+[a_n]s^n} \tag{6.25}$$

where $m \leq n$ and

$$[\mathbf{a}] = \begin{pmatrix} [a_0] \\ [a_1] \\ [a_2] \\ \vdots \\ [a_n] \end{pmatrix} \quad [\mathbf{b}] = \begin{pmatrix} [b_0] \\ [b_1] \\ [b_2] \\ \vdots \\ [b_m] \end{pmatrix}$$

Similar to the design problem presented in the previous section, the main objective is to design robust controller for which robust performances hold for any system $G(s)$ part of the family of systems defined by $[G](s,[\mathbf{a}],[\mathbf{b}])$. Also, in addition to the desired performance specifications of the closed-loop system, it is often desired to design low-order controllers for simplicity of implementation, especially for embedded systems. For that, a fixed structure of the controller can be *a priori* imposed as follows

$$[C](s,[\theta]) = \frac{[d_0]+[d_1]s+[d_2]s^2+\ldots+[d_l]s^l}{[c_0]+[c_1]s+[c_2]s^2+\cdots+[c_k]s^k} \tag{6.26}$$

where $[\theta] = ([c_0],\cdots,[c_k],[d_0],\ldots,[d_l])^T$ is an unknown vector of interval parameters and $l \leq k$ to have the causality of the controller.

The issue is to find the set (or subset) of the suitable values of the controller parameters so that the closed-loop system respects some given performances despite the parametric uncertainties considered in the transfer function of the controlled system. For that, the controller parameters can be adjusted using H_∞-criterion. Such

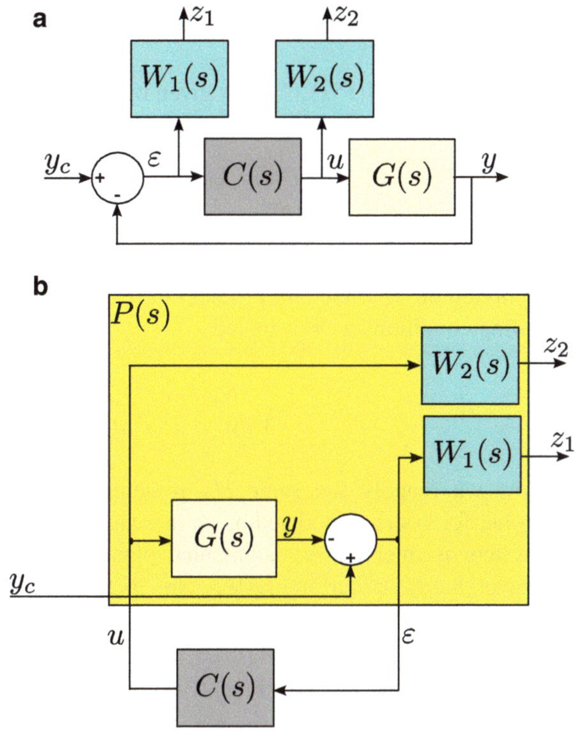

Fig. 6.2 Standard H_∞ control scheme (**a**): the weighted closed-loop block-diagram. (**b**): the corresponding standard form

criterion is defined as the H_∞-norm of some weighted transfer functions of the closed-loop to be less than or equal to one.

Let's remind the H_∞-standard principle that considers the tracking performances and the input control limitation [3, 18]. It is based on the standard block pictured in Fig. 6.2b where $P(s)$ is called the augmented system. This standard scheme is derived from the weighted closed-loop in Fig. 6.2a. While the weighting $W_1(s)$ is used to transcribe the tracking performances, the weighting $W_2(s)$ is used to transcribe the input control limitation.

The H_∞ problem is to find a controller stabilizing the closed-loop system and achieving the following H_∞-criterion

$$\|F_l(P(s),C(s))\|_\infty \leq \gamma \tag{6.27}$$

where γ is a positive scalar. If $\gamma \leq 1$, the nominal (specified) performances are achieved.

The linear fractional transformation $F_l(P(s),C(s))$ is the transfer between the weighted outputs and the exogenous inputs of Fig. 6.2b. It is defined as follows

$$F_l(P(s),C(s)) = z(s)\, y_c^{-1}(s) \tag{6.28}$$

with $z = \begin{pmatrix} z_1 \\ z_2 \end{pmatrix}$

From Fig. 6.2a $F_l(P(s), C(s))$ is given by

$$F_l(P(s), C(s)) = \begin{pmatrix} W_1(s)S(s) \\ W_2(s)C(s)S(s) \end{pmatrix} \tag{6.29}$$

where $S(s) = (1 + C(s)G(s))^{-1}$ is the sensivity function.

Applying the H_∞ standard problem in (6.27) to (6.28) and (6.29), we obtain the following conditions to be satisfied

$$\begin{cases} \|W_1(s)S(s)\|_\infty \le \gamma \\ \|W_2(s)C(s)S(s)\|_\infty \le \gamma \end{cases} \tag{6.30}$$

Now we reapply the same H_∞ principle presented above to design robust controller for systems modeled by an interval transfer function $[G](s, [\mathbf{a}], [\mathbf{b}])$. Since the system is interval, the augmented plant will also be interval, $[P](s, [\mathbf{a}], [\mathbf{b}])$. Moreover, the H_∞-criterion $\|F_l([P](s, [\mathbf{a}], [\mathbf{b}]), [C](s, [\theta]))\|_\infty \le \gamma$ is given by

$$\begin{cases} \|W_1(s)[S](s)\|_\infty \le \gamma \\ \|W_2(s)[C](s, [\theta])[S](s)\|_\infty \le \gamma \end{cases} \tag{6.31}$$

In this case, if $\gamma \le 1$, the robust performances are achieved.

Let's denote by Θ the set of the suitable values corresponding to the controller parameters that ensures the requirements. Based on the H_∞ principle above, the design problem can be formulated as follows [7, 9, 10].

Find the set Θ so that H_∞ performance holds for any positive number $\gamma \le 1$, i.e.,

$$\Theta = \left\{ \theta \in \mathscr{D} \,\middle|\, \begin{cases} \|W_1(s)[S](s)\|_\infty \le \gamma \\ \|W_2(s)[C](s, [\theta])[S](s)\|_\infty \le \gamma \end{cases} \right\} \tag{6.32}$$

where $\mathscr{D}$ is the definition domain of θ. The interval sensivity function $[S](s)$ is defined as follows

$$[S](s) = \frac{1}{1 + [C](s, [\theta])[G](s, [\mathbf{a}], [\mathbf{b}])} \tag{6.33}$$

However, the resolution of the problem (6.32) requires the computation of the H_∞-norm of certain interval transfers. This computation can be done by applying the following theorems which are due to the results in [1, 2, 17].

Theorem 4.1. *Consider an interval system $[G](s, [\mathbf{a}], [\mathbf{b}])$ defined as in (6.25). The H_∞-norm of $[G]$ is the maximal among the H_∞-norm of the sixteen transfers,*

$$\|[G]\|_\infty = \max_{i=1 \to 16} \left\| G^{(i)} \right\|_\infty \tag{6.34}$$

where $G^{(i)}$, for $i = 1,2,\ldots,16$ are sixteen (point) systems based on the eight Kharitonov vertex polynomials corresponding to the numerator and denominator of the interval system, i.e., sixteen transfer functions formed by combining the four Kharitonov vertex polynomials of the numerator of $[G](s,[\mathbf{a}],[\mathbf{b}])$ and the four Kharitonov vertex polynomials of its denominator.

Proof. see [1,2].

When the interval system $[G]$ is weighted by a weighting function (not interval transfer) $W(s)$, it is not advised to compute the multiplication $W[G]$ first and then compute the H_∞-norm of the resulting interval plant afterwards. Indeed, developing the multiplication of the intervals polynomials produces a multi-occurrence of the parameters and therefore a overestimation of the resulting intervals. Thus, the H_∞-norm of $W[G]$ is defined as follows [1,2]

$$\|W[G]\|_\infty = \max_{i=1\to 16} \left\| W G^{(i)} \right\|_\infty \tag{6.35}$$

Also, in this control approach, we need to compute the H_∞-norm of the sensivity function of an interval system $[G](s,[\mathbf{a}],[\mathbf{b}])$. This has been addressed in the following theorem proposed by *Long-Wang* [17].

Theorem 4.2. *Consider an interval system $[G](s,[\mathbf{a}],[\mathbf{b}])$ and its sensivity function $[S] = \frac{1}{1+[G]} = \frac{[D]}{[N]+[D]}$, where $[N]$ and $[D]$ are the numerator and denominator polynomials of $[G]$. The H_∞-norm of the sensivity $[S]$ is defined by the maximal among the H_∞-norm of twelve vertex systems out of sixteen vertex systems,*

$$\|[S]\|_\infty = \left\| \frac{[D]}{[N]+[D]} \right\|_\infty = \max_{i=1\to 12} \left\| S^{(i)} \right\|_\infty \tag{6.36}$$

Proof. see [17].

The computation of $\|W_1(s)[S](s)\|_\infty$ and $\|W_1(s)[C](s,[\theta])[S](s)\|_\infty$ given in (6.32) can be easily carried out by applying the above theorems.

$$\begin{aligned} \|W_2[C][S]\|_\infty &= \max_{i=1\to 16} \left\| W_2 M^{(i)} \right\|_\infty \\ \|W_1[S]\|_\infty &= \max_{i=1\to 12} \left\| W_1 S^{(i)} \right\|_\infty \end{aligned} \tag{6.37}$$

where $[M] = [C][S]$ and $M^{(i)}$ $(i = 1,2,\ldots,16)$ are the sixteen vertex of $[M]$.

The problem given in (6.32) is known as a set-inversion problem which can be solved using set inversion algorithms. By using SIVIA algorithm [5,6], it is possible to approximate the set solution Θ corresponding to the controller parameters for which the problem (6.32) is fulfilled. In fact, testing the existing or not of a solution (existing of a candidate controller) for the problem (6.32) requires to have a knowledge on the minimum and the maximum values of the H_∞-norm of the involved interval transfers. However, Theorems 4.1 and 4.2 allow only to evaluate the maximum value of the H_∞-norm of interval transfers. For that, we present in Fig. 6.3, a flow chart describing the recursive SIVIA algorithm allowing to solve the above design problem (6.32). The controller computation requires a search box $[\theta_0]$ also called initial box. The subpaving Θ is initially empty. ε represents the wanted accuracy of computation. Note that, contrary to the standard H_∞ problem (for point systems) where the optimal value of γ is found by dichotomy, its value here is directly set to one, $\gamma = 1$. The objective is to find directly the controller parameters for which the specified performances are met.

Remark 4. The controller computation based on the algorithm shown in Fig. 6.3 takes more time due to the high number of bisections carried out on the domain of the parameters θ.

6.5 *A Posteriori* Performances Analysis Using Standard H_∞ and Interval Tools Combined

Contrary to the problem presented in the two last sections where the objective was to design robust controller for interval systems, in this part, we deal with the inverse problem. This latter is as follows.

Consider an uncertain system modeled by an interval transfer $[G](s, [\mathbf{a}], [\mathbf{b}])$ and controlled by a controller $C(s)$ (see Fig. 6.4) to ensure for the closed-loop system a more desirable behavior.

Assume that a candidate controller $C^*(s)$ is available (for example, computed using the method presented in Sect. 6.3), then the natural question: How can one check if a such controller $C^*(s)$ achieves the required performance specifications for the closed-loop system? This point can be carried out by means of H_∞ approach combined with interval analysis.

The principle of H_∞ synthesis combined with interval analysis discussed in Sect. 6.4 consists first in transcribing during the synthesis, the requirements into weighting functions (see Fig. 6.5), then computing a controller for which a H_∞ criterion holds,

$$\|F_l([P](s, [\mathbf{a}], [\mathbf{b}]), C(s)\|_\infty \leq 1 \tag{6.38}$$

where $F_l(P(s), C(s))$ is the transfer of the interconnection between $C(s)$ and the augmented plant $[P](s, [\mathbf{a}], [\mathbf{b}])$.

In our case the controller $C^*(s)$ is known, so we need to check the fulfillment of the condition (6.38) for the controller $C(s) = C^*(s)$. From Fig. 6.5, the H_∞ criterion becomes,

$$\begin{cases} \|W_1(s)[S](s)\|_\infty \leq \gamma \\ \|W_2(s)C(s)[S](s)\|_\infty \leq \gamma \end{cases} \tag{6.39}$$

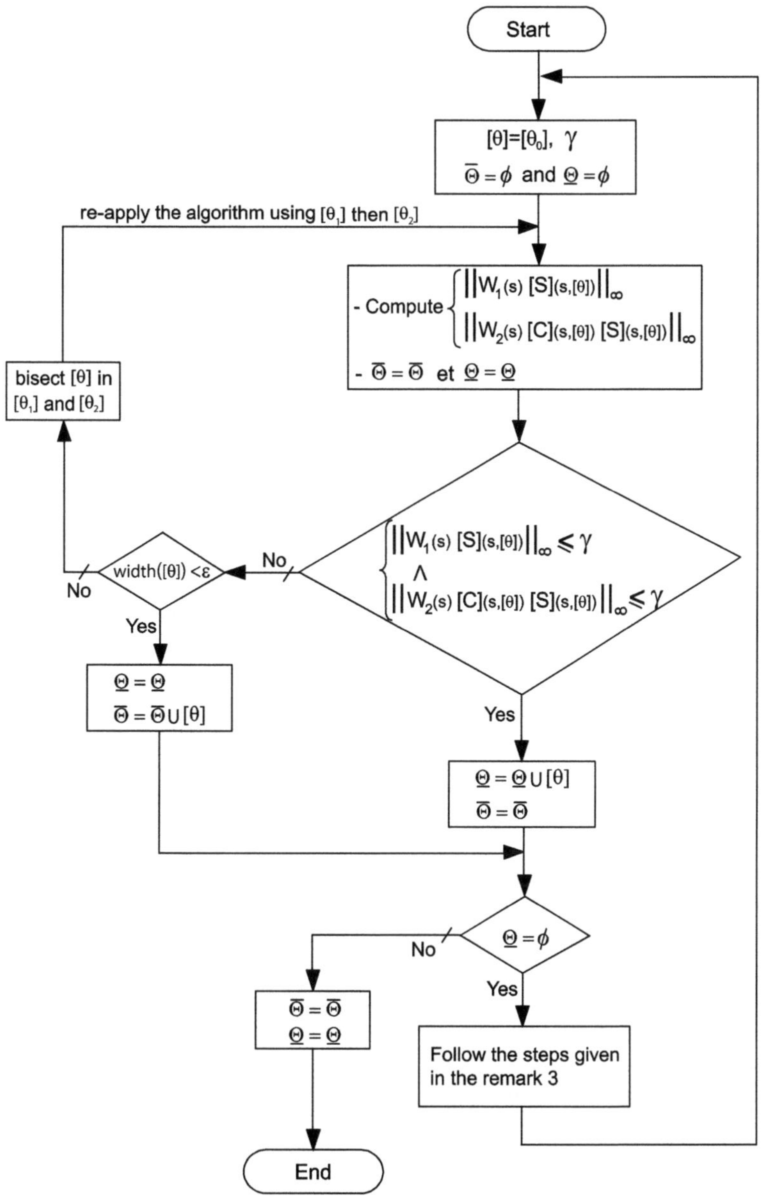

Fig. 6.3 Flow chart corresponding to the SIVIA algorithm used for solving the problem (6.32)

Fig. 6.4 Closed-loop control system

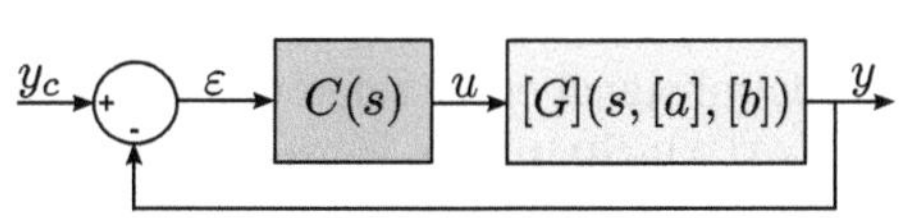

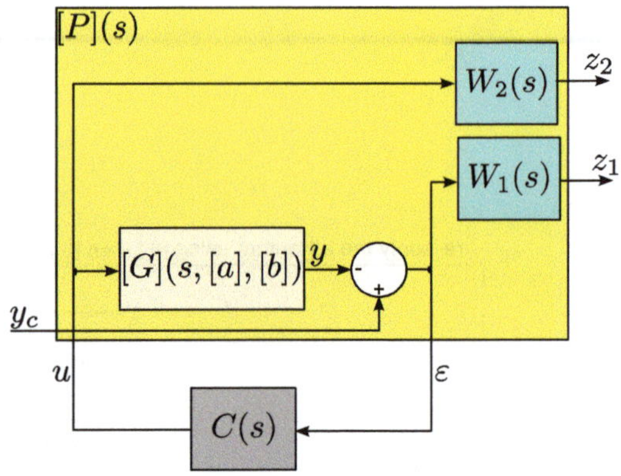

Fig. 6.5 H_∞-standard problem

then satisfying the conditions defined in (6.39) for $C(s) = C^*(s)$, means that the controller $C^*(s)$ guarantees robust performances for any $G(s)$ within the interval system $[G](s)$. The computation of the maximal H_∞ norm of the interval transfers given in (6.39) can be carried out by applying the Theorems 4.1 and 4.2.

Remark 5. The H_∞ conditions given in (6.39) are only sufficient, so if these constraints are not satisfied, then no conclusion on the achievement of the desired performances can be done.

6.6 Application to Piezocantilevers and Experimental Results

The aim of this section is to apply the interval control methods previously presented to control piezoelectric microactuators used in microgrippers. In fact, a piezoelectric microgripper is composed of two piezoelectric cantilevers (microactuators) generally with rectangular section. Figure 6.6 pictures a microgripper made at the AS2M department of FEMTO-ST Institute manipulating a small gear.

In general, one of the two actuators that compose the microgripper is used for the precise positioning while the second actuator is used to measure or control the manipulation force. In this application, we are interested by the modeling and control of the positioning. The actuator used is a unimorph cantilever made up of one piezoelectric layer (PZT material) and one passive layer (Copper material). Figure 6.7 presents the setup used for the rest of the chapter which includes,

- The piezoelectric actuator itself.
- A computer and a dSPACE board for the data acquisition, for generating the control signal or the reference signal and for the controller implementation. The MATLAB-SIMULINK is used for the implementation and the sampling time is set equal to $T_s = 0.2\,\text{ms}$.

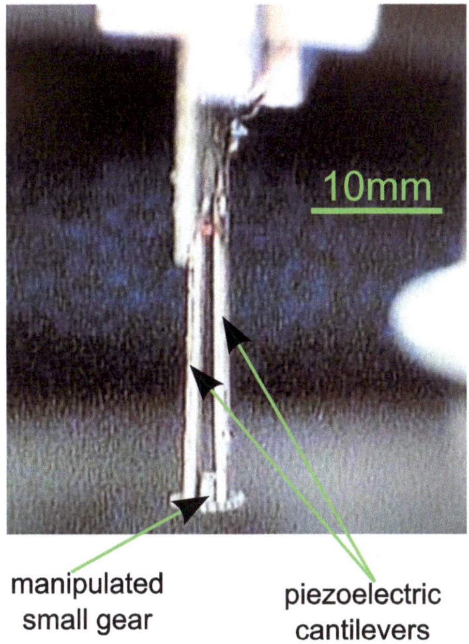

Fig. 6.6 A piezoelectric microgripper manipulating a small gear

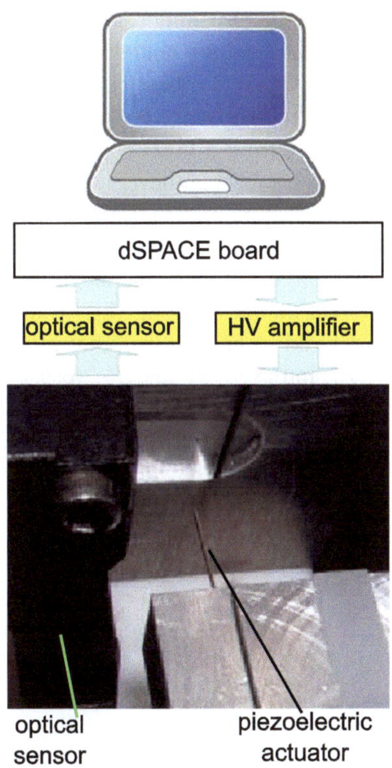

Fig. 6.7 Setup used for the experiments

- An optical sensor (*Keyence LC-2420*) which is set to have a resolution of 50 nm.
- A high-voltage (HV) amplifier ($\pm$200 V).

Modeling and identification of microsystems are very delicate because of their small sizes, their fragility, and the lack of convenient (accurate and high bandwidth) sensors to report precise measurements. These systems are also very sensitive to environmental disturbances (temperature, vibrations, manipulated objects, etc.). As a result, their behavior parameters may change during their functioning or during the tasks and therefore the wanted performances or even the stability may be lost. In this application, we bound the uncertain parameters of piezocantilever models by intervals that are able to account the above complex characteristics. Afterwards, control design approaches presented previously can be easily applied to improve the performances of piezocantilevers.

6.6.1 *Interval Model Derivation*

The models of piezocantilevers are often subjected to variation due to the environment (small thermal variation, manipulated object, etc.). In fact, these characteristics stem from the relatively small sizes of the piezoelectric actuators used in micromanipulation and micropositioning applications which finally make them very sensitive to any minor variation. The model parameters can be considered as uncertain and thus bounded by intervals within its range of variation in order to further design a robust controller. However, for an ease of identification in this application, we will not characterize the parameter variations of the piezoelectric actuator during a micropositioning or a micromanipulation task. We will use two unimorph piezocantilevers denoted by P_1 and P_2. The first piezocantilever P_1 has the dimensions $length \times width \times thickness = 16\,\text{mm} \times 1\,\text{mm} \times 0.45\,\text{mm}$, while the second one P_2 has dimensions of $14\,\text{mm} \times 1\,\text{mm} \times 0.45\,\text{mm}$. The difference in their length generate nonnegligible difference on their model parameters. The interval model $[G](s,[\mathbf{a}],[\mathbf{b}])$ which represents a family of piezocantilever models is derived using the two point models $G_1(s)$ and $G_2(s)$ corresponding to the used piezocantilevers P_1 and P_2, respectively, where the models $G_1(s)$ and $G_2(s)$ are identified without performing the above tasks. After a frequency identification for each piezocantilever and performing some computation, we obtain the following interval model $[G](s,[\mathbf{a}],[\mathbf{b}])$,

$$[G](s,[\mathbf{a}],[\mathbf{b}]) = \frac{[b_2]s^2 + [b_1]s + [b_0]}{[a_2]s^2 + [a_1]s + [a_0]} \tag{6.40}$$

where

$$
\begin{aligned}
[b_2] &= [7.042, 8.0313] \times 10^{-8} \\
[b_1] &= [1.808, 1.809] \times 10^{-4} \\
[b_0] &= 1 \\
[a_2] &= [8.802, 9.794] \times 10^{-8} \\
[a_1] &= [5.24, 5.364] \times 10^{-6} \\
[a_0] &= [1.291, 1.44]
\end{aligned}
$$

In order to increase the stability margin and to ensure that the interval model really contains the models of the two piezocantilevers, we propose to expand by 10 % the interval width of each parameter of the model $[G](s, [\mathbf{a}], [\mathbf{b}])$. This choice is a compromise. If the widths are too large, it is difficult to find a controller that respects both the stability and performances for the closed-loop. Finally, the extended interval model that will be used for the computation of a controller is as follows

$$
[G](s, [\mathbf{a}], [\mathbf{b}]) = \frac{[6.992, 8.08] \times 10^{-8} s^2 + [1.807, 1.809] \times 10^{-4} s + 1}{[8.753, 9.844] \times 10^{-8} s^2 + [5.234, 5.37] \times 10^{-6} s + [1.283, 1.448]} \tag{6.41}
$$

6.6.2 *Specifications and Controller Structure*

Piezocantilevers are very resonant (more than 60 % of overshoot). Such overshoot is not desirable in micromanipulation and microassembly tasks. The following specifications are therefore considered for the closed-loop,

- Zero or very small overshoot.
- Settling time $tr_{5\%} \leq 8\,\text{ms}$.
- Static error $|\varepsilon| \leq 1\,\%$.

These specifications often correspond to the requirement in micropositioning tasks for microassembly and micromanipulation that use piezoelectric microgrippers.

To ensure the above requirements, the control design approach does not require any specified structure for the controller. So, any structure can be chosen for the controller $[C](s)$ as long as Remark 2 is satisfied. In this example, we consider a *PI* (Proportional–Integral) structure because of its low-order (two parameters) and its wide use in the industry

$$
[C](s, [K_p], [K_i]) = \frac{[K_p]s + [K_i]}{s} \tag{6.42}
$$

where $[K_p]$ and $[K_i]$ are the proportional and integral gains, respectively.

Next, the both proposed control approaches will be applied to achieve these requirements.

6.6.3 PI Controller Computation Using PIT Approach

Based on the interval model in (6.41) and the interval controller in (6.42), the general model of the closed-loop can be expressed as follows

$$[H_{\mathrm{cl}}](s,[\mathbf{p}],[\mathbf{q}]) = \frac{[q_3]s^3+[q_2]s^2+[q_1]s+1}{[p_3]s^3+[p_2]s^2+[p_1]s+[p_0]} \tag{6.43}$$

where $[q_3]=\dfrac{[K_p][b_2]}{[K_i]}$, $[q_2]=\dfrac{[K_p][b_1]}{[K_i]}+[b_2]$, $[q_1]=\dfrac{[K_p]}{[K_i]}+[b_1]$,

$[p_3]=\dfrac{[a_2]+[K_p][b_2]}{[K_i]}$, $[p_2]=\dfrac{[a_1]+[K_p][b_1]}{[K_i]}+[b_2]$, $[p_1]=\dfrac{[a_0]+[K_p]}{[K_i]}+[b_1]$

and $[p_0]=1$.

Concerning the reference model, its computation is carried out according to the closed-loop (6.43) and to the required specifications. According to the specifications (see Sect. 6.6.2), a first order model can be used for the reference model.

$$[H](s,[K],[\tau]) = \frac{[K]}{[\tau]s+1} \tag{6.44}$$

where the parameters $[K]$ and $[\tau]$ define the static error and settling time, respectively:

- $[K]=1+\varepsilon=[0.99,1.01]$.
- $[\tau]=\dfrac{[tr_{5\%}]}{3}=[0,2.66\,\mathrm{ms}]$.

However, it is necessary that the interval reference model has the same structure than that of the closed-loop in order to apply the parameter by parameter inclusion as required in (6.24). Thus we add some poles and zeros far from the imaginary axis to (6.44)

$$[H](s,[K],[\tau]) = \frac{[K]\left(\frac{[\tau]}{10}s+1\right)^3}{([\tau]s+1).\left(\frac{[\tau]}{10}s+1\right)^2} \tag{6.45}$$

which can also be rewritten as follows:

$$[H](s,[\overline{\mathbf{p}}],[\overline{\mathbf{q}}]) = \frac{[\overline{q}_3]s^3+[\overline{q}_2]s^2+[\overline{q}_1]s+1}{[\overline{p}_3]s^3+[\overline{p}_2]s^2+[\overline{p}_1]s+[\overline{p}_0]} \tag{6.46}$$

where $[\overline{q}_3]=0.001[\tau]^3$, $[\overline{q}_2]=0.03[\tau]^2$, $[\overline{q}_1]=0.3[\tau]$, $[\overline{p}_3]=0.01\dfrac{[\tau]^3}{[K]}$,

$[\overline{p}_2]=0.21\dfrac{[\tau]^2}{[K]}$, $[\overline{p}_1]=1.2\dfrac{[\tau]}{[K]}$ and $[\overline{p}_0]=\dfrac{1}{[K]}$.

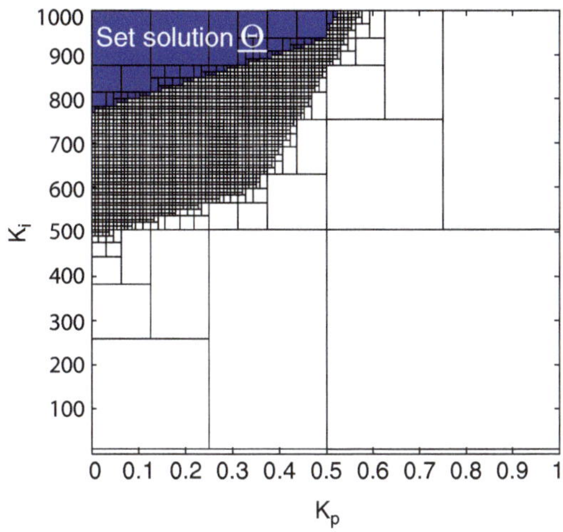

Fig. 6.8 Set solution of the parameters $[K_p]$ and $[K_i]$ ensuring the wanted performances

According to the control method discussed in Sect. 6.3, the admissible values of the PI controller parameters that guarantee the required specifications for the interval model (6.41) can be obtained by solving the following system of inclusions.

$$\begin{aligned}
&\frac{[K_p][b_2]}{[K_i]} \subseteq 0.001[\tau]^3 && \frac{[a_1]+[K_p][b_1]}{[K_i]}+[b_2] \subseteq \frac{0.21[\tau]^2}{[K]} \\
&\frac{[K_p][b_1]}{[K_i]}+[b_2] \subseteq 0.03[\tau]^2 && \frac{[a_0]+[K_p]}{[K_i]}+[b_1] \subseteq \frac{1.2[\tau]}{[K]} \\
&\frac{[K_p]}{[K_i]}+[b_1] \subseteq 0.3[\tau] && 1 \subseteq \frac{1}{[K]} \\
&\frac{[a_2]+[K_p][b_2]}{[K_i]} \subseteq \frac{0.01[\tau]^3}{[K]}
\end{aligned} \tag{6.47}$$

The application of the SIVIA algorithm implemented in the Matlab-Software, with an initial box $[K_{p0}] \times [K_{i0}] = [0,1] \times [0.1, 1000]$, provides the subpaving shown in Fig. 6.8. The dark colored subpaving ($\underline{\Theta}$) corresponds to the inner approximation, i.e., the set parameters $[K_p]$ and $[K_i]$ of the controller (6.42) that ensures the above inclusions and consequently that meets the performances for the interval model.

The controller $[C](s,[K_p],[K_i])$ is an interval and is not directly implementable. Point parameters K_p and K_i within the set solution $\underline{\Theta}$ must be chosen and the corresponding point controller $C(s,K_p,K_i) = C(s)$ has to be implemented. In this example, we test the following PI controller

$$C(s) = \frac{0.1s+900}{s} \tag{6.48}$$

This controller has been independently tested on the both piezocantilevers P_1 and P_2. A step response analysis is performed on each closed-loop by applying a step

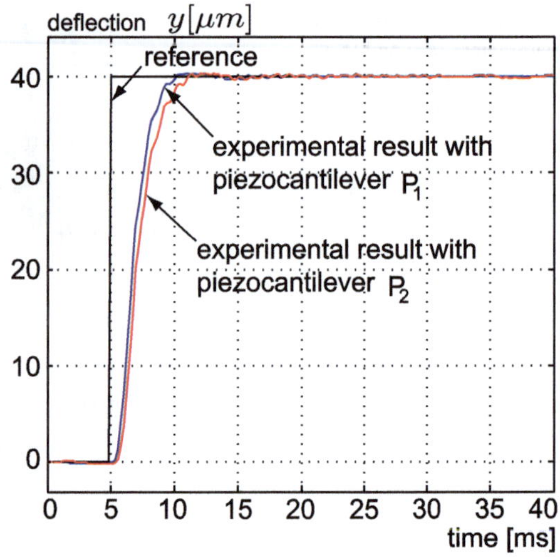

Fig. 6.9 Experimental step responses when testing the implemented controller $C(s)$

reference of amplitude 40 µm. The application of the implemented controller $C(s)$ to the both piezocantilevers leads to the experimental results shown in Fig. 6.9.

As shown in Fig. 6.9, the implemented controller has played its role and achieved the wanted performances for the closed-loops. Indeed, the experimental settling times are about $tr_1 = 4\,\text{ms}$ and $tr_2 = 4.7\,\text{ms}$ with the piezocantilever P_1 and P_2, respectively. Moreover, the obtained behaviors are with very small overshoot and the experimental static errors are neglected and belong to the required interval $|\varepsilon| \leq 1\,\%$.

6.6.4 PI Controller Computation by Combining the H_∞ Technique with Interval Analysis

In this section, we apply the robust control approach proposed in Sect. 6.4 to control the deflection (position) of piezocantilevers having model inside the interval model $[G](s, [\mathbf{a}], [\mathbf{b}])$ defined in (6.41). The same requirements presented in Sect. 6.6.2 are considered here. Moreover, it is necessary to limit the applied voltage in order to avoid any damage of the actuators. For that, we add a condition on the amplitude of the input voltage U applied to the piezocantilever. We particularly choose a maximal voltage $U^{\max} = 2.5\,\text{V}$ for each 1 µm of reference. Also, without loss of generality, we consider the design of the previous PI (Proportional–Integral) controller structure (6.42).

Figure 6.10a presents the closed-loop scheme for the controller design, where the weighting function $W_1(s)$ is added to transcribe the tracking performances and $W_2(s)$ for the input control limitation.

The weighting functions $W_1(s)$ and $W_2(s)$ were chosen according to the required performances. We choose

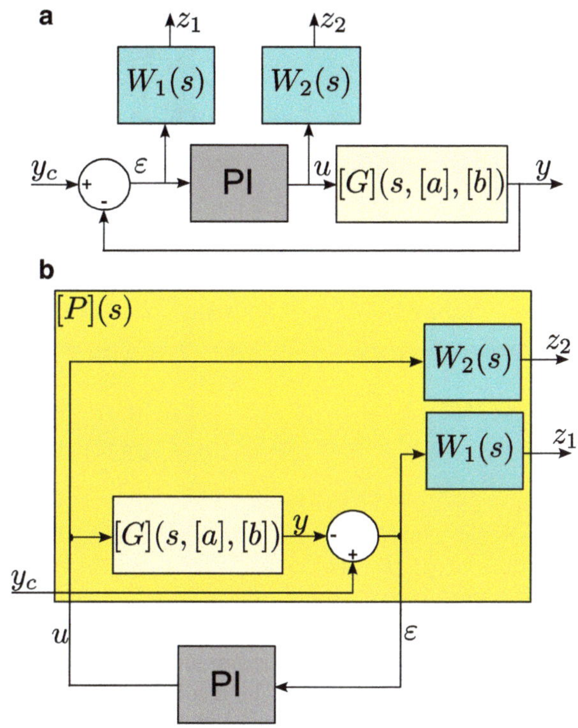

Fig. 6.10 (**a**) The closed-loop scheme with the weighting functions. (**b**) The H_∞-standard scheme

$$\begin{cases} W_1(s) = \dfrac{0.002667s+1}{0.002667s+0.01} \\ \\ W_2(s) = \dfrac{1}{2.5} \end{cases} \tag{6.49}$$

We aim to find the set-solution Θ of the PI controller parameters that ensures H_∞ performance for $\gamma = 1$, i.e.,

$$\Theta = \left\{ \theta \in [\theta] \,\middle|\, \left\{ \begin{array}{l} \|W_1(s)[S](s)\|_\infty \le \gamma \\ \|W_2(s)[C](s,[\theta])[S](s)\|_\infty \le \gamma \end{array} \right. \right\} \tag{6.50}$$

where $[S](s) = (1+[C](s,[\theta])[G](s,[\mathbf{a}],[\mathbf{b}]))^{-1}$ is the sensivity function defined as follows:

$$[S](s) = \frac{\frac{[a_2]}{[K]}s^3 + \frac{[a_1]}{[K]}s^2 + \frac{1}{[K]}s}{\frac{[a_2]}{[K]}s^3 + \frac{[a_1]}{[K]}s^2 + \frac{1}{[K]}s + [K_p][b_2]s^3 + ([K_i][b_2]+[K_p][b_1])s^2 + ([K_p]+[K_i][b_1])s + [K_i]} \tag{6.51}$$

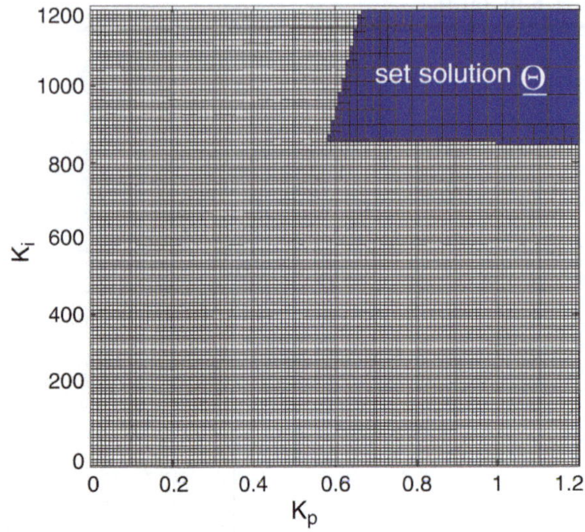

Fig. 6.11 Set-solution of the parameters $[K_p]$ and $[K_i]$ ensuring the wanted performances

Now, we solve the set-inversion problem in (6.50) using the recursive algorithm presented in Fig. 6.3. We choose an initial box for the controller parameters $[K_{p0}] \times [K_{i0}] = [0, 1.2] \times [0.1, 1200]$. The resulting subpaving is presented in Fig. 6.11. The dark colored subpaving $\underline{\Theta}$ corresponds to the set parameters $[K_p]$ and $[K_i]$ of the PI controller (6.42) that ensures the performances defined by the H_∞-criterion (6.50).

Note that any choice of the parameters $[K_p]$ and $[K_i]$ within the dark colored subpaving $\underline{\Theta}$ (see Fig. 6.11) satisfies the conditions (6.50) and consequently ensures the required performances. In the case where the problem (6.50) is not feasible (with the imposed controller), i.e., $\Theta = \emptyset$, the initial box of the parameters $[K_{p0}] \times [K_{i0}]$ must be changed and/or the structure of the controller must be modified (increase the order, for example) and/or the specifications must be modified (degrade the specifications).

Similar to the previous case, the controller $C(s)$ to be implemented is chosen by taking any point parameters K_p and K_i within the set-solution $\underline{\Theta}$ in Fig. 6.11. In this example, we test the following controller:

$$C(s) = \frac{s + 1,000}{s} \tag{6.52}$$

In order to prove that the inequalities (6.50) are satisfied, the magnitudes of the bounds $\left|\frac{1}{W_1(s)}\right|$ and $\left|\frac{1}{W_2(s)}\right|$ are compared to the magnitudes of the sensivity function $|[S](s)|$ and of the transfer $|C(s)[S](s)|$, respectively, when using the implemented controller (6.52). This comparison is given by Fig. 6.12.

The obtained results in Fig. 6.12 prove that the magnitudes of $[S](s)$ and $C(s)[S](s)$ are effectively bounded by that of $\frac{1}{W_1(s)}$ and $\frac{1}{W_2(s)}$, respectively, when using the computed controller $C(s)$. This fact confirms that the specified performances are effectively ensured.

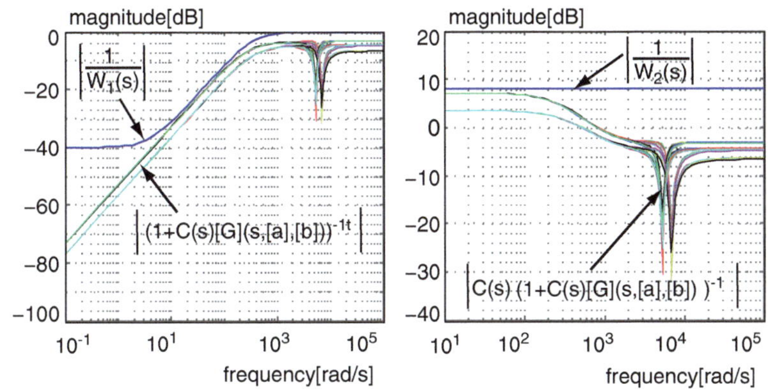

Fig. 6.12 Magnitudes of the bounds compared to the sensivity $[S](s)$ and to the input transfer $C(s)[S](s)$

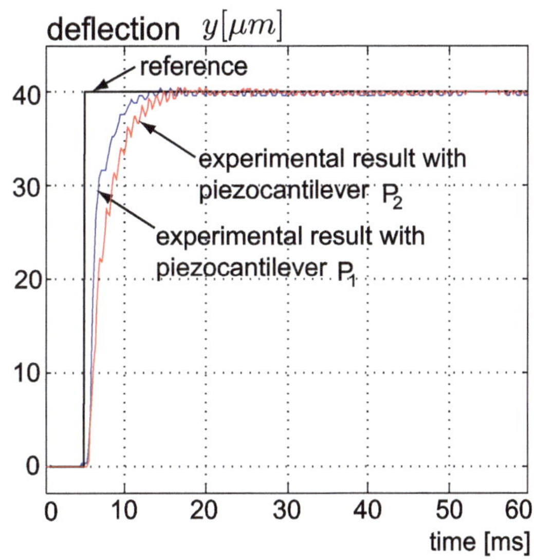

Fig. 6.13 Experimental step responses of the piezocantilevers when using $C(s)$

Now, we implement the computed controller $C(s)$ using the first piezocantilever with length $l = 16\,\text{mm}$ then the second one with length $l = 14\,\text{mm}$. Figure 6.13 shows the experimental results when a step reference of $40\,\mu\text{m}$ is applied. As shown on the Fig. 6.13, the implemented controller (6.52) has played its role since the closed-loop piezocantilevers satisfy the wanted specifications. Indeed, experimental settling times obtained with the piezocantilevers P_1 and P_2 are about $tr_1 = 5.2\,\text{ms}$ and $tr_2 = 7\,\text{ms}$, respectively. The overshoots and static errors are neglected ($D_{1,2} \approx 0$, $\varepsilon_{1,2} \approx 0 < 1\,\%$). Furthermore, the maximal voltages U applied to the both piezocantilevers are less than $40 \times 2.5 = 100\,\text{V}$, which should be the limit for a displacement of $40\,\mu\text{m}$. Indeed, the experiments show that the maximal input voltage is $U^{\max} = 97\,\text{V}$.

6.7 Conclusion

This chapter presents the modeling and robust control of piezoelectric microactuators. These latters are characterized by models with uncertain parameters and need convenient modeling and robust control laws. The challenge in micromanipulation, microassembly, and micropositioning application is to ensure robust performances despite the variation in model parameters. For that, interval analysis has been introduced to describe uncertain parameters in the models of microactuators. The main advantage of a such description by interval is the ease and natural way to bound these uncertainties. Moreover, interval techniques can be used to solve many engineering problems, such as control system problems. In the second part of this chapter, two control design approaches for interval systems have been proposed. The first control design method is based on the inclusion of interval transfers and their time and frequency responses, while the second one combines the H_∞-standard method with interval techniques. The main advantage of the proposed approaches is that they can provide low-order controllers that are able to ensure robust performances for uncertain systems and that are convenient for real-time embedded systems. It has been noted that these proposed control methods are based on some sufficient conditions. This is one limitation of these proposed methods, since sometimes the fulfillment of the constrained conditions does not hold for a given controller, however the required performances measures can be met with this latter. Also, based on the principle of the second control approach, it has been shown that it is possible to perform *a posteriori* analysis of the performances of interval closed-loop system when the controller is assumed to be known. At the end of this chapter, the proposed control design methods have been applied to control the deflection of piezocantilevers which are typically uncertain systems. The derived controllers were with very low-order (first order) that are suitable for embedded systems. The obtained experimental results confirmed the robustness of the implemented controllers and also the efficiency of the proposed control approaches. As a conclusion interval analysis can be viewed as a guaranteed and powerful tool to represent uncertainties in real systems. Moreover, it can be introduced also to formulate and solve many engineering problems.

References

1. S.-J. An, L. Huang, E. Wang, On the parametric H_∞ problems of weighted interval plants, IEEE Trans. Automat. Contr. **45**, 332–335 (2000)
2. S. J. An, X. Hu, B. Vucetic, W. Liu, Vertex results for parametric shifted H_∞ performance of weighted interval plants. IEEE Conf. Decis. Contr. **5**, 4195–4196 (2000)
3. G.J. Balas, J.C. Doyle, K. Glover, A. Packard, R. Smith, μ-synthesis and synthesis toolbox for use with Matlab. The Mathworks (2001)
4. L. Iorga, H. Baruh, I. Ursu, A review of H_∞ robust control of piezoelectric smart structures. Appl. Mech. Rev. **61**(4), 04082-1–04082-15 (2008)

5. L. Jaulin, E. Walter, Set inversion via interval analysis for nonlinear bounded-error estimation. Automatica. **29**(4), 1053–1064 (1993)
6. L. Jaulin, M. Kieffer, O. Didrit, E. Walter, *Applied Interval Analysis* (Springer, London, 2001)
7. S. Khadraoui, Calcul par intervalles et outils de l'automatique permettant la micromanipulation prcision qualifie pour le microassemblage, Thse de Doctorat, Universit de Franche-Comt Besanon, 2012
8. S. Khadraoui, M. Rakotondrabe, P. Lutz, PID-structured controller design for interval systems: application to piezoelectric microactuators, in *American Control Conference (ACC)*, San Francisco, CA, USA, 2011, pp. 3477–3482
9. S. Khadraoui, M. Rakotondrabe, P. Lutz, Combining H_∞ and interval techniques to design robust low order controllers: application to piezoelectric actuators, in *American Control Conference (ACC)*, Montral, Canada, 2012
10. S. Khadraoui, M. Rakotondrabe, P. Lutz, Combining H_∞ approach and interval tools to design a low order and robust controller for systems with parametric uncertainties: application to piezoelectric actuators. Int. J. Contr. **85**(3), 251–259 (2012)
11. S. Khadraoui, M. Rakotondrabe, P. Lutz, Interval modeling and robust control of piezoelectric microactuators. IEEE Trans. Contr. Syst. Technol. **20**(2),486–494 (2012)
12. R.E. Moore, *Interval Analysis* (Prentice-Hall, Englewood Cliffs, 1966)
13. M. Rakotondrabe, Performances inclusion for stable interval systems, in *American Control Conference (ACC)*, San Francisco, CA, USA, June–July 2011, pp. 4367–4372
14. M. Rakotondrabe, C. Clevy, P. Lutz, Modelling and robust position/force control of a piezoelectric microgripper, in *IEEE - International Conference on Automation Science and Engineering (CASE)*, Scottsdale, AZ, USA, 2007, pp. 39–44
15. M. Rakotondrabe, Y. Haddab, P. Lutz, Quadrilateral modelling and robust control of a nonlinear piezoelectric cantilever. IEEE Trans. Contr. Syst. Technol. **17**(3), 528–539 (2009)
16. A. Sebastian, A. Pantazi, S.O.R. Moheimani, H. Pozidis, E. Eleftheriou, Achieving sub-nanometer precision in a MEMS-based storage device during self-servo write process. IEEE Trans. Nanotechnol. **7**(5), 586–595 (2008)
17. L. Wang, H_∞ performance of interval systems. eprint arXiv:math/0211013 **1**, 1–8 (2002)
18. K. Zhou, J. Doyle, K. Glover, *Robust and Optimal Control* (Prentice-Hall, Englewood Cliffs, New Jersey 07632, Upper Saddle River, 1996)

Chapter 7
Kalman Filtering and State-Feedback Control of a Nonlinear Piezoelectric Cantilevered Actuator

Micky Rakotondrabe, Juan-Antonio Escareno, Didace Habineza, and Sergio Lescano

Abstract This chapter deals with the state estimation with noise rejection in a piezoelectric cantilevered actuator and its state-feedback control. The noises which come from the sensor used, strain gage, are important and should be filtered. For that, we employ the classical Kalman filtering for their rejection and for the state estimation and we apply afterwards a state-feedback control with integral action to improve the general performances of the actuator. However, as the actuator exhibits hysteresis nonlinearity, we propose first its linearization thanks to a feedforward control before application of the above filtering and feedback control. The experimental results confirm the efficiency of the approach and demonstrate the interest of the method for precise positioning such as in micropositioning applications.

7.1 Introduction

Most of smart and active materials used as actuators in micro/nanopositioning systems exhibit nonlinearities like hysteresis. Piezoelectric ceramics-based actuators are in this case. When working at large deformation, the hysteresis that typifies them becomes non-negligible and leads to a loss of precision of the whole system even if they have a high resolution. For an enhancement of the performances, closed-loop control techniques have been widely used [1–10], Chaps. 5 and 10. In fact, these techniques seem to be the best way to reach overall substantial performances such as accuracy, repeatability, disturbances and coupling rejection,

M. Rakotondrabe (✉) • J.-A. Escareno • D. Habineza • S. Lescano
Automatic Control and Micro-Mechatronic Systems Department, AS2M,
FEMTO-ST Institute, 24 rue Alain Savary, Besançon 25000, France
e-mail: mrakoton@femto-st.fr; juanantonio.escareno@femto-st.fr; didace.habineza@femto-st.fr; sergio.lescano@femto-st.fr

M. Rakotondrabe (ed.), *Smart Materials-Based Actuators at the Micro/Nano-Scale: Characterization, Control, and Applications*, DOI 10.1007/978-1-4614-6684-0_7,
© Springer Science+Business Media New York 2013

bandwidth augmentation, etc. In addition to that, closed-loop control techniques permit a robustification of the closed-loop scheme in case of model uncertainties. Unfortunately, closed-loop techniques are not convenient for most of applications at the micro/nano-world. Indeed, there is a lack of convenient sensors usable for feedback at this scale, in particular for piezoelectric-based micro/nanopositioning systems. In fact, these systems require high-bandwidth, high-resolution, and high-precision sensors but the actual and existing sensors that could provide such performances are bulky (optical sensors) and are not embeddable. Hence, they are not suitable for more complex structured and/or batch fabricated small systems, additionally to their expensive cost. On the other hand, sensors that are embeddable such as strain gage do not offer the required performances. As their sizes are small, most of them are highly sensitive to (thermal, surrounding, etc.) noises and then provide high noise to signal ratio. Alternative techniques used to enhance the performances of piezoelectric actuators were therefore the open-loop (feedforward) control techniques [10–18], Chaps. 6, 8 and 9. These techniques allow a very high packageability and a low cost of the systems since no sensors are used to control them. Nevertheless, they cannot provide enough robustness as closed-loop control techniques can provide, and then when non-perfect systems models are used, a loss of precision appears. Finally, other approaches to control piezoelectric micro/nanopositioning systems are the self-sensing techniques. Initially used for vibration control and AFM microscope systems [19–23], these techniques were recently extended to work in micro/nanopositioning tasks that require long-term measurement of constant displacement or force [10, 24–26]. Although very interesting, self-sensing techniques for micro/nanopositioning applications with feedback control objectives are still under development. In particular, the integration of the hysteresis behavior in self-sensing is still under research. This chapter proposes to use smaller and embeddable sensors to measure the displacement in piezoelectric actuators and then integrate the signals in a feedback control. Knowing that these embeddable sensors are noisy, we propose to employ a Kalman filtering to improve the measured signals. Furthermore, this Kalman filtering also permits the estimation of the state vector of the actuator. This state vector is afterwards used in a state-feedback control technique for a general performance enhancement. As the actuator exhibits hysteresis nonlinearity, a hysteresis compensator (feedforward control) is first implemented before the Kalman filtering and the feedback control. The main advantage of the proposed approach is the enhancement of the performances of the actuator with robust feedback controller using an embeddable sensor.

The chapter is organized as follows. We present in Sect. 7.2 the piezoelectric actuator to be controlled and the setup used for the experiments. Section 7.3 is devoted to the linearization by feedforward control of the actuator. In Sect. 7.4, we present the Kalman filtering in order to estimate its state and to minimize the measurement noise that affects the displacement signal provided by the sensor used. Finally, the state-feedback control of the actuator is presented in Sect. 7.5.

7.2 Presentation of the Setup to be Controlled

In the sequel, we use a unimorph piezoelectric cantilever (piezocantilever). A unimorph piezocantilever is a cantilever with rectangular section and made up of two layers: a piezoelectric layer and a passive layer glued themselves. When a voltage is applied to the piezoelectric layer, it contracts/expands which results in a global deflection of the cantilever (see Fig. 7.1). It is also possible to obtain the deflection of the actuator when more than two layers compose it.

In this chapter, we employ a PZT material (lead zirconate titanate) for the piezoelectric layer and nickel for the passive layer. The actuator has dimensions of 15 mm × 2 mm × 0.3 mm, where the thickness of the piezoelectric layer is 0.2 mm and that of the passive layer is 0.1 mm. The experimental setup, pictured in Fig. 7.2, is composed of:

- The unimorph piezoelectric actuator.
- A strain gauge embedded on its surface and used to measure the deflection. The calibration of the strain gauge has been carried out *a priori*.
- An electronic conditioner for the strain gauge.
- A computer and a dSPACE-board that supply the control signal and acquire the measurement. They are cadanced at a fresh time of 0.2 ms. The Matlab-Simulink software is used to implement the signal generators, the Kalman filtering, and the controllers.
- A high-voltage amplifier that amplifies the control voltage from the computer/dSPACE.

Remind that we also use an optical sensor (from *Keyence*) to measure the deflection of the cantilever. This optical sensor is only used to characterize the actuator's hysteresis in the next subsection and to validate the different results during the linearization and the Kalman filtering.

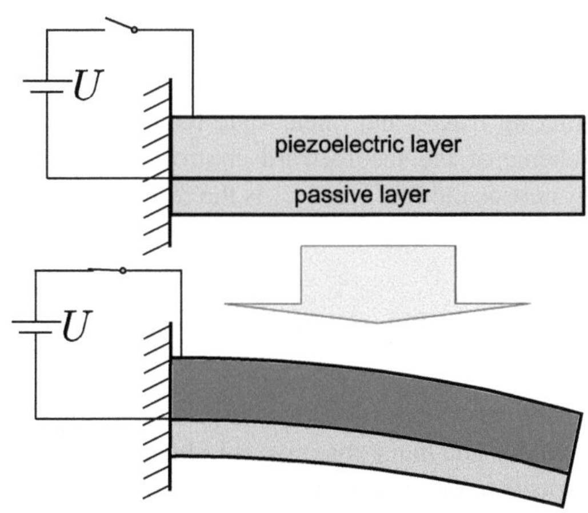

Fig. 7.1 A unimorph piezoelectric cantilever

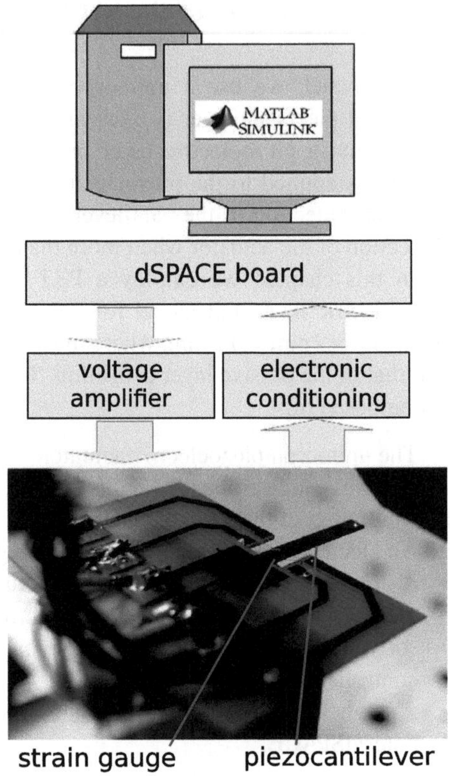

Fig. 7.2 The experimental setup

7.3 Linearization by Feedforward Control of the Piezoelectric Actuator

This section is devoted to the characterization, modeling, and feedforward control of the hysteresis that typifies the piezoelectric actuator. The aim of the control is to obtain a new linear system to which we can afterwards apply the linear Kalman filtering (LKF) and linear state-feedback control. Figure 7.3 depicts the principle scheme of the feedforward control. In the figure, Γ^{-1} is an inverse model of a hysteresis model Γ and y_{rh} is the reference input of hysteresis controlled system.

7.3.1 Hysteresis Characterization

In this part, we characterize the piezocantilever's behaviors, in particular in the static domain. The main objective is to have an idea about the hysteresis nonlinearity that appears in the piezocantilever behavior. For that, we apply a sine input voltage U to the actuator and measure the resulting deflection y at its tip. The amplitude U_A of the

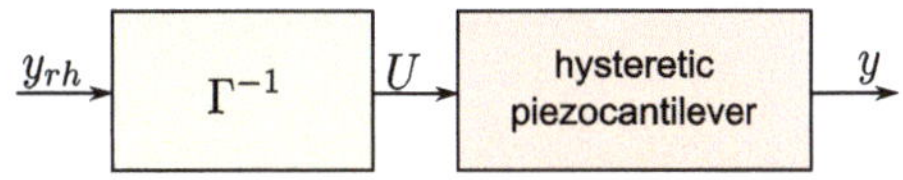

Fig. 7.3 Principle scheme of a feedforward control of the hysteresis

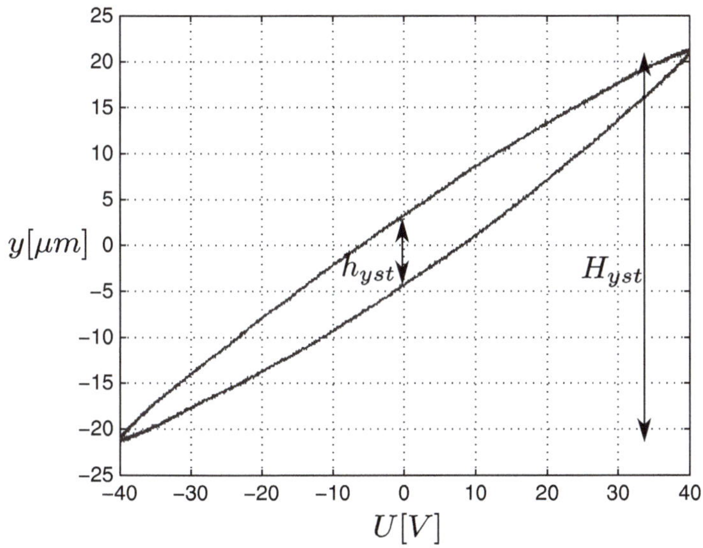

Fig. 7.4 The hysteresis of the actuator

voltage is chosen to cover the range of operation of the actuator. For that we choose $U_A = 40\,\text{V}$. The frequency is chosen to be low enough in order to avoid the phase-lag, i.e., to avoid the influence of the dynamics of the actuator on the shape of the plotted hysteresis [2]. Different tests show that a frequency of 0.1 Hz is convenient for our setup. Afterwards, the input–output (U, y) map is plotted. Figure 7.4 depicts the results which shows a hysteresis of $\frac{h_{yst}}{H_{yst}} \approx 19\,\%$.

7.3.2 Hysteresis Modeling

Different approaches have been used to model and compensate the hysteresis phenomenon in piezoelectric actuators. A small survey can be found in Chap. 9. In this chapter, we propose to use the classical Prandtl–Ishlinskii approach which offers a high precision and a convenience for real-time applications. In the approach, a complex hysteresis can be modeled by the superposition of several elementary hysteresis called hysterons. The hysteron itself, called play-operator or backlash operator and denoted γ, has the following equation:

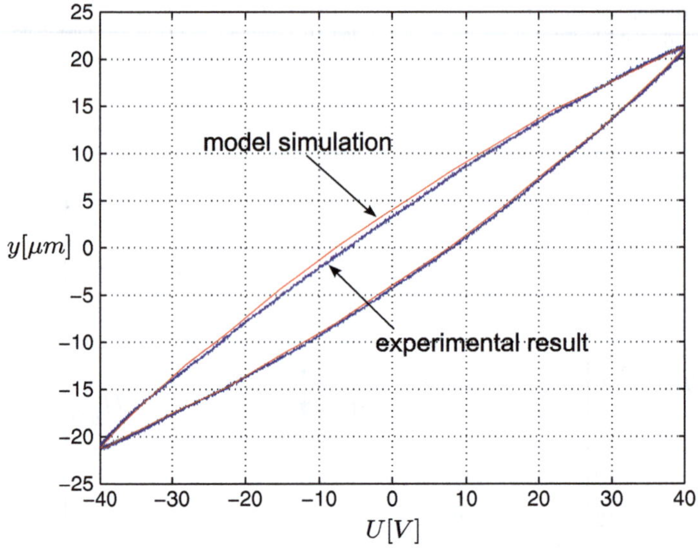

Fig. 7.5 Hysteresis curve: experimental result and simulation of the identified model

$$\begin{cases} y(t) = \gamma(U(t), y(t-T)) = \max\{U(t) - r, \min\{U(t) + r, y(t-T)\}\} \\ y(0) = y_0 \end{cases} \tag{7.1}$$

where r is called threshold of the backlash operator, T is the sampling time, and $y(t-T)$ is the output displacement measured at the previous time.

Hence, the complex hysteresis, denoted Γ, is written as follows [27]:

$$\begin{cases} y(t) = \Gamma = \sum_{i=1}^{n_h} w_i \cdot \max\{U(t) - r_i, \min\{U(t) + r_i, y_{ei}(t-T)\}\} \\ y(0) = y_0 \end{cases} \tag{7.2}$$

where n_h is the number of superposed hysterons, r_i is the backlash, and y_{ei} is the elementary output of the ith backlash operator. Finally the gain w_i is used to weight the corresponding operator.

Following the procedure proposed in [13], the parameters r_i and w_i of the hysteresis of Fig. 7.4 were identified. Figure 7.5 pictures the comparison of the experimental result and of the model simulation which shows a good adequacy between the identified model and the real hysteresis.

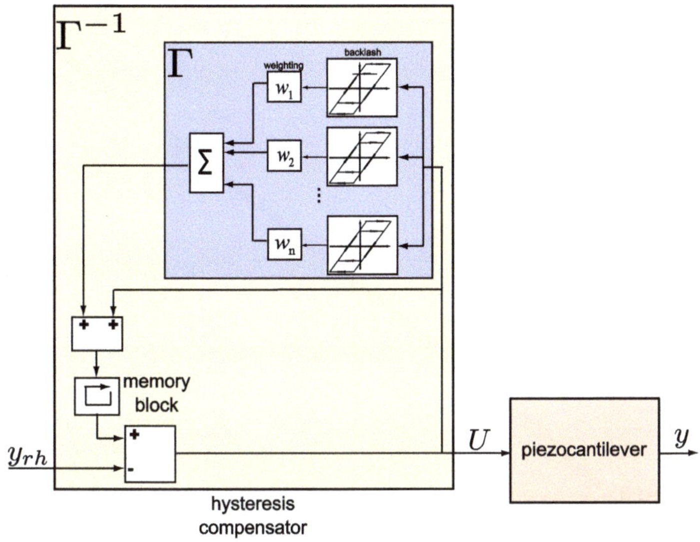

Fig. 7.6 Implementation of hysteresis compensator based on the inverse multiplicative inverse structure [17]

7.3.3 Feedforward Control

To compensate a hysteresis modeled with a classical Prandtl–Ishlinskii approach, another classical Prandtl–Ishlinskii model can be used as the feedforward controller (or compensator) [13, 28]. This requires an additional calculation of the parameters of the controller however. Another compensation technique is to employ the same model Γ as base of the compensator but with a slightly modified structure, more precisely with the inverse multiplicative structure. This technique was proposed in [17] and its main advantage is the non-requirement of additional calculation of the compensator's parameters. Indeed, the initial model Γ with its parameters is directly used in the compensator. We use this technique which is reminded by the following theorem.

Theorem 3.1. *Reconsider the PI hysteresis model in Eq. (7.2). If the compensator is defined by*

$$U(t) = \sum_{i=0}^{n} w_i \cdot \max\{U(t-T) - r_i, \min\{U(t-T) + r_i, y_{ei}(t-2T)\}\} - y_{rh}(t)$$

then the hysteresis will be compensated.

Proof. See [17]

The implementation of the compensator in Theorem 3.1 is pictured in Fig. 7.6.

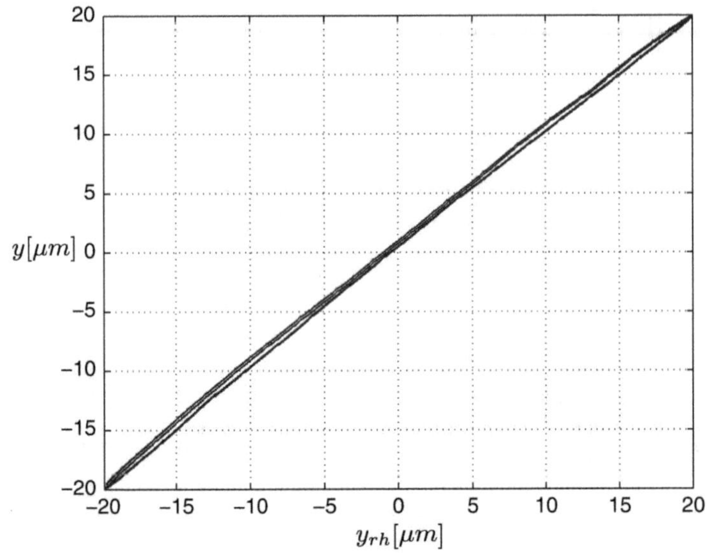

Fig. 7.7 Experimental results with the hysteresis compensator

The compensator was implemented in Simulink following the block diagram of Fig. 7.6. To verify its efficiency, a sine input reference y_{rh} is applied. The amplitude should be less or equal to the output amplitude of y obtained during the identification, i.e., less than the output in Fig. 7.5. We choose an amplitude of $y_{rhA} = 20\,\mu\text{m}$. A frequency of 0.1 Hz is also used for the sine input. In general, a frequency (much) higher than that used during the identification would not work. Indeed, the hysteresis model used here (the classical Prandtl–Ishlinskii model) is a static hysteresis model. Consequently, the model is valid for the frequency (and for low frequencies) with which it has been identified. To be efficient at high frequencies, a dynamics such as a transfer function should be combined with the static hysteresis model. Such dynamics will be introduced in the next sections. Figure 7.7 pictures the experimental results with the hysteresis compensator. It shows that the new system is linear.

7.3.4 The New System

Having controlled the piezoelectric cantilever thanks to a hysteresis feedforward compensator, we obtain a new linear system. Remind that the compensator only removed the static nonlinearity and then allowed to derive a (linear) static gain. Combining this linear static gain with a linear dynamics (transfer function, differential equation), the complete model (static and dynamic model) is yielded. This new system is pictured in Fig. 7.8.

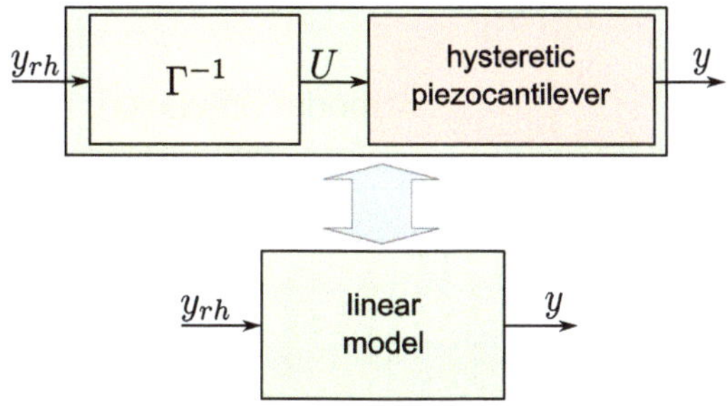

Fig. 7.8 The new system

In the sequel, we will use a second-order dynamic model for the linearized actuator. The advantage of such a model is its low order but at the same time its ability to account the vibration, if the latter exists. We use:

$$a\ddot{y} + b\dot{y} + y = ky_{rh} \Leftrightarrow \frac{y(s)}{y_{rh}(s)} = \frac{k}{as^2+bs+1} \tag{7.3}$$

where $k = \partial y/\partial y_{rh}$ is the static gain identified from Fig. 7.7, and a and b are the coefficients of the dynamics of the new system. The parameters a and b can be determined by identifying the linearized system. For that, a step reference input y_{rh} is applied to the system, then system identification techniques with Matlab[1] are used. We have $k \approx 1$, $a = 4.5 \times 10^{-8}$, and $b = 4.2 \times 10^{-6}$. Figure 7.9 depicts the response of the linearized system (new system) when a step input of $y_{rh} = 20\,\mu\text{m}$ is applied. In the same figure, we also plot the step response of the model (Eq. (7.3)) with the identified parameters. Remind that the experimental response in Fig. 7.9 was measured with the optical sensor.

7.4 Kalman Filtering Applied to the Linearized Piezoelectric Cantilever

In this section, we use the strain gauge as displacement sensor for the piezocantilever. Its main advantage with respect to the optical sensor used so far is its embeddability onto the actuator (see Fig. 7.2). Unfortunately, the strain gauge provides noisy signal which may, among others, alter the performances of the system if a closed-loop control is employed. To filter the noise, we propose to use a Kalman filtering. As we have now a linearized system, we use the LKF. In addition to the noise filtering, this also enables us to estimate the state of the piezocantilever. This estimate state can be afterwards used in a state-feedback control scheme.

[1] Identification Matlab Toolbox.

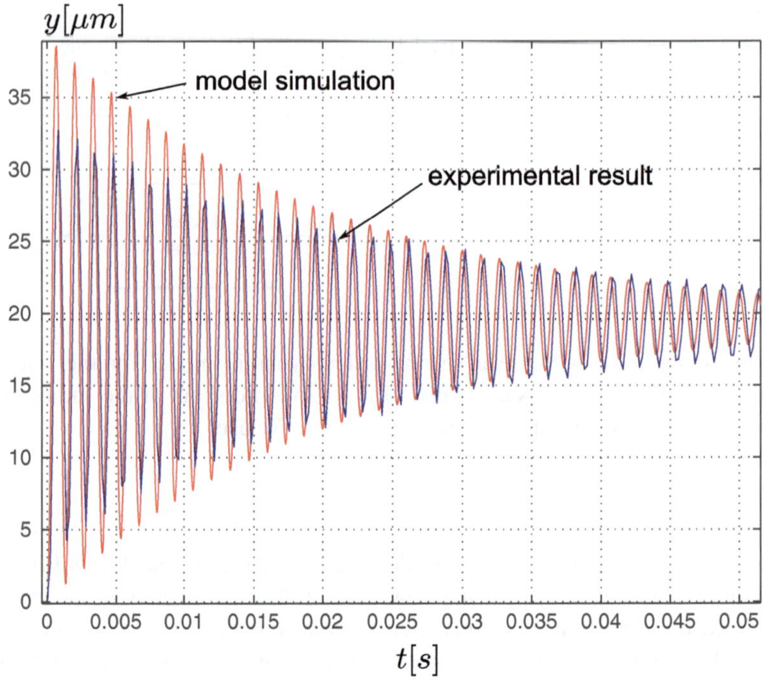

Fig. 7.9 Step response of the linearized system

7.4.1 *The Linear Kalman Filtering*

When introducing the noise, the state-space representation of the overall dynamic equation in (7.3) becomes

$$\frac{\mathrm{d}X}{\mathrm{d}t} = AX + By_{rh} + \eta_p \mathbf{w} \tag{7.4}$$

where $X = (y, (\mathrm{d}y/\mathrm{d}t))^T$ and $\mathbf{w} = (w_{vel}, w_{acc})^T$ represent the state vector and the process noise vector, respectively. The matrices

$$A = \begin{pmatrix} 0 & 1 \\ -\frac{1}{a} & -\frac{b}{a} \end{pmatrix},\ B = \begin{pmatrix} 0 \\ \frac{k}{a} \end{pmatrix} \text{ and } \eta_p = \begin{pmatrix} 0\ 0 \\ 0\ 1 \end{pmatrix} \tag{7.5}$$

represent the state transition matrix, the input control distribution matrix, and the process-noise distribution matrix, respectively. The noise vector $\mathbf{w}$ is modeled as random walk process dictated by white Gaussian noise corresponding to modeling inaccuracies. Such a noise process is described by a continuous covariance matrix Q as follows:

$$Q = \begin{pmatrix} \sigma_{vel}^2 & 0 \\ 0 & \sigma_{acc}^2 \end{pmatrix} \tag{7.6}$$

Likewise the measurement equation is

$$y = CX + \mathbf{v} \tag{7.7}$$

where $\mathbf{v}$ corresponds to the noise that affects the current measurement. The matrix

$$C = \begin{pmatrix} 1 & 0 \end{pmatrix} \tag{7.8}$$

represents the output-measurement distribution matrix. The noise vector $\mathbf{v}$ is described by the variance of the displacement measurement $R = \sigma_{\text{meas}}^2$, with $R = 0.7$ in our case.

Hence, the LKF for the continuous model in Eqs. (7.4) and (7.7), also called Kalman–Bulcy filtering, is given by [29]:

$$\frac{\mathrm{d}\hat{X}}{\mathrm{d}t} = A\hat{X} + By_{rh} + K_f\left(y - C\hat{X}\right) \tag{7.9}$$

where K_f is the gain of the filter. This gain K_f is calculated as follows:

$$K_f(t) = P_f(t)C^T\left(R\right)^{-1} \tag{7.10}$$

where $P_f(t)$ is the solution of the following differential Riccati equation:

$$\frac{\mathrm{d}P_f}{\mathrm{d}t} = \eta_p Q\left(\eta_p\right)^T + AP_f + P_f A^T - P_f C^T\left(R\right)^{-1} CP_f \tag{7.11}$$

Remark 1. The inputs of the LKF comprise the actual deflection measurement (provided by the strain gauge) as well as the input y_{rh} of the hysteresis compensator, as pictured in Fig. 7.10. In the figure, the state noise $\mathbf{w}$ is not represented. The output $\hat{X}$ of the LKF represents the filtered estimate state vector.

7.4.2 LKF Implementation

The LKF equation in Eq. (7.9) was implemented in the Simulink. Figure 7.11 depicts the estimate and filtered output deflection $\hat{y}$ provided by the LKF and the measurement y from the optical sensor in order to validate the efficiency of the strain gauge sensor combined with the LKF. Notice that y is the first element of the vector state X, i.e., $y = X(1)$ and $\hat{y} = \hat{X}(1)$.

Concerning the second element of the state vector, i.e., the velocity $X(2) = \mathrm{d}y/\mathrm{d}t$, Fig. 7.12a shows the estimate $\hat{X}(2)$ provided by the LKF. Figure 7.12b depicts the

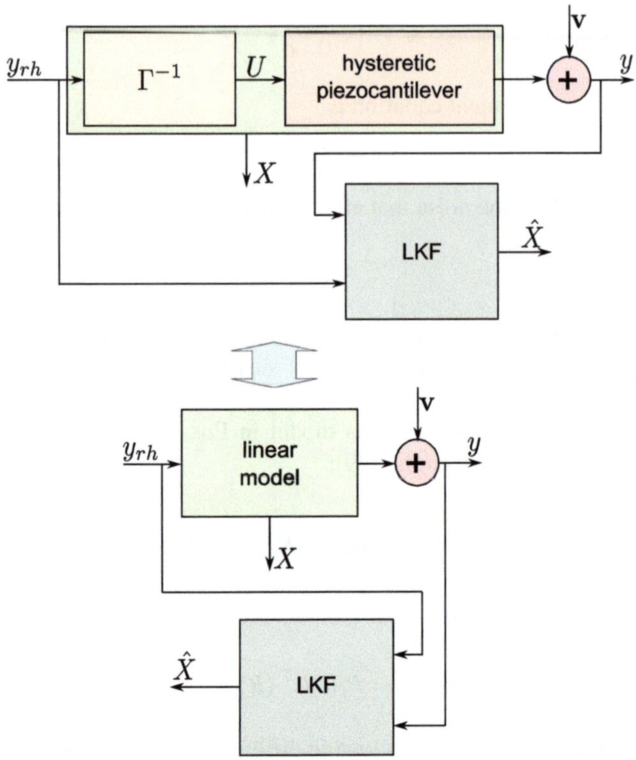

Fig. 7.10 Block diagram of the linearized system with the LKF for state estimation and noise rejection

estimate $\hat{X}(2)$ obtained with a numerical derivative of the above estimate $\hat{X}(1) = \hat{y}$ of the bending. Finally Fig. 7.12c depicts the estimate $\hat{X}(2)$ obtained with a numerical derivative of the measurement y. As we can see from these figures, the estimate $\hat{X}(2)$ from the LKF presents the most clear signal as it is almost without noise.

7.5 State-Feedback Control with Integral Action

Having now the state available, we can perform a state-feedback control of the linearized system. In this section, such a feedback control scheme is proposed. The pole assignment is used to adjust and to enhance the dynamics of the closed loop while an integral action is introduced to cancel the steady-state error and then to increase the accuracy.

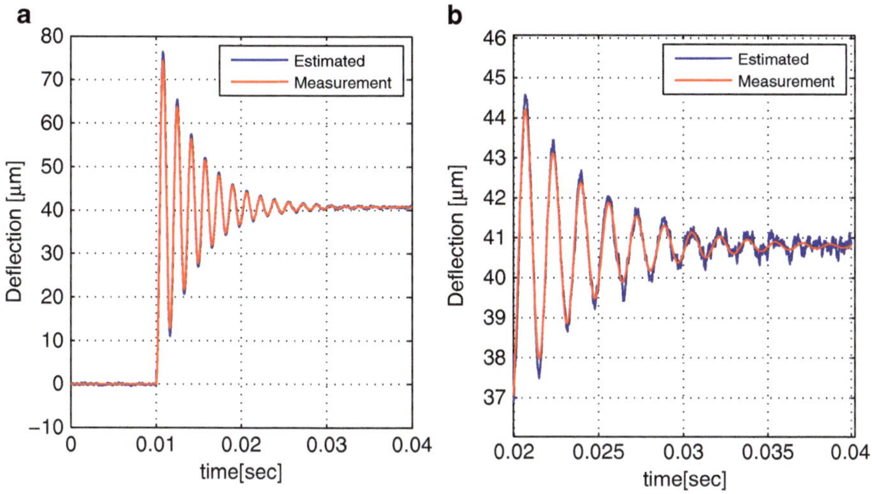

Fig. 7.11 (**a**) Measured bending y and estimate bending $\hat{y}$ of the piezocantilever and (**b**) zoom

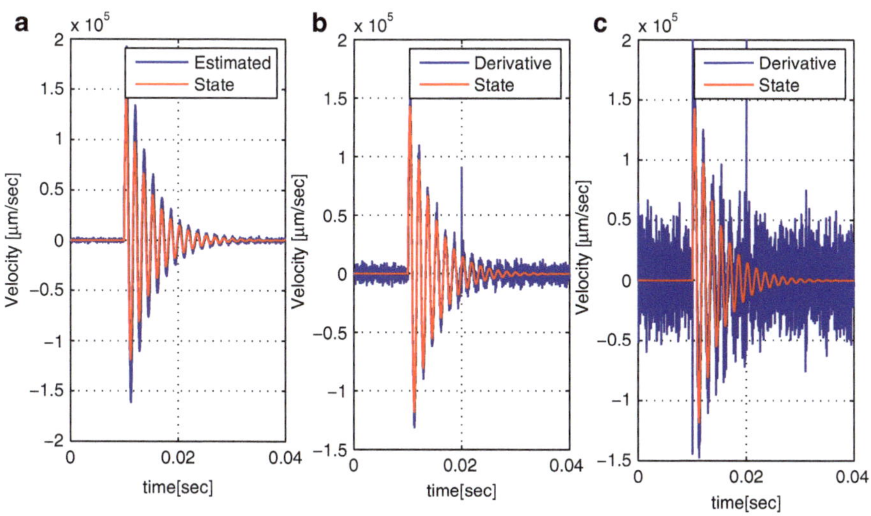

Fig. 7.12 (**a**) Estimate $\hat{X}(2)$ obtained with the LKF. (**b**) Estimate $\hat{X}(2)$ obtained with a numerical derivative of the previous estimate $\hat{X}(1)$. (**c**) Estimate $\hat{X}(2)$ obtained with a numerical derivative of the measurement y

7.5.1 Remind of the New System to be Controlled and Principle Scheme

The system to be controlled is composed of:

- The piezocantilever with hysteresis.
- The hysteresis compensator.
- The strain gauge sensor to derive the measurement of the bending y. Remind that the measurement from this sensor is noisy.
- A LKF that provides an estimate of the internal state of the system and that filters the noise.

The scheme of the system to be controlled is pictured in Fig. 7.10. Assuming that the observer, i.e., the Kalman filtering, is highly quick relative to the further controlled system (closed loop), we can write the following equality for the sequel:

$$\hat{X} = X \Leftrightarrow C\hat{X} = CX \Leftrightarrow \hat{y} = y \tag{7.12}$$

for a control synthesis point of view.

This assumption is equivalent to saying that the transient part time of the estimation/observation is very low and negligible relative to the time characteristics of the control. Hence, the system to be controlled, in its state-space representation, can be written as follows:

$$\begin{cases} \frac{\mathrm{d}X}{\mathrm{d}t} = AX + By_{rh} \\ y = CX \end{cases} \tag{7.13}$$

instead of

$$\begin{cases} \frac{\mathrm{d}\hat{X}}{\mathrm{d}t} = A\hat{X} + By_{rh} \\ y = C\hat{X} \end{cases} \tag{7.14}$$

Notice that the model in Eq. (7.13) is similar to the model in Eqs. (7.4) and (7.7) except the absence of the noises thanks to the Kalman filtering.

7.5.2 Scheme of the Closed Loop

Let Fig. 7.13a be the scheme of the closed loop. In this, the controller is based on a state-feedback gain K_c and an integral gain K_i. The gain K_c is used to enhance the dynamics of the closed loop while the gain K_i is employed to suppress the static error. The main advantage of using an integral in the loop for the static error suppression, instead of employing a prefilter gain, is its robustness. The reason for which we use the estimate output $\hat{y} = C\hat{X}$ as external feedback loop instead

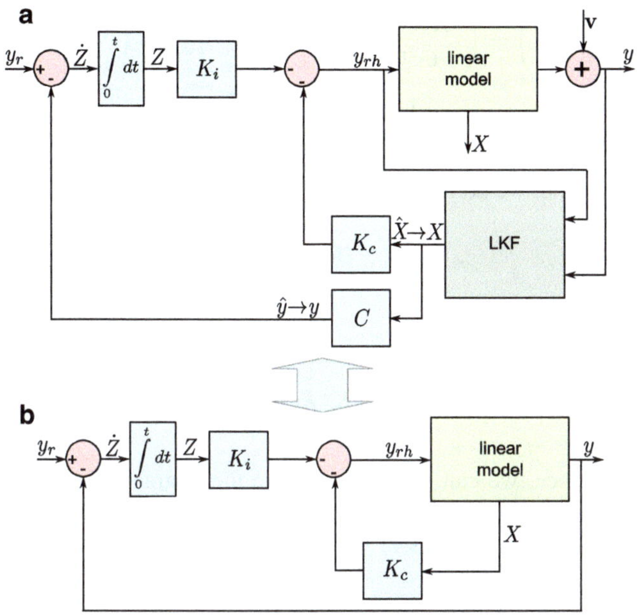

Fig. 7.13 (**a**): Scheme of the state-feedback control with integral action; (**b**): the simplified equivalent scheme

of the measured output y is because $\hat{y}$ is with minimized noise. Let Fig. 7.13b be the simplified equivalent scheme of Fig. 7.13a. In the figure, we have

- y_r as the reference input of the closed loop.
- Z as a new variable that we introduce such that $\frac{\mathrm{d}Z}{\mathrm{d}t} = y_{y_r} - y$ is the error.

7.5.3 *Equations of the Closed-Loop*

From Eq. (7.13) and Fig. 7.13, we have:

$$\begin{cases} \frac{\mathrm{d}X}{\mathrm{d}t} = AX + By_{rh} \\ y = CX \\ \frac{\mathrm{d}Z}{\mathrm{d}t} = y_r - CX \end{cases} \tag{7.15}$$

Considering a new vector state $\begin{pmatrix} X \\ Z \end{pmatrix}$, we obtain a new state-space representation, called augmented state-space, of the closed loop as follows:

$$\begin{cases} \frac{\mathrm{d}}{\mathrm{d}t}\begin{pmatrix} X \\ Z \end{pmatrix} = \begin{pmatrix} A & [0] \\ -C & [0] \end{pmatrix}\begin{pmatrix} X \\ Z \end{pmatrix} + \begin{pmatrix} B \\ [0] \end{pmatrix} y_{rh} + \begin{pmatrix} [0] \\ I \end{pmatrix} y_r \\ y = \begin{pmatrix} C & [0] \end{pmatrix}\begin{pmatrix} X \\ Z \end{pmatrix} \end{cases} \tag{7.16}$$

On the other hand, we derive from Fig. 7.13 that:

$$y_{rh} = -(K_c X + K_i Z) \tag{7.17}$$

which is equivalent to

$$y_{rh} = -\begin{pmatrix} K_c & K_i \end{pmatrix}\begin{pmatrix} X \\ Z \end{pmatrix} \tag{7.18}$$

As a consequence, we can say that we have the augmented state-space model given by Eq. (7.16) which is feedback controlled by an augmented-state-feedback gain $K = \begin{pmatrix} K_c & K_i \end{pmatrix}$. The matrices of the augmented state-space model (Eq. (7.16)) are:

The augmented state matrix

$$A_{\mathrm{aug}} = \begin{pmatrix} A & [0] \\ -C & [0] \end{pmatrix} \tag{7.19}$$

The augmented input matrix

$$B_{\mathrm{aug}} = \begin{pmatrix} B \\ [0] \end{pmatrix} \tag{7.20}$$

The augmented output matrix

$$C_{\mathrm{aug}} = \begin{pmatrix} C & [0] \end{pmatrix} \tag{7.21}$$

Finally, from all these remarks, the closed-loop controlled scheme of Fig. 7.13 has the following state-space representation:

$$\begin{cases} \frac{\mathrm{d}}{\mathrm{d}t}\begin{pmatrix} X \\ Z \end{pmatrix} = \begin{pmatrix} A & [0] \\ -C & [0] \end{pmatrix}\begin{pmatrix} X \\ Z \end{pmatrix} - \begin{pmatrix} B \\ [0] \end{pmatrix}\begin{pmatrix} K_c & K_i \end{pmatrix}\begin{pmatrix} X \\ Z \end{pmatrix} + \begin{pmatrix} [0] \\ I \end{pmatrix} y_r \\ y = \begin{pmatrix} C & [0] \end{pmatrix}\begin{pmatrix} X \\ Z \end{pmatrix} \end{cases} \tag{7.22}$$

which is equivalent to

$$\begin{cases} \frac{\mathrm{d}}{\mathrm{d}t}\begin{pmatrix} X \\ Z \end{pmatrix} = \begin{pmatrix} A-BK_c & -BK_i \\ -C & [0] \end{pmatrix}\begin{pmatrix} X \\ Z \end{pmatrix} + \begin{pmatrix} [0] \\ I \end{pmatrix} y_r \\ y = \begin{pmatrix} C & [0] \end{pmatrix}\begin{pmatrix} X \\ Z \end{pmatrix} \end{cases} \tag{7.23}$$

The aim is therefore to find the augmented feedback gain $K = \begin{pmatrix} K_c & K_i \end{pmatrix}$ such that we have convenient eigenvalues of the augmented closed-loop state matrix $\mathbb{A}$:

$$\mathbb{A} = \begin{pmatrix} A-BK_c & -BK_i \\ -C & [0] \end{pmatrix} \tag{7.24}$$

The order of $\mathbb{A}$ is $n+1=3$, where $n=2$ is the order of A. We mention that the static error is still rejected. Indeed, letting $\frac{\mathrm{d}}{\mathrm{d}t}\begin{pmatrix} X \\ Z \end{pmatrix} = [0]$ (at steady-state regime) in Eq. (7.24) yields

$$\begin{cases} (A-BK_c)X = BK_iZ \\ CX = y_r \\ y = CX \end{cases} \tag{7.25}$$

The two last equation of Eq. (7.25) yields

$$y = y_r \tag{7.26}$$

which means that the static error is always null.

7.5.4 Calculations of the Controller Gains

The calculation of the augmented feedback gain $K = \begin{pmatrix} K_c & K_i \end{pmatrix}$ is based on the following requirements [10]:

- Stability condition—the matrix $\mathbb{A}$ should be HURWITZ, i.e., it should have eigenvalues with negative real parts.
- Performances condition—the closed-loop system should have better performance than that of the initial system. In other words, the eigenvalues of $\mathbb{A}$ should be “more performant” than the eigenvalues of A_{aug}.

The second condition requires information about the eigenvalues of A_{aug}. From the model (Eq. (7.5)) and the numerical values of its parameters, we yield the eigenvalues:

$$\text{eig}\left(A_{\text{aug}}\right) = \begin{pmatrix} 0 \\ -46.7 + 4713i \\ -46.7 - 4713i \end{pmatrix} \tag{7.27}$$

with $i^2 = -1$. It is worth to notice that the presence of the imaginary parts indicates the vibration that we can see in the step response in Fig. 7.9.

By assigning eigenvalues better than Eq. (7.27) to $\mathbb{A}$, we can derive the gain $K = \left(K_c \; K_i\right)$. Such a technique is called synthesis by eigenvalues or pole assignment. Different values of eigenvalues of $\mathbb{A}$ have been tested. We present here the results with the following chosen eigenvalues:

$$\text{eig}\left(\mathbb{A}\right) = \mathbb{P} = \begin{pmatrix} -200 \\ -200 + 1000i \\ -200 - 1000i \end{pmatrix} \tag{7.28}$$

For the calculation of $K = \left(K_c \; K_i\right)$, the following pole assignment function of Matlab is used:

$$K = \text{place}\left(A_{\text{aug}}, B_{\text{aug}}, \mathbb{P}\right) \tag{7.29}$$

We obtain

$$K = \left(K_c \; K_i\right) = \left(-21102222 \; 507 \; -207999999\right) \tag{7.30}$$

from which we extract the feedback gain K_c and the integral gain K_i:

$$\begin{cases} K_c = \left(-21102222 \; 507\right) \\ K_i = -207999999 \end{cases} \tag{7.31}$$

7.5.5 Controller Implementation

The controller gains in Eq. (7.31) were implemented in Simulink additionally to the feedforward control in Sect. 7.3 and to the LKF designed in Sect. 7.4. Figure 7.14a depicts the block diagram of the whole implementation and of the piezoelectric actuator with hysteresis. Experimental tests with different values of step input references y_r were carried out. Figure 7.14b shows the step response which demonstrates the efficiency of the designed controller in terms of oscillation diminution and rapidity augmentation. Indeed, the important overshoot seen in the step response of the initial system (see Fig. 7.9) is highly reduced by the proposed closed-loop control scheme. Furthermore the settling time is less than 11 ms with the closed-loop controller while it was more than 50 ms for the initial system according to Fig. 7.9.

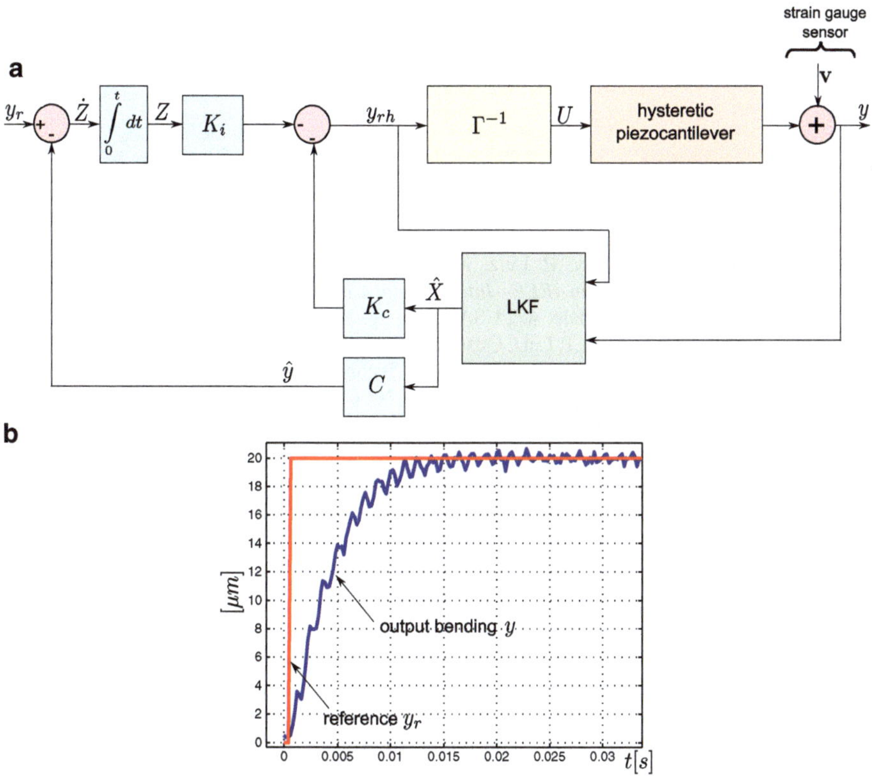

Fig. 7.14 (**a**) Block diagram of the hysteresis feedforward controller, of the LKF, and of the state-feedback controller. (**b**) Step response of the closed loop

7.6 Conclusion

This chapter presented the use of Kalman filtering and the application of a state-feedback control with integral action in piezoelectric cantilevered actuators. The main objective was to employ embeddable sensors (strain gage) for the output measurement and then for a feedback control of the actuator in order to enhance its general performances. As the sensor provided noisy signals, the Kalman filtering allowed the noise rejection additionally to the state estimation. The feedback control is based on a state-feedback scheme with integral action in the loop. Its main advantage is the guaranteed derivation of a zero static error. Moreover, it also provides an improvement of the dynamics of the whole system. As the piezoelectric actuator exhibits hysteresis, we proposed first a feedforward controller to compensate this before applying the Kalman filtering and the state-feedback control. The experimental results confirmed the efficiency of the approach and demonstrated its interest for micropositioning applications.

Acknowledgements This work is supported by the national ANR-Emergence MYMESYS-project (ANR-11-EMMA-006: High Performances Embedded Measurement Systems for multi-Degrees of Freedom Microsystems) and partially by the MIM-Hac project.

References

1. M. Rakotondrabe, C. Clévy, P. Lutz, Modelling and robust position/force control of a piezoelectric microgripper, in *IEEE—International Conference on Automation Science and Engineering (CASE)*, Scottsdale, AZ, USA, 2007, pp. 39–44
2. M. Rakotondrabe, Y. Haddab, P. Lutz, Quadrilateral modelling and robust control of a nonlinear piezoelectric cantilever. IEEE Trans. Contr. Syst. Technol. (T-CST) **17**(3), 528–539 (2009)
3. M. Rakotondrabe, K. Rabenorosoa, J. Agnus, N. Chaillet, Robust feedforward-feedback control of a nonlinear and oscillating 2-dof piezocantilever. IEEE Trans. Autom. Sci. Eng. (T-ASE) **8**(3), 506–519 (2011)
4. S. Khadraoui, M. Rakotondrabe, P. Lutz, Interval modeling and robust control of piezoelectric microactuators. IEEE Trans. Contr. Syst. Technol. (T-CST) **20**(2), 486–494 (2012)
5. Y. Shan, K.K. Leang, Accounting for hysteresis in repetitive control design. Automatica **48**(8), 1751–1758 (2012)
6. A. Sebastian, A. Gannepalli, M.V. Salapaka, A review of the systems approach to the analysis of dynamic-mode atomic force microscopy. IEEE Trans. Contr. Syst. Technol. **15**(5), 952–959 (2007)
7. Q. Xu, Y. Li, Model predictive discrete-time sliding mode control of a nanopositioning piezostage without modeling hysteresis. IEEE Trans. Contr. Syst. Technol. **20**(4), 983–994 (2012)
8. A. Bazaei, Y.K. Yong, S.O.R. Moheimani, A. Sebastian, Tracking of triangular references using signal transformation for control of a novel AFM scanner stage. IEEE Trans. Contr. Syst. Technol. **20**(2), 453–464 (2012)
9. S. Devasia, E.E. Eleftheriou, R. Moheimani, A survey of control issues in nanopositioning. IEEE Trans. Contr. Syst. Technol. **15**(5), 802–823 (2007)
10. M. Rakotondrabe, *Piezoelectric Cantilevered Structures: Modeling, Control and Measurement/Estimation Aspects* (Springer, Berlin, 2013)
11. D. Croft, G. Shed, S. Devasia, Creep, hysteresis and vibration compensation for piezoactuators: atomic force microscopy application. ASME J. Dyn. Syst. Meas. Contr. **123**(1), 35–43 (2001)
12. A. Dubra, J. Massa, C.l. Paterson, Preisach classical and nonlinear modeling of hysteresis in piezoceramic deformable mirrors. Opt. Express **13**(22), 9062–9070 (2005)
13. M. Rakotondrabe, C. Clévy, P. Lutz, Complete open loop control of hysteretic, creeped and oscillating piezoelectric cantilever. IEEE Trans. Autom. Sci. Eng. (TASE) **7**(3), 440–450 (2010)
14. W.T. Ang, P.K. Kholsa, C.N. Riviere, Feedforward controller with inverse rate-dependent model for piezoelectric actuators in trajectory-tracking applications. IEEE/ASME Trans. Mechatron. **12**(2), 134–142 (2007)
15. B. Mokaberi, A.A.G. Requicha, Compensation of scanner creep and hysteresis for AFM nanomanipulation. IEEE Trans. ASE **5**(2), 197–208 (2008)
16. M. Al Janaideh, P. Krejci, Inverse rate-dependent Prandtl–Ishlinskii model for feedforward compensation of hysteresis in a piezomicropositioning actuator. IEEE/ASME Trans. Mechatron. (2012). doi:10.1109/TMECH.2012.2205265
17. M. Rakotondrabe, Classical Prandtl–Ishlinskii modeling and inverse multiplicative structure to compensate hysteresis in piezoactuators, in *ACC (American Control Conference)*, Montréal, Canada, June 2012, pp. 1646–1651
18. M. Rakotondrabe, Bouc–Wen modeling and inverse multiplicative structure to compensate hysteresis nonlinearity in piezoelectric actuators. IEEE Trans. ASE **8**(2), 428–431 (2011)

19. J.J. Dosch, D.J. Inman, E. Garcia, A self-sensing piezoelectric actuator for collocated control. J. Intell. Mater. Syst. Struct. **3**(1), 166–185 (1992)
20. W.W. Law, W.-H. Liao, J. Huang, Vibration control of structures with self-sensing piezoelectric actuators incorporating adaptive mechanisms. Smart Mater. Struct. **12**, 720–730 (2003)
21. C.K. Pang, G. Guo, B.M. Chen, T.H. Lee, Self-sensing actuation for nanopositioning and active-mode damping in dual-stage HDDs. IEEE/ASME Trans. Mechatron. (T-mech) **11**(3), 328–338 (2006)
22. S.O.R. Moheimani, Y.K. Yong, Simultaneous sensing and actuation with a piezoelectric tube scanner. Rev. Sci. Instrum. **79**, 073702 (2008)
23. S.O.R. Moheimani, Invited review article: accurate and fast nanopositioning with piezoelectric tube scanners: emerging trends and future challenges. Rev. Sci. Instrum. **79**, 071101 (2008)
24. A. Ivan, M. Rakotondrabe, P. Lutz, N. Chaillet, Quasi-static displacement self-sensing method for cantilevered piezoelectric actuators. Rev. Sci. Instrum. (RSI) **80**(6), 065102 (2009)
25. A. Ivan, M. Rakotondrabe, P. Lutz, N. Chaillet, Current integration force and displacement self-sensing method for cantilevered piezoelectric actuators. Rev. Sci. Instrum. (RSI) **80**(12), 2126103 (2009)
26. M. Rakotondrabe, I.A. Ivan, S. Khadraoui, C. Clevy, P. Lutz, N. Chaillet, Dynamic displacement self-sensing and robust control of canxtilevered piezoelectric actuators dedicated to microassembly tasks, in *IEEE/ASME—AIM (International Conference on Intelligent Materials)*, Montréal, Canada, July 2010, pp. 557–562
27. M.A. Krasnosel'skii, A.V. Pokrovskii, *Systems with Hysteresis* (Springer, Berlin, 1989)
28. K. Kuhnen, H. Janocha, Complex hysteresis model of a broad class of hysteretic nonlinearities, in *Proceedings of 8th International Conference on New Actuators*, Bremen, pp. 688–691 (2002)
29. R.G. Brown, P.Y.C. Hwang, *Introduction to Random Signals and Applied Kalman Filtering*, 3rd edn. (Wiley, New York, 1997)

Chapter 8
Intelligent Hysteresis Modeling and Control of Piezoelectric Actuators

Qingsong Xu

Abstract This chapter presents a new approach to hysteresis modeling and compensation of piezoelectric actuators by resorting to an intelligent model. A least squares support vector machine (LSSVM)-based hysteresis model is developed and used for the purposes of both hysteresis characterization and compensation. By this way, the hysteresis inverse is not needed in the feedforward hysteresis compensator since only the hysteresis model is used. The effectiveness of the presented approach is validated by experimental studies on a piezoactuated system. Experimental results confirm that this approach is superior to Bouc–Wen model in terms of both hysteresis modeling and compensation.

8.1 Introduction

Smart materials-based actuators are popularly employed for actuation in various precision engineering applications such as micropositioning, micromanipulation, and microassembly. As a typical smart actuator, piezoelectric actuators are particularly attractive owing to the merits of subnanometer positioning resolution and rapid response speed [14]. Even though intensive works have concentrated on the research and applications of piezoelectric actuators, the nonlinear piezoelectric effects, especially the hysteresis, still pose big challenges to precise positioning nowadays.

To fulfill the requirements of ultrahigh-precision positioning, the piezoelectric hysteresis behavior has to be suppressed by developing a suitable control strategy [20]. It has been shown that the hysteresis can be greatly alleviated by using a charge-driven approach or a capacitor insertion method [12]; it is however

Q. Xu (✉)
Department of Electromechanical Engineering, University of Macau,
Av. Padre Tomás Pereira Taipa, Macao, China
e-mail: qsxu@umac.mo

M. Rakotondrabe (ed.), *Smart Materials-Based Actuators at the Micro/Nano-Scale: Characterization, Control, and Applications*, DOI 10.1007/978-1-4614-6684-0_8,
© Springer Science+Business Media New York 2013

at the cost of stroke reduction. Thus, voltage actuation is still widely adopted in practice. Generally, the existing control schemes fall into two categories of hysteresis model-based and hysteresis model-free methods. In the first approach, a hysteresis model (e.g., Preisach model) is generated and used to construct an inverse-based feedforward compensator [1, 2]. It is known that the inverse model-based compensation is capable of achieving an accurate positioning, whereas the result is very sensitive to the model accuracy [5, 6]. In order to suppress the residual hysteresis as well as creep effects, a combination of feedforward with feedback control can be adopted [3, 15]. The major advantage of the second approach is that no hysteresis model is required. Instead, the unmodeled hysteresis is considered as an uncertainty or a disturbance [19] to the nominal system, which is tolerated by a robust or adaptive control technique. For instance, the sliding mode control [19], H_∞ robust control [18], fuzzy logic control [4], and neural network control [10] have been applied in recent works.

Even though it is possible to compensate the hysteresis nonlinearity by designing an advanced feedback controller without modeling the hysteresis effect, it is true that a feedforward controller by resorting to a simple hysteresis model in combination with a simple feedback (e.g., PID) controller makes it more feasible to suppress the hysteresis effect. The reason lies in that the latter approach allows the relief of burden to develop complicated modern controllers. For example, it has been shown that by modeling the hysteresis with Prandtl–Ishlinskii model [9], a feedforward compensator combined with a PID feedback controller is capable of effectively compensating the nonlinear hysteresis. Nevertheless, the majority of existing works on model-based hysteresis compensation employ an inverse hysteresis model. Hence, both a forward model and an inverse hysteresis model are required for the purposes of hysteresis characterization and compensation. Recently, it has been shown that it is possible to compensate the hysteresis by using a simple Bouc–Wen hysteresis model without adopting the hysteresis inverse [7, 13] .

The purpose of the current research is to characterize and compensate the piezoelectric hysteresis by only using an intelligent hysteresis model without solving the hysteresis inverse. It has been shown that support vector machine (SVM) is superior to artificial neural networks (ANN) in terms of global optimization and higher generalization capability, hence SVM is widely employed to estimate nonlinear system models accurately [16]. Specifically, in the current research, a least squares support vector machine (LSSVM)-based hysteresis model is established and a feedforward compensator is developed without modeling the hysteresis inverse. It is shown that the LSSVM model is more effective than Bouc–Wen model in terms of both hysteresis modeling and hysteresis compensation.

8.2 Modeling of Dynamics with Hysteresis Behavior

The Bouc–Wen model has been extensively applied in piezoelectric hysteresis modeling. It has been shown that the entire dynamic model of the piezoactuated system can be established as follows [7, 11]:

$$M\ddot{y}(t)+B\dot{y}(t)+Ky(t)=K[Du(t)-H(t)] \tag{8.1}$$

$$\dot{H}(t)=\alpha D\dot{u}(t)-\beta|\dot{u}(t)|H(t)-\gamma\dot{u}(t)|H(t)| \tag{8.2}$$

where t is the time variable, parameters M, B, K, and y represent the mass, damping coefficient, stiffness, and displacement response of the piezoactuated system, respectively; D is the piezoelectric coefficient, u denotes the input voltage, and H indicates the hysteretic loop in terms of displacement whose magnitude and shape are determined by parameters α, β, and γ.

The hysteresis model describes the relationship between the input voltage and output displacement of the piezostage system. On the other hand, the input voltage used to produce a desired displacement value is obtained by solving (8.1).

$$u(t)=\frac{1}{KD}[M\ddot{y}(t)+B\dot{y}(t)+Ky(t)+KH(t)]. \tag{8.3}$$

It is observed that the feedforward control signal (8.3) is generated by using the hysteresis term $H(t)$ without solving the inverse hysteresis model. A block diagram constructed with Matlab and Simulink software is given in [7].

Motivated by the hysteresis compensation using Bouc–Wen model which does not solve the hysteresis inverse, an intelligent model based on LSSVM is proposed below.

The dynamic model (8.1) of the system is written into the form

$$\ddot{y}(t)+2\xi\omega_n\dot{y}(t)+\omega_n^2y(t)=du(t)+h(t) \tag{8.4}$$

where ξ and ω_n denote the damping ratio and natural frequency of the piezoactuated system, respectively, d is a positive parameter, and h represents the hysteresis effect in terms of acceleration.

If the hysteresis model h is established, a feedforward hysteresis compensator can be constructed.

$$u(t)=\frac{1}{d}[\ddot{y}(t)+2\xi\omega_n\dot{y}(t)+\omega_n^2y(t)-h(t)] \tag{8.5}$$

which uses the hysteresis model directly without solving the inverse hysteresis model. The hysteresis model based on LSSVM is established in the following discussions.

8.3 Hysteresis Modeling Using LSSVM

It is well known that the hysteresis effect is rate-dependent. Specifically, the hysteresis behavior is dependent not only on the amplitude but also on the frequency of input voltage signals. Moreover, due to the hysteresis nonlinearity, an input voltage corresponds to multiple position outputs. Typically, LSSVM only treats the problem of single-valued mapping between the input and output. Hence, one of the challenges in modeling the hysteresis behavior with LSSVM lies in how to convert the multivalued mapping problem into a single-valued one.

In previous work [21] of the authors, a one-to-one mapping is constructed by introducing the current input value and input variation rate as one data set. Nevertheless, in the case that the input data are accompanied with noises, the variation rate is not smooth and will induce modeling error. Moreover, it is unknown how many orders of the variation rate are sufficient to establish the mapping. Later, a regression model of hysteresis is developed in [17] by employing the current and previous inputs and previous outputs as exogenous inputs to transform the multivalued mapping is into a single-valued one. Yet, the above work takes the voltage and position as the input and output variables, respectively, which is different from the situation in the current research as shown below.

8.3.1 Regression Model Development

In the current research, the hysteresis term is expressed by taking into account (8.4):

$$h(t) = \ddot{y}(t) + 2\xi\omega_n\dot{y}(t) + \omega_n^2 y(t) - du(t). \tag{8.6}$$

It is observed from (8.6) that the output variable is the hysteresis term h, whereas both input voltage u and output position y appear as input variables. It is different from the scenario in [17].

Using LSSVM, a nonlinear regression model is formulated to capture the hysteresis behavior

$$\hat{h}(t) = f(\mathbf{x}_k) \tag{8.7}$$

with

$$\mathbf{x}_k = [u_k, \ldots, u_{k-m}, y_k, \ldots, y_{k-n}, h_{k-1}, \ldots, h_{k-l}] \tag{8.8}$$

where $\hat{h}_k$ denotes the predicted hysteresis term by LSSVM at the current time instant k. u_{k-1}, y_{k-1}, and h_{k-1} represent the input voltage, output position, and hysteresis term at previous time instant $k-1$, respectively. In addition, $m \geq 0$, $n \geq 0$, and $l \geq 1$ define the order of the model.

8.3.2 Modeling with LSSVM

It is known that the LSSVM maps the input data into a high-dimensional feature space and constructs a linear regression function therein [17, 21]. The unknown hysteresis function is approximated by the equation

$$h(\mathbf{x}) = \mathbf{w}^T \varphi(\mathbf{x}) + b \tag{8.9}$$

with the given training data set $\{\mathbf{x}_k, h_k\}_{k=1}^N$ where N represents the number of training data set, $\mathbf{x}_k \in \mathbf{R}^{m+n+l+2}$ is an input vector as shown in (8.8), and $h_k \in \mathbf{R}$ are the output data. Additionally, $\mathbf{w}$ is a weight vector, $\varphi(\cdot)$ denotes a nonlinear mapping from the input space to a higher-dimensional feature space, and b is the bias.

The LSSVM approach formulates the regression as an optimization problem in the primal weight space. Then, the conditions for optimality are obtained by solving a series of partial derivatives, which are used to construct the dual formulation, i.e.,

$$\begin{bmatrix} 0 & \mathbf{1}_N^T \\ \mathbf{1}_N & \Omega + \gamma^{-1}\mathbf{I}_N \end{bmatrix} \begin{bmatrix} b \\ \alpha \end{bmatrix} = \begin{bmatrix} 0 \\ \mathbf{h} \end{bmatrix} \tag{8.10}$$

where $\alpha = [\alpha_1, \alpha_2, \ldots, \alpha_N]^T$ is called the support vector. The support values are $\alpha_k = \gamma e_k$ with $\gamma \in \mathbb{R}$ denoting the regularization factor. In addition, $\mathbf{1}_N = [1, 1, \ldots, 1]^T$, $\mathbf{h} = [h_1, h_2, \ldots, h_N]^T$, and $\mathbf{I}_N$ is an identity matrix. Besides, the kernel trick is employed to derive that

$$\Omega_{kj} = \varphi(\mathbf{x}_k)^T \varphi(\mathbf{x}_j) = K(\mathbf{x}_k, \mathbf{x}_j), \quad k, j = 1, 2, \ldots, N \tag{8.11}$$

where K is a predefined kernel function. The purpose of introducing the kernel function is to avoid the explicit computation of the map $\varphi(\cdot)$ in dealing with the high-dimensional feature space.

After calculating b and α from (8.10), one can obtain the solution for the regression problem

$$h(\mathbf{x}) = \sum_{k=1}^{N} \alpha_k K(\mathbf{x}, \mathbf{x}_k) + b \tag{8.12}$$

where K is the kernel function satisfying Mercer's condition, $\mathbf{x}_k$ is the training data, and $\mathbf{x}$ denotes the new input data.

By adopting the radial basis function (RBF) as kernel function,

$$K(\mathbf{x}, \mathbf{x}_k) = \exp\left(-\frac{\|\mathbf{x} - \mathbf{x}_k\|^2}{\sigma^2}\right) \tag{8.13}$$

with $\sigma > 0$ denoting the width parameter (which specifies the kernel sample variance σ^2) and $\|\cdot\|$ representing the Euclidean distance, the LSSVM model for the hysteresis model estimation becomes

$$h(\mathbf{x}) = \sum_{k=1}^{N} \alpha_k \exp\left(-\frac{\|\mathbf{x} - \mathbf{x}_k\|^2}{\sigma^2}\right) + b. \tag{8.14}$$

With the assigned regularization parameter γ and kernel parameter σ, the purpose of the training process is to determine the support values α_k and the bias b. A good generalization ability of the LSSVM model relies on an appropriate tuning of the two hyperparameters (γ and σ). In the current research, the leave-one-out cross-validation approach is adopted to infer the values of the hyperparameters.

8.4 Experimental Investigations on Hysteresis Modeling

In this section, the hysteresis modeling based on Bouc–Wen model and LSSVM approach is carried out by experimental studies.

8.4.1 Experimental Setup

Figure 8.1 depicts the experimental setup, where a four-layer piezoelectric bimorph actuator with the dimension of $28 \times 5 \times 0.86\,\text{mm}^3$ is driven by a high-voltage amplifier. A USB-6259 board (from National Instruments Corp.) with 16-bit DAC and ADC channels is adopted to produce an analog voltage, which is then amplified by a high-voltage amplifier (model: EPA-104 from Piezo Systems, Inc.) with an adjustable gain of 10 to provide a high voltage for driving the piezo-actuator. The output displacement at the end point of piezo-bimorph is measured by a laser displacement sensor (model: LK-H055, from Keyence Corp.). The analog voltage output of the sensor signal conditioner is acquired by a PC through one ADC channel of the USB-6259 board. Moreover, LabVIEW software is employed to implement a real-time control of the piezoactuated system.

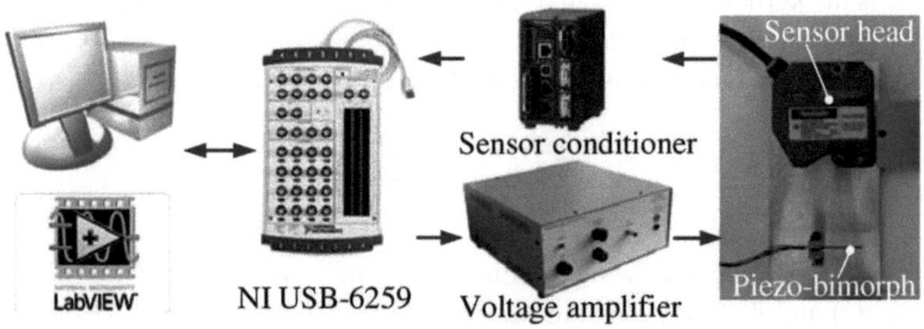

Fig. 8.1 Experimental setup of a piezoactuated system

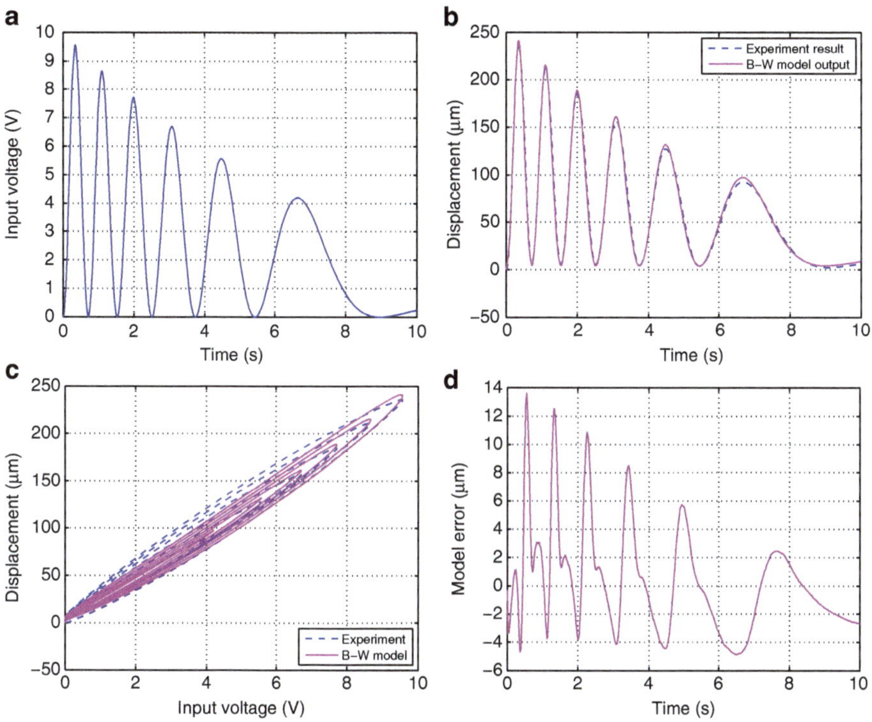

Fig. 8.2 Results of the identified Bouc–Wen model. (**a**) Input voltage; (**b**) experimental result and Bouc–Wen model output; (**c**) displacement–voltage hysteresis loops; (**d**) Bouc–Wen model output errors

8.4.2 Bouc–Wen Model Results

8.4.2.1 Bouc–Wen Model Identification

To identify the hysteresis model, the input voltage signal is chosen as shown in Fig. 8.2a, and the corresponding output data are depicted in Fig. 8.2b.

It has been shown that the seven parameters (M, B, K, D, α, β, and γ) of the Bouc–Wen model can be identified by minimizing the following fitness function [7]:

$$f(M,B,K,D,\alpha,\beta,\gamma) = \frac{1}{N}\sum_{i=1}^{N}\left(y_i - y_i^{\mathrm{BW}}\right)^2 \tag{8.15}$$

where N denotes the total number of samples and $y_i - y_i^{\mathrm{BW}}$ represents the error of the ith sample which is calculated as the deviation of Bouc–Wen model output (y_i^{BW}) from experimental result (y_i).

Table 8.1 Parameters of the Bouc–Wen hysteresis model

Parameter	Search space	Value
M	–	1
B	$[10^3, 10^6]$	7.8486×10^5
K	$[10^4, 10^9]$	5.3959×10^7
D	[0, 1]	0.0234
α	[0, 1]	0.9991
β	[0, 1]	6.2399×10^{-9}
γ	[0, 1]	4.2805×10^{-5}

By setting a time interval of 0.02 s, 500 training data sets are obtained as shown in Fig. 8.2a, b. Note that the voltage signal is applied to the high-voltage amplifier. Then, the Bouc–Wen model is identified by optimizing the seven parameters to minimize the fitness function (8.15). Specifically, the particle swarm optimization (PSO) is adopted, and the identified model parameters are shown in Table 8.1. It is noticeable that the mass M is normalized as 1 in order to reduce the number of the model parameters.

8.4.2.2 Modeling Results

The experimental output and the Bouc–Wen model output are plotted in Fig. 8.2b–d. It is found that the Bouc–Wen model cannot exactly represent the complicated hysteresis of the piezostage system. A relatively large error exists between the identified model output and the experimental result as shown in Fig. 8.2d. Specifically, the maximum model error is 13.63 μm, which accounts for 5.8% of the travel range of the piezo-actuator. The root-mean-square error (RMSE) is 3.71 μm, which accounts for 1.6% of the travel range. It is observed that smaller model error is obtained when the input has lower magnitude and frequency. Hence, the model errors vary greatly at different amplitudes and frequencies of the input signal, which indicates that the Bouc–Wen model cannot capture the rate dependency of the hysteresis precisely.

8.4.2.3 Generalization Study

To test the generalization of the Bouc–Wen model, a new input signal is selected as shown in Fig. 8.3a. The model output is depicted in Fig. 8.3b, c. The model error with respect to the actual output (y_d) obtained by experiments is illustrated in Fig. 8.3d. It is observed that the Bouc–Wen model produces an RMSE of 2.86 μm, which accounts for 2.1% of the motion range.

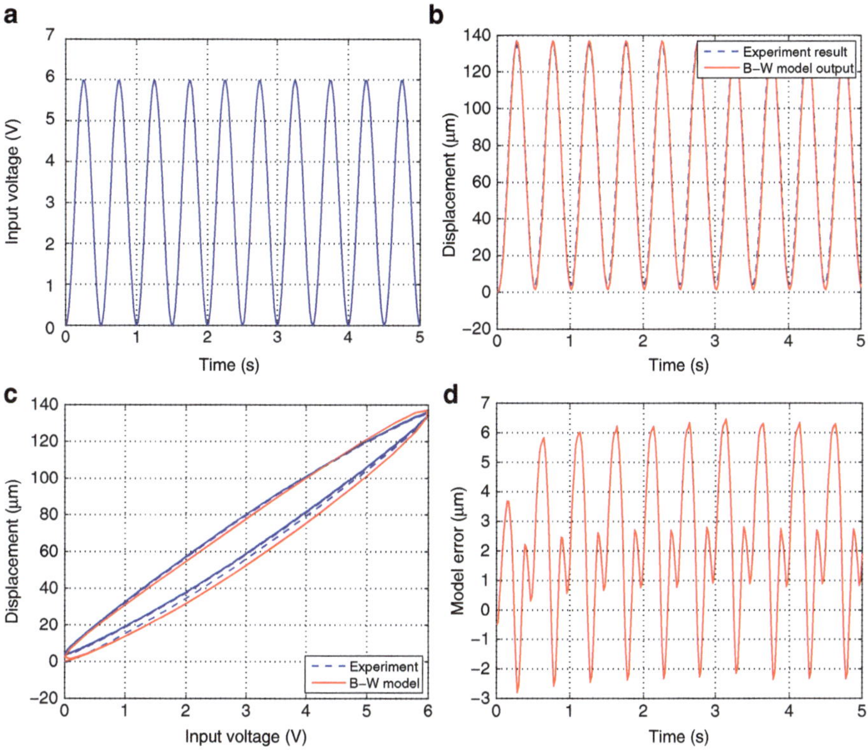

Fig. 8.3 Generalization test results of the Bouc–Wen model. (**a**) Input voltage; (**b**) experimental result and Bouc–Wen model output; (**c**) displacement-voltage hysteresis loops; (**d**) Bouc–Wen model output error

8.4.3 LSSVM Model Results

8.4.3.1 Dynamic Model Identification

Before developing the LSSVM model, a linear dynamic model of the system plant is identified by the swept-sine approach. The magnitudes of frequency responses obtained from the experimental data and the identified model are compared in Fig. 8.4. It is found that the first resonant mode occurs around 404 Hz, and the identified second-order model matches the system dynamics well in the frequencies below 600 Hz. The identified transfer function is

$$G_c(s) = \frac{1.247 \times 10^8}{s^2 + 1.847s + 6.477 \times 10^6} \tag{8.16}$$

which is employed to demonstrate the effectiveness of the proposed hysteresis modeling and compensation scheme.

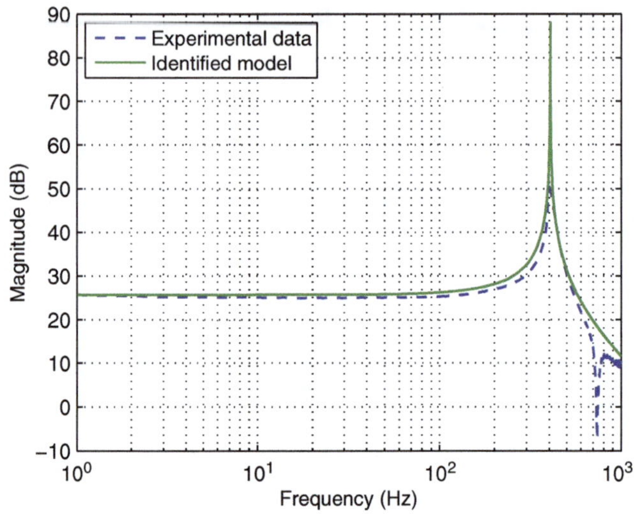

Fig. 8.4 Magnitude of frequency response of the system plant

By comparing the parameters of (8.4) and (8.16), one can deduce that $\omega = 2.5450 \times 10^3$ rad/s, $\xi = 3.6287 \times 10^{-4}$, and $d = 1.247 \times 10^8\,\mu\text{m/s}^2$-V.

8.4.3.2 LSSVM Modeling and Testing

To train the LSSVM model, the same exciting voltage signal as shown in Fig. 8.5a is employed. In order to accurately capture the hysteresis behavior based on LSSVM model, a suitable input vector (8.8) is required to be determined. Without loss of generality, the position and hysteresis terms are considered as input variables in the current research.

By setting $n = 2$ and $l = 2$, the LSSVM model is trained by using the corresponding input variables and the output variables, i.e., the hysteresis term h as shown in Fig. 8.5b–d which is generated by resorting to (8.6). The training results of the LSSVM model are shown in Fig. 8.6. By employing the input signal (see Fig. 8.3a), the testing results are illustrated in Fig. 8.7.

It is observed that the LSSVM model is trained to produce an RMSE of 0.01% for the hysteresis term h, which leads to a percent RMSE of 1.05% for the output position x. With the new testing signal, the LSSVM model generates an RMSE of 0.77% for h, which results in 1.42% RMSE for the output position x.

It is evident that the LSSVM model has reduced the testing error of output position by 32% in comparison with the Bouc–Wen model. Therefore, the effectiveness of LSSVM model is confirmed by the hysteresis modeling results.

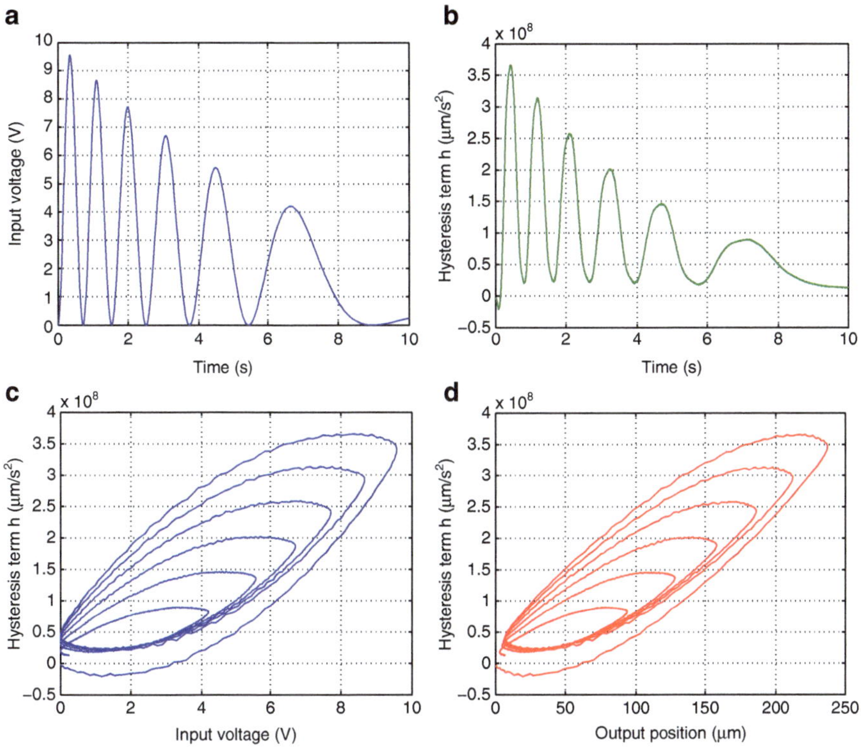

Fig. 8.5 The hysteresis term *h*. (**a**) Time history of input voltage; (**b**) time history of *h*; (**c**) *h* versus input voltage; (**d**) *h* versus output position

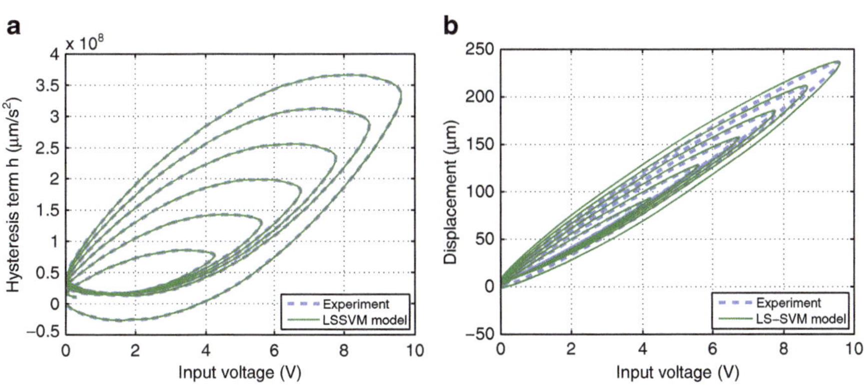

Fig. 8.6 Results of the trained LSSVM model. (**a**) Hysteresis term versus voltage; (**b**) displacement–voltage hysteresis loops

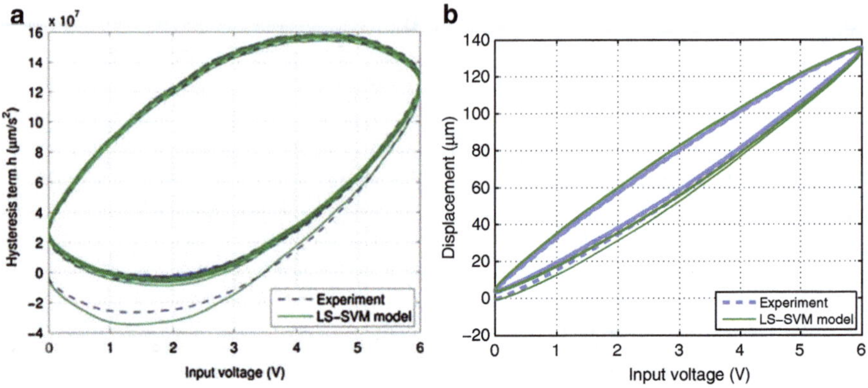

Fig. 8.7 Generalization test results of the LSSVM model. (**a**) Hysteresis term versus voltage; (**b**) displacement–voltage hysteresis loops

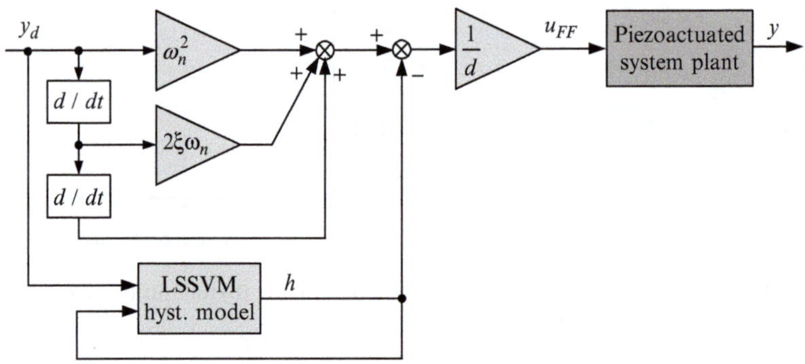

Fig. 8.8 Block diagram of LSSVM model-based feedforward control scheme

8.5 Feedforward Hysteresis Compensation and Results

In this section, the feedforward control schemes based on the developed Bouc–Wen and LSSVM models are realized to compensate for the hysteresis effect.

To suppress the hysteresis nonlinearity, a feedforward (FF) control (8.3) based on the Bouc–Wen model is first implemented. Moreover, the block diagram of LSSVM model-based FF control is depicted in Fig. 8.8. It is obvious that only the hysteresis model is needed whereas no inverse hysteresis model is required in the implementation of the feedforward compensation.

In order to demonstrate the superiority of LSSVM over Bouc–Wen model for hysteresis compensation, several experimental studies have been carried out. For instance, concerning a desired position trajectory as shown in Fig. 8.9a, the FF control results of Bouc–Wen model and LSSVM model are illustrated in Fig. 8.9a, c, and the tracking errors are compared in Fig. 8.9b. It is observed that the Bouc–Wen

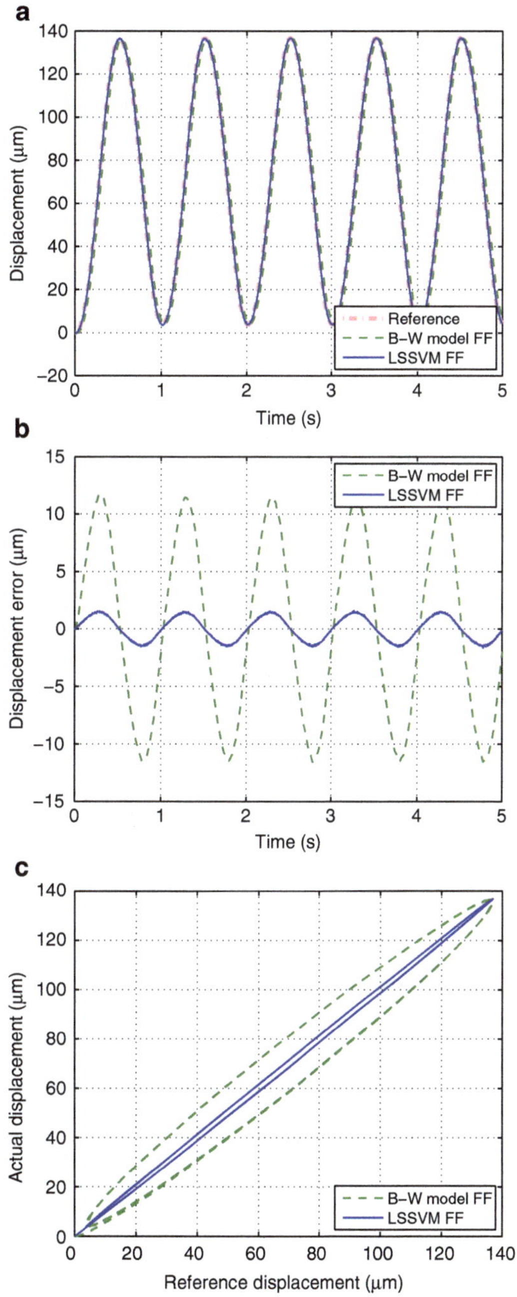

Fig. 8.9 Feedforward compensation results. (**a**) Reference and actual positions; (**b**) displacement errors of Bouc–Wen and LSSVM models-based FF compensation; (**c**) actual versus reference positions

model gives an RMSE of 7.92 μm (i.e., 5.8% of motion range), and the LSSVM model produces a 1.02-μm RMSE (i.e., 0.7% of motion range). As compared with Bouc–Wen model, the LSSVM is capable of suppressing the tracking error more by 88%, which leads to a negligible hysteresis as depicted in Fig. 8.9c. Thus, the effectiveness of LSSVM in the hysteresis compensation is demonstrated by the experimental results.

Although some degree of compensation error exists as evident from Fig. 8.9b, it can be easily reduced by combining a feedback controller as shown in [8].

8.6 Conclusion

This chapter is dedicated to hysteresis modeling and compensation of piezoelectric actuators. It is shown that the nonlinear hysteresis can be well compensated by resorting to a trained intelligent hysteresis model without modeling the hysteresis inverse. That means that the established hysteresis model can be used for both hysteresis characterization and compensation, which is more computationally effective than most of existing approaches where both a hysteresis model and an inverse hysteresis model are employed. Experimental results confirm that the LSSVM model is superior to the Bouc–Wen model in terms of hysteresis modeling accuracy and hysteresis compensation effectiveness. Furthermore, the proposed approach applies to hysteresis modeling and compensation of other smart materials-based actuators as well.

Acknowledgments This work was supported by the Macao Science and Technology Development Fund (grant no. 024/2011/A) and the Research Committee of the University of Macau (grant no. MYRG083(Y1-L2)-FST12-XQS).

References

1. M. Al Janaideh, S. Rakheja, C. Su, An analytical generalized Prandtl–Ishlinskii model inversion for hysteresis compensation in micropositioning control. IEEE/ASME Trans. Mechatron. **16**(4), 734–744 (2011)
2. W.T. Ang, P.K. Khosla, C.N. Riviere, Feedforward controller with inverse rate-dependent model for piezoelectric actuators in trajectory-tracking applications. IEEE/ASME Trans. Mechatron. **12**(2), 134–142 (2007)
3. P. Ge, M. Jouaneh, Generalized Preisach model for hysteresis nonlinearity of piezoceramic actuators. Precis. Eng. **20**(2), 99–111 (1997)
4. C.L. Hwang, Microprocessor-based fuzzy decentralized control of 2-D piezo-driven systems. IEEE Trans. Ind. Electron. **55**(3), 1411–1420 (2008)
5. S.W. John, G. Alici, C.D. Cook, Inversion-based feedforward control of polypyrrole trilayer bender actuators. IEEE/ASME Trans. Mechatron. **15**(1), 149–156 (2010)
6. K.K. Leang, Q. Zou, S. Devasia, Feedforward control of piezoactuators in atomic force microscope systems: inversion-based compensation for dynamics and hysteresis. IEEE Contr. Syst. Mag. **29**(1), 70–82 (2009)

7. Y. Li, Q. Xu, Adaptive sliding mode control with perturbation estimation and PID sliding surface for motion tracking of a piezo-driven micromanipulator. IEEE Trans. Contr. Syst. Technol. **18**(4), 798–810 (2010)
8. Y. Li, Q. Xu, A novel piezoactuated XY stage with parallel, decoupled, and stacked flexure structure for micro-/nanopositioning. IEEE Trans. Ind. Electron. **58**(8), 3601–3615 (2011)
9. Y. Li, Q. Xu, A totally decoupled piezo-driven XYZ flexure parallel micropositioning stage for micro/nanomanipulation. IEEE Trans. Autom. Sci. Eng. **8**(2), 265–279 (2011)
10. H.C. Liaw, B. Shirinzadeh, Neural network motion tracking control of piezo-actuated flexure-based mechanisms for micro-/nanomanipulation. IEEE/ASME Trans. Mechatron. **15**(4), 517–527 (2009)
11. C.J. Lin, S.Y. Chen, Evolutionary algorithm based feedforward control for contouring of a biaxial piezo-actuated stage. Mechatronics **19**(6), 829–839 (2009)
12. J. Minase, T.F. Lu, B. Cazzolato, S. Grainger, A review, supported by experimental results, of voltage, charge and capacitor insertion method for driving piezoelectric actuators. Precis. Eng. **34**(10), 692–700 (2010)
13. M. Rakotondrabe, Bouc–Wen modeling and inverse multiplicative structure to compensate hysteresis nonlinearity in piezoelectric actuators. IEEE Trans. Automat. Sci. Eng. **8**(2), 428–431 (2011)
14. M. Rakotondrabe, C. Clévy, P. Lutz, Complete open loop control of hysteretic, creeped and oscillating piezoelectric cantilever. IEEE Trans. Automat. Sci. Eng. **7**(3), 440–450 (2010)
15. J.C. Shen, W.Y. Jywe, H.K. Chiang, Y.L. Shu, Precision tracking control of a piezoelectric-actuated system. Precis. Eng. **32**(2), 71–78 (2008)
16. J.A.K. Suykens, T.V. Gestel, J.D. Brabanter, B.D. Moor, J. Vandewalle, *Least Squares Support Vector Machines* (World Scientific, Singapore, 2002)
17. P.K. Wong, Q. Xu, C.M. Vong, H.C. Wong, Rate-dependent hysteresis modeling and control of a piezostage using online support vector machine and relevance vector machine. IEEE Trans. Ind. Electron. **59**(4), 988–2001 (2012)
18. Y. Wu, Q. Zou, Robust-inversion-based 2-DOF control design for output tracking: piezoelectric actuator example. IEEE Trans. Contr. Syst. Technol. **17**(5), 1069–1082 (2009)
19. Q. Xu, Y. Li, Micro-/nanopositioning using model predictive output integral discrete sliding mode control. IEEE Trans. Ind. Electron. **59**(2), 1161–1170 (2012)
20. Q. Xu, Y. Li, Model predictive discrete-time sliding mode control of a nanopositioning piezostage without modeling hysteresis. IEEE Trans. Contr. Syst. Technol. **20**(4), 983–994 (2012)
21. Q. Xu, P.K. Wong, Hysteresis modeling and compensation of a piezostage using least squares support vector machines. Mechatronics **21**(7), 1239–1251 (2011)

Chapter 9
Compensation of Rate-Dependent Hysteresis in a Piezomicropositioning Actuator

Mohammad Al Janaideh

Abstract Piezomicropositioning actuators exhibit strong rate-dependent hysteresis nonlinearities that affect the accuracy of these micropositioning systems in open-loop system and may even lead to system instability of the closed-loop control system. Compensation of rate-dependent hysteresis effects using inverse rate-independent hysteresis models may yield high compensation error at high-excitation frequencies since these hysteresis effects increase as the excitation frequency of the input voltage increases. The inverse rate-dependent Prandtl–Ishlinskii model is utilized for compensation of the rate-dependent hysteresis nonlinearities in a piezomicropositioning stage. The exact inversion of the rate-dependent model is on hold under the condition that the distances between the thresholds do not decrease in time. The inverse of the rate-dependent model is applied as a feedforward compensator to compensate for the rate-dependent hysteresis nonlinearities of a piezomicropositioning actuator at different excitation frequencies between 0.1 and 50 Hz. The results show that the inverse compensator suppresses the hysteresis percent and the maximum positioning error in the output displacement of the piezomicropositioning actuator at different excitation frequencies, respectively.

9.1 Introduction

Piezomicropositioning actuators are increasingly used in micro and nano-positioning applications because of their advantages which include nanometer resolution, high stiffness, and fast response [1–9]. However, piezomicropositioning

M. Al Janaideh (✉)
Department of Mechatronics Engineering, The University of Jordan, Amman 11942, Jordan

Department of Aerospace Engineering, The University of Michigan, Ann Arbor, MI 48109, USA
e-mail: aljanaideh@gmail.com

M. Rakotondrabe (ed.), *Smart Materials-Based Actuators at the Micro/Nano-Scale: Characterization, Control, and Applications*, DOI 10.1007/978-1-4614-6684-0_9,

© Springer Science+Business Media New York 2013

actuators show hysteresis nonlinearities between the applied input voltage and output displacement. These nonlinearities have been associated with oscillations in the open-loop system's responses, and poor tracking performance and potential instabilities of the closed-loop system [10]. Different piezomicropositioning actuators show obvious increase in hysteresis nonlinearities when the excitation frequency of the applied input voltage increases in a nonlinear manner [11–15]. The inverse-based control methods generally employ a cascade of a rate-independent hysteresis model and its inverse together with a controller to compensate for the error of the compensation in piezomicropositioning actuators; see for example [16]. These methods, however, necessitate the formulation of the hysteresis model inverse, which is often a challenging task.

Different closed-loop control systems, however, have been proposed with inverse rate-independent hysteresis models to compensate for hysteresis nonlinearities at different excitation frequencies. Ge and Jouaneh [16] used inverse Preisach model, which was obtained using a numerical algorithm, as a feedforward compensator with PID feedback control system. Hu et al. [17] applied inverse Preisach model formulated with a dynamic density function in a closed-loop control system. In a similar manner, Song et al. [18] applied the inverse Preisach model with PD-lag and PD-lead controllers in a closed-loop control system. Esbrook et al. [19] applied a servocompensator with inverse Prandtl–Ishlinskii model in a closed-loop control system. Shan and Leang[20] applied discrete-time repetitive controller combined with an inverse hysteresis compensator based on the Prandtl–Ishlinskii model. Feedback control techniques could compensate for the rate-dependent hysteresis in peizomicropositioning actuators. However, the accurate and large bandwidth sensors as well as the feedback control techniques inserted in the closed-loop control systems may limit the use of the piezomicropositioning systems in different industrial applications. Ang et al. [21] applied the inverse modified Prandtl–Ishlinskii model as a feedforward compensator to compensate for hysteresis nonlinearities under different excitation frequencies.

In this chapter, the open-loop control technique is applied to compensate for the rate-dependent hysteresis nonlinearities over different excitation frequencies. The rate-dependent hysteresis nonlinearities are characterized using the rate-dependent Prandtl–Ishlinskii model. The analytical exact inverse of the rate-dependent Prandtl–Ishlinskii model is formulated and applied as a feedforward compensator to compensate for the rate-dependent hysteresis nonlinearities in a piezomicropositioning actuator. The main advantage of the rate-dependent Prandtl–Ishlinskii model over the other hysteresis models used in the literature is that its inverse can be attained analytically, and it can be implemented as a feedforward compensator to control the piezomicropositioning actuator over different excitation frequencies without inserting feedback control techniques.

In [7, 22], the analytical inverse of the Prandtl–Ishlinskii model is constructed with dynamic thresholds. The explicit inversion formula for the Prandtl–Ishlinskii model presented in [23] remains applicable also for the case of time-dependent thresholds, provided the distances between them do not decrease in time. This inverse is applied in this chapter as a feedforward controller in order to compensate for the rate-dependent hysteresis nonlinearities. Such compensations are experimentally illustrated by a piezomicropositioning actuator.

9.2 Background

The Prandtl–Ishlinskii model integrates play operators Γ_r with different thresholds r and with positive weights in order to characterize hysteresis nonlinearities in actuators and materials; see [23, 24]. For $t \in [0,T]$, when an input $u(t) \in C[0,T]$ is applied, where $C[0,T]$ is the space of continuous functions on the time interval $[0,T]$, the output of the Prandtl–Ishlinskii model for $i = 1,\ldots,n$, where n is an integer represents the number of the play operators, is given, according to [23], by the formula

$$\Pi[u](t) = a_0 u(t) + \sum_{i=1}^{n} a_i \Gamma_{r_i}[u,x_i](t), \tag{9.1}$$

where a_0 and a_i are positive weights.

The Prandtl–Ishlinskii model (9.1) is a rate-independent hysteresis model, attributed to the time-independent play operator that the model employs. This model has a unique advantage since it admits an exact inverse, which has been established in [25]. In [23], the output of the inverse rate-independent independent Prandtl–Ishlinskii model is written in the form

$$\Pi^{-1}[u](t) = b_0 u(t) + \sum_{i=1}^{n} b_i \Gamma_{s_i}[u,y_i](t). \tag{9.2}$$

This inverse has been widely applied as a feedforward controller to compensate for hysteresis nonlinearities in smart-material actuators at low-excitation frequencies of the applied input. However, these actuators exhibit hysteresis nonlinearities that are strongly rate-dependent. Consequently, the use of the inverse rate-independent Prandtl–Ishlinskii model as a feedforward compensator may cause considerable errors in the output displacement. It is therefore necessary to design a model and a compensator capable of incorporating rate-dependent hysteresis effects. This can be accomplished by adding a viscosity term in the constitutive relation [7].

9.3 Rate-Dependent Prandtl–Ishlinskii Model and Its Inverse

In this section the rate-dependent Prandtl–Ishlinskii model and its inverse are presented. The rate-dependent Prandtl–Ishlinskii model is employed in this study to characterize the rate-dependent hysteresis nonlinearities in a piezomicropositioning stage. This model is formulated in [22] using a play operator of time-dependent (dynamic) threshold.

9.3.1 *The Rate-Dependent Prandtl–Ishlinskii Model*

Let $AC(0,T)$ the space of real absolutely continuous functions defined on the interval $[0,T]$. For an input $u(t) \in AC(0,T)$, the output of the rate-dependent Prandtl–Ishlinskii model, constructed based on the rate of the applied input $\dot{u}(t)$, is given by the formula

$$\Psi[u](t) = a_0 u(t) + \sum_{i=1}^{n} a_i \Phi_{r_i(\dot{u}(t))}[u, x_i](t). \tag{9.3}$$

The output of the rate-dependent play operator is denoted as

$$z_i(t) = \Phi_{r_i(\dot{u}(t))}[u, x_i](t). \tag{9.4}$$

Let x_i be given initial conditions for $i = 1, 2, \ldots, n$ such that for $i = 1, \ldots, n-1$

$$\begin{aligned} |x_1| &\leq r_1(\dot{u}(0)), \\ |x_{i+1} - x_i| &\leq r_{i+1}(\dot{u}(0)) - r_i(\dot{u}(0)). \end{aligned} \tag{9.5}$$

The dynamic thresholds $r_i(t)$ are defined for $t \in [0,T]$ as

$$0 \leq r_1(\dot{u}(t)) \leq r_2(\dot{u}(t)) \leq \cdots \leq r_n(\dot{u}(t)). \tag{9.6}$$

As shown in [11], the rate-dependent Prandtl–Ishlinskii model can characterize the rate-dependent hysteresis nonlinearities in piezomicropositioning actuators over different excitation frequencies. The inverse of the rate-dependent Prandtl–Ishlinskii model is achievable and can be applied as a feedforward compensator to compensate for the rate-dependent hysteresis nonlinearities in real-time systems.

9.3.2 *Inverse Rate-Dependent Prandtl–Ishlinskii Model*

The concept of the open-loop control system is used to obtain identity mapping between the input $u(t)$ and the output $v(t)$ such that $u(t) = v(t)$. When the output of the exact inverse of the rate-dependent Prandtl–Ishlinskii model $\Psi^{-1}[u](t)$ is applied as a feedforward controller to compensate for the rate-dependent hysteresis nonlinearities presented by the rate-dependent Prandtl–Ishlinskii model $\Psi[u](t)$, the output of the compensation is expressed as

$$v(t) = \Psi \circ \Psi^{-1}[u](t). \tag{9.7}$$

The exact inversion is on hold under the condition that the distances between the dynamic thresholds $r_i(\dot{u}(t))$ do not decrease in time [22]. Analytically for $\forall i = 1, \ldots, n-1$

$$\frac{\mathrm{d}}{\mathrm{d}t}\left(r_{i+1}(\dot{u}(t)) - r_i(\dot{u}(t))\right) \geq 0. \tag{9.8}$$

The proof of the inversion formula is based in a substantial way on the so-called Brokate formula for the superposition of play operators with different thresholds. It was established for constant thresholds in [26], and the extension to moving thresholds has been done in [22]. The inverse of the rate-dependent Prandtl–Ishlinskii is also a rate-dependent Prandtl–Ishlinskii model. The output of the inverse model is expressed as

$$\Psi^{-1}[u](t) = b_0, u(t) + \sum_{i=1}^{n} b_i \Phi_{s_i(\dot{u}(t))}[u, y_i](t). \tag{9.9}$$

Let the output of the rate-dependent play operator of the inverse model is

$$d_i(t) = \Phi_{s_i(\dot{u}(t))}[u, y_i](t). \tag{9.10}$$

The thresholds of the inverse model are

$$\begin{aligned} s_1(\dot{u}(t)) &= a_0 r_1(\dot{u}(t)), \\ s_{i+1}(\dot{u}(t)) - s_i(\dot{u}(t)) &= \left(\sum_{j=0}^{i} a_j\right)\left(r_{i+1}(\dot{u}(t)) - r_i(\dot{u}(t))\right). \end{aligned} \tag{9.11}$$

The weights of the inverse model $b_0, b_1, \ldots, b_n$ are

$$\begin{aligned} b_0 &= \frac{1}{a_0}, \\ b_i &= \frac{1}{\sum_{j=0}^{i} a_j} - \frac{1}{\sum_{j=0}^{i-1} a_j}. \end{aligned} \tag{9.12}$$

The initial conditions of the inverse model $y_1, y_2, \ldots, y_n$ are

$$\begin{aligned} y_1 &= a_1 x_1, \\ y_{i+1} - y_i &= \left(\sum_{j=0}^{i} a_j\right)(x_{i+1} - x_i). \end{aligned} \tag{9.13}$$

9.3.3 The Dynamic Threshold

The following dynamic thresholds are used:

$$r_i(\dot{u}(t)) = \alpha_i + g(\dot{u}(t)). \tag{9.14}$$

This can be shown to be mathematically equivalent to modeling hysteresis and creep by means of an analogical model with elastic, plastic, and viscous elements as in [7,27]. Then

$$\alpha_i - \alpha_{i-1} \geq \sigma, \tag{9.15}$$

where σ is a positive constant. The constants α_i in (9.14) represent the rate-independent hysteresis effects, while the function $g(\dot{u}(t))$ is proposed to characterize the rate-dependent hysteresis effects. With this choice

$$r_{i+1}(\dot{u}(t)) - r_i(\dot{u}(t)) \geq 0 \tag{9.16}$$

and

$$\frac{\mathrm{d}}{\mathrm{d}t}\left(r_{i+1}(\dot{u}(t)) - r_i(\dot{u}(t))\right) = 0. \tag{9.17}$$

From these equations it can be concluded that the exact inversion formula for the rate-dependent Prandtl–Ishlinskii model holds. It should be mentioned that the proposed formulation for the dynamic threshold reduces the rate-dependent Prandtl–Ishlinskii model $\Psi[u](t)$ into the rate-independent Prandtl–Ishlinskii model $\Pi[u](t)$ for $g(\dot{u}(t)) = 0$.

9.3.4 Numerical Implementation

The numerical implementation of the rate-dependent Prandtl–Ishlinskii model and its inverse is presented for an input $u(t)$ with h sampling time. The rate of the applied input $\dot{u}(t)$ can be estimated for $k = 1,2,3,\ldots,K = T/h$ as

$$u_s(k) = \frac{u(k) - u(k-1)}{h}, \tag{9.18}$$

where

$$h = t(k) - t(k-1). \tag{9.19}$$

The discrete dynamic threshold is presented as

$$r_i(u_s(k)) = \alpha_i + g(u_s(k)). \tag{9.20}$$

The discrete output of the rate-dependent play operator is expressed as

$$z_i(k) = \max(u(k) - r_i(u_s(k)), \min(u(k) + r_i(u_s(k)), z_i(k-1))). \tag{9.21}$$

The discrete output of rate-dependent Prandtl–Ishlinskii model can thus be derived as

$$\Psi[u](k) = a_0 u(k) + \sum_{i=1}^{n} a_i \Phi_{r_i(u_s(k))}[u](k). \tag{9.22}$$

The discrete output of the inverse rate-dependent Prandtl–Ishlinskii model can thus be expressed as

$$\Psi^{-1}[u](k) = b_0 u(k) + \sum_{i=1}^{n} b_i \Phi_{s_i(u_s(k))}[u](k). \tag{9.23}$$

9.4 Experimental Results and Hysteresis Modeling

9.4.1 Experimental Results

The experiments were performed on a piezomicropositioning actuator (P-753.31C) from Physik Instrumente Company. The actuator provided maximum displacement of 100 μm from its static equilibrium position, and it integrates a capacitive sensor (sensitivity = 1 μm/V; resolution 0.1 nm) for measurement of stage displacement response. The excitation module compromises a voltage amplifier (LVPZT, E-505) with a fixed gain of 10, which provides the excitation voltage to the actuator. The actuator displacement response signal was acquired by a DSpace DS1104 controller board. The measurements with the piezomicropositioning stage were performed under a harmonic input of $u(t) = 40\cos(2f\ \pi\ t)$ at seven excitation frequencies (1, 5, 10, 20, 30, 40, and 50 Hz). The input voltage and output displacement signals were acquired at a sampling frequency of 10 KHz.

The measured data were further analyzed to quantify hysteresis and displacement attenuation as a function of the applied excitation frequency. The resulting hysteresis loops relating displacement responses to the input voltage are shown at excitation frequencies in Fig. 9.1. The results show that hysteresis nonlinearities increase with increasing the excitation frequency of the applied input voltage. It is evident that the micropositioning stage exhibits highly rate-dependent hysteresis effects between the input voltage and the output displacement.

9.4.2 Parameter Identification

Parameter identifications of the rate-dependent Prandtl–Ishlinskii model are presented in this section. Measured rate-dependent hysteresis loops between the applied input voltage and the output displacement are used to identify the parameters of the rate-dependent Prandtl–Ishlinskii model and its inverse. Let $g(\dot{u}(t)) = \beta(\dot{u}(t))^2$.

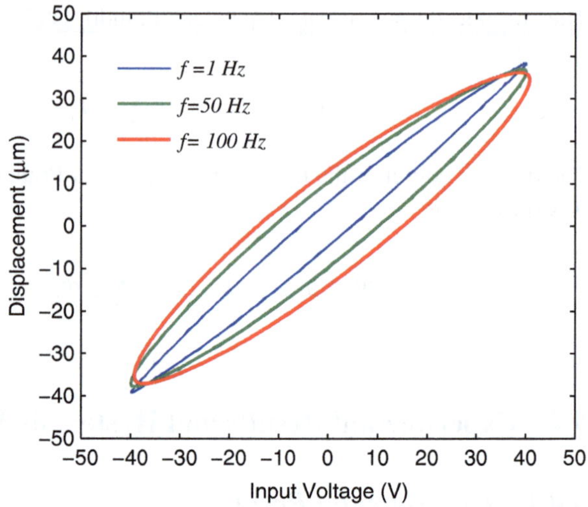

Fig. 9.1 Hysteresis loops at different excitation frequencies

The parameter vector $X = \{\beta, \zeta, a_0, a_1, a_2, \ldots, a_n\}$ was identified through minimization of the an error function over different excitation frequencies. This function is given by

$$Q(X) = \Theta(k). \tag{9.24}$$

The model error function Θ is used to identify the parameters of the rate-dependent Prandtl–Ishlinskii model $\Psi[v](t)$. The error function Θ is expressed as

$$\Theta(k) = \sum_{p=1}^{P} \sum_{k=1}^{K} A_p(\Psi[v](k) - y_m(k))^2, \tag{9.25}$$

where $\Psi[v](k)$ is the displacement response of the model corresponding to a particular excitation frequency and $y_m(k)$ is the measured displacement under the same excitation frequency. The model error function is constructed through summation of squared errors over a range of input frequencies, denoted by p ($p = 1, 2, \ldots, P$). The index k ($k = 1, \ldots, K$) refers to the number of data points considered to compute the error function Q for one complete hysteresis loop. Two hundred data points ($K = 200$) were available for each measured hysteresis loop. Seven excitation frequencies ($P = 7$) of 1, 5, 10, 20, 30, 40, and 50 Hz are used. Owing to the higher error at excitation frequencies, a weighting constant A_p was introduced to emphasize the error minimization at higher excitation frequencies. The weights A_p for $p = 1, 2, \ldots, 7$ are selected based on the hysteresis percent as

$$A_p = \frac{H_p}{H_1}, \tag{9.26}$$

where H_p represents the hysteresis percent for the p excitation frequency. The weights A_p are obtained as: 1, 1.18, 1.38, 1.57, 1.78, and 1.96 for 1, 5, 10, 20, 30, 40,

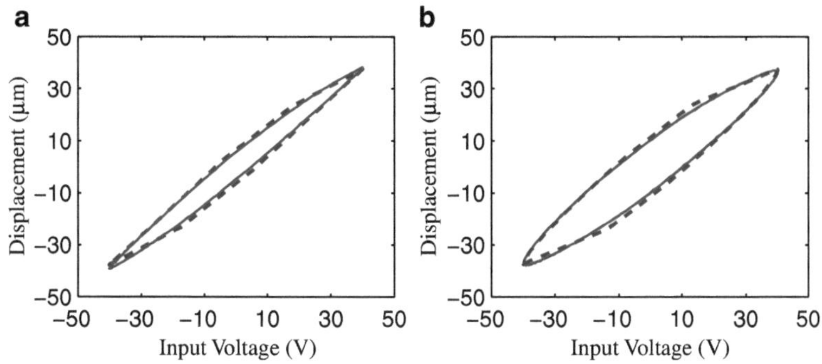

Fig. 9.2 Comparison between the output of the rate-dependent Prandtl–Ishlinskii model (*solid line*) and the measured hysteresis nonlinearities (*dashed line*) in the piezomicropositioning actuator at: (**a**) 1 Hz and (**b**) 50 Hz

and 50 Hz, respectively. The error minimization is performed using the MATLAB constrained optimization toolbox subject to the following constraints:

$$(\beta, \zeta, a_0, a_1, a_2) > 0, \zeta \gg \beta. \tag{9.27}$$

The rate-dependent Prandtl–Ishlinskii model is used to characterize the rate-dependent hysteresis nonlinearities of the piezomicropositioning actuator between 1 and 50 Hz. The results propose two rate-dependent play hysteresis operators ($n = 2$) to characterize the rate-dependent hysteresis nonlinearities.

9.4.3 Hysteresis Modeling

The validity of the model was examined by comparing the model displacement responses to the measured data. The results suggest that the model can effectively predict the hysteresis properties of the piezomicropositioning actuator at different excitation frequencies between 1 and 50 Hz. Figure 9.2 shows the capability of the model to characterize the rate-dependent hysteresis nonlinearities at 1 and 50 Hz.

9.5 Feedforward Compensation of Rate-Dependent and Rate-Independent Hysteresis Nonlinearities

In this section the inverse rate-dependent Prandtl–Ishlinskii model is applied as a feedforward compensator to compensate for hysteresis nonlinearities under different excitation frequencies.

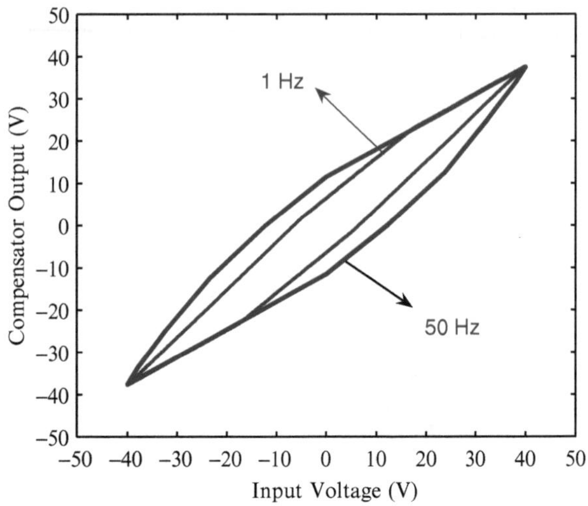

Fig. 9.3 The input–output characteristics of the inverse rate-dependent Prandtl–Ishlinskii model at excitation frequencies of 1 and 50 Hz

9.5.1 The Inverse Compensator

The input–output characteristics of the inverse rate-dependent Prandtl–Ishlinskii model at excitation frequencies of 1, 20, and 50 Hz are shown in Fig. 9.3. This figure shows that hysteresis nonlinearities in the output of the inverse model increase as the excitation frequency of the input voltage increases.

9.5.2 Compensation of Rate-Dependent Hysteresis

The inverse rate-dependent Prandtl–Ishlinskii model obtained in the previous section is applied as a feedforward compensator to compensate for the rate-dependent hysteresis nonlinearities between 1 and 50 Hz. The measured output–input characteristics of the piezomicropositioning stage with inverse compensator are illustrated in Fig. 9.4 at excitation frequencies of 1, 20, and 50 Hz. The results show that the inverse rate-dependent model can effectively compensate the hysteresis effects at different excitation frequencies. Figure 9.5 shows the time history of the input voltage and the measured displacement with and without the inverse compensator at excitation frequency of 50 Hz.

The positioning error is computed as the deviation between measured displacement and the input voltage, which represents the desired displacement, at different excitation frequencies (Fig. 9.6). Figure 9.7 illustrates comparison of the maximum positioning errors with and without the inverse compensator. Without the inverse compensator, the maximum positioning errors are between 6.5 and 13.8 μm, while the measured responses with the inverse compensator show maximum positioning errors between 2.3 and 3.7 μm across the entire 1–50 Hz band. In a similar manner,

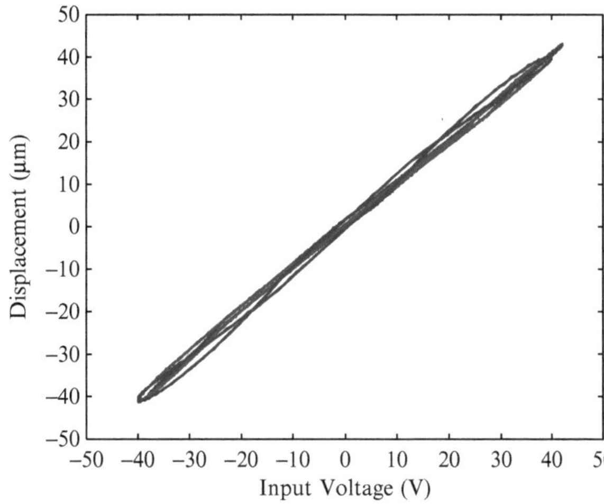

Fig. 9.4 Measured output displacement of the piezomicropositioning stage when the inverse rate-dependent Prandtl–Ishlinskii model is applied as a feed-forward compensator at different excitation frequencies between 1 and 50 Hz

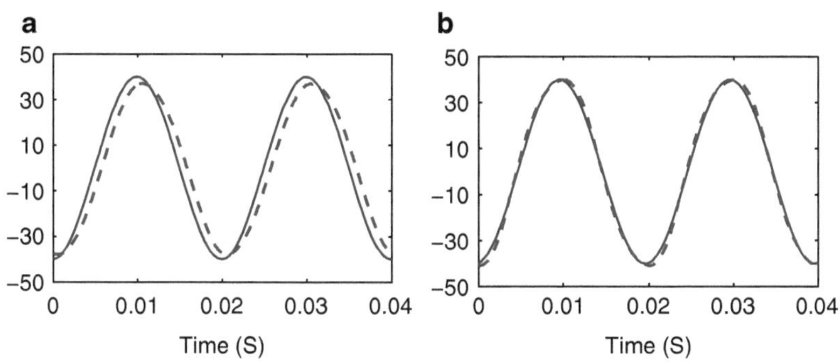

Fig. 9.5 The time history of the input voltage (*solid line*) and the measured displacement (*dashed line*) at excitation frequency of 50 Hz: (**a**) without the inverse compensator and (**b**) with the inverse compensator

Fig. 9.8 shows that the inverse compensator decreases the hysteresis percent to 3.5. It is obvious that the inverse rate-dependent Prandtl–Ishlinskii model suppresses the error due to rate-dependent hysteresis disregarding the excitation frequency of the input voltage and the tracking accuracy remains consistent.

In Fig. 9.18, the compensation effectiveness of the inverse rate-dependent Prandtl–Ishlinskii model is further evaluated by comparing the time history of the measured displacement responses of the piezomicropositioning stage with and without the inverse compensator at 20 and 50 Hz frequencies. The results show the effectiveness of the inverse rate-dependent model under low- and high-excitation frequencies.

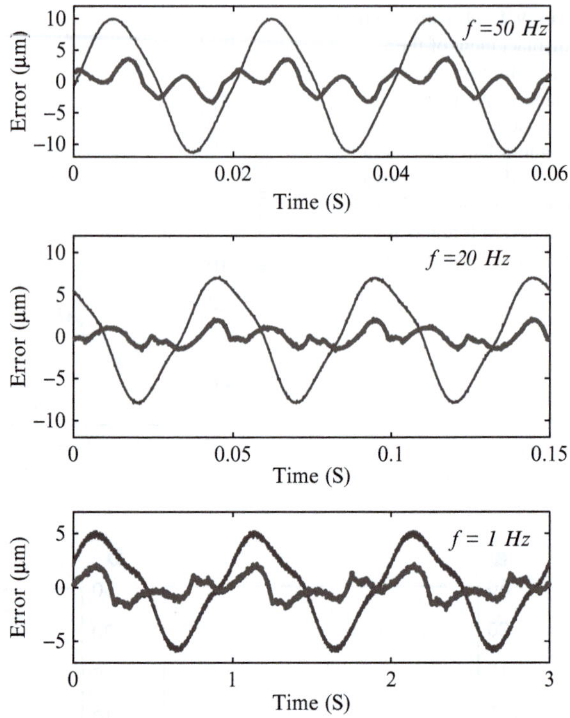

Fig. 9.6 Comparison of the positioning error with the inverse compensator (*red line*) and without the inverse compensator (*blue line*)

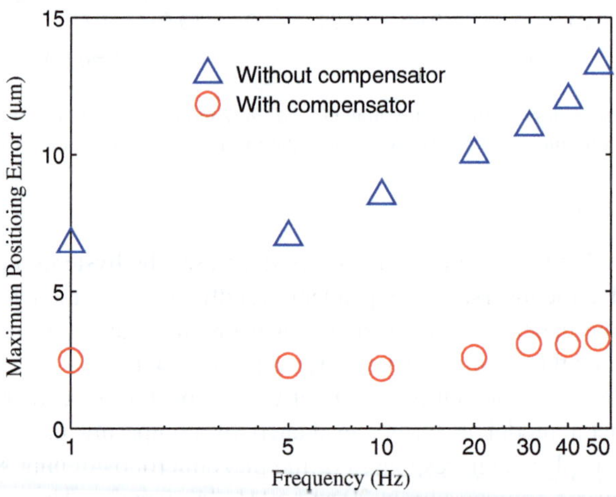

Fig. 9.7 Comparison between the maximum positioning error with and without inverse rate-dependent Prandtl–Ishlinskii model

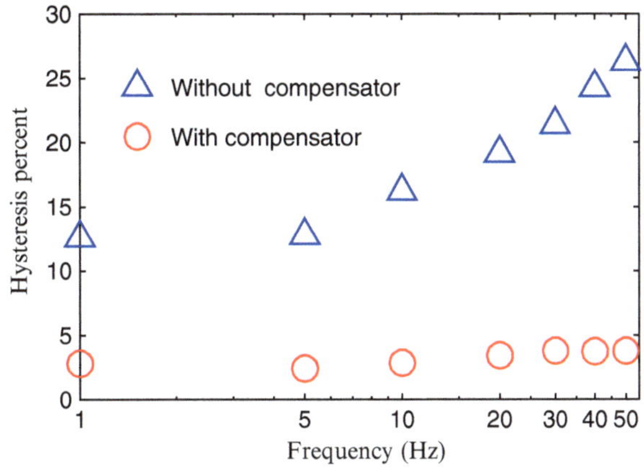

Fig. 9.8 Comparison between the percent of hysteresis nonlinearities with and without inverse rate-dependent Prandtl–Ishlinskii model

9.5.3 Compensation of Hysteresis Nonlinearities at Low-Excitation Frequencies

At low-excitation frequencies, the rate-dependent Prandtl–Ishlinskii model (9.3) and its inverse (9.9) constructed based on the dynamic threshold (9.20) are reduced to the rate-independent Prandtl–Ishlinskii model and its inverse. Analytically, the dynamic thresholds are reduced to

$$r(\dot{u}(t)) \approx \zeta i. \tag{9.28}$$

Then

$$\Pi[v](t) \approx \Psi[v](t). \tag{9.29}$$

Further investigation shows that the performance of the inverse rate-dependent Prandtl–Ishlinskii model still remains at low-excitation frequencies. In other words, the inverse rate-dependent Prandtl–Ishlinskii model can be applied to compensate for rate-independent hysteresis nonlinearities. As shown in Fig. 9.9, the inverse rate-dependent Prandtl–Ishlinskii model compensates for the hysteresis nonlinearities at excitation frequencies of 0.1 and 0.5 Hz. The time history for the positioning error is presented in Fig. 9.10. It can be concluded that the inverse rate-dependent Prandtl–Ishlinskii can be used as a feedforward compensator also at low-excitation frequencies.

9.5.4 Triangular Waveform

A triangular waveform of amplitude 40 at excitation frequencies of 10 and 20 Hz is applied as an input voltage. The experimental results show that the percent of the hysteresis nonlinearities are 13.54 and 15.62 for 10 Hz and 20 Hz, respectively. As shown in Fig. 9.11, the inverse model compensates for the hysteresis nonlinearities

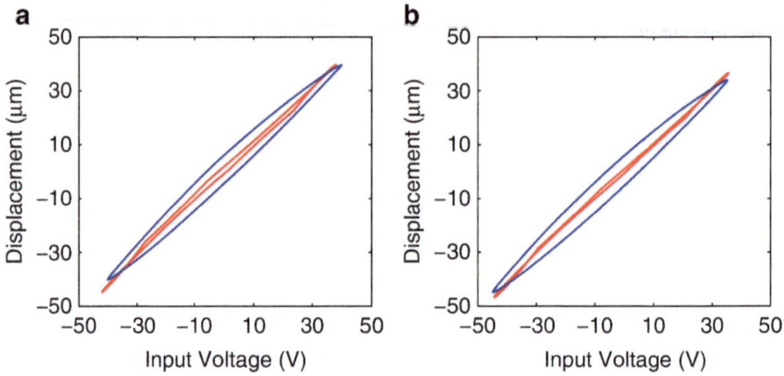

Fig. 9.9 The input–output of the piezomicropositioning actuator with (*red line*) and without (*blue line*) the inverse compensator at (**a**) $f = 0.10$ Hz and (**b**) $f = 0.50$ Hz

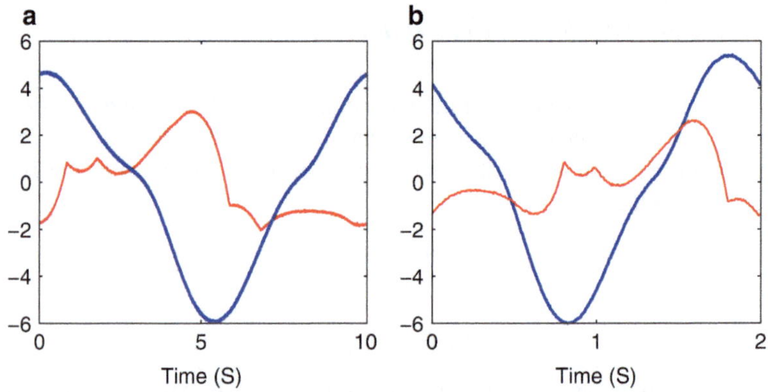

Fig. 9.10 The time history of the positioning error of the piezomicropositioning actuator with (*red line*) and without (*blue line*) the inverse compensator at (**a**) $f = 0.10$ Hz and (**b**) $f = 0.5$ Hz

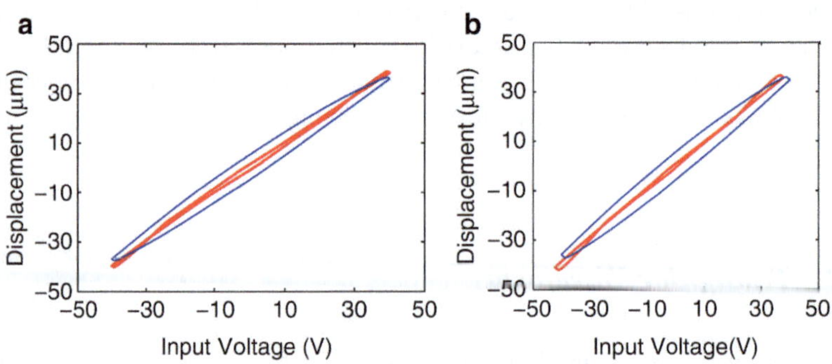

Fig. 9.11 The output of the piezomicropositioning stage with (*red line*) and without (*blue line*) the inverse compensator when a triangular waveform is applied at: (**a**) $f = 10$ Hz and (**b**) $f = 20$ Hz

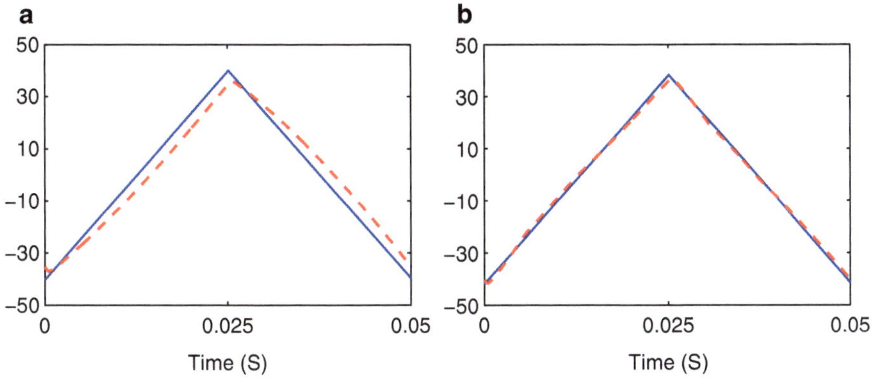

Fig. 9.12 The time history of the triangular input voltage (*blue line*) and the measured displacement (*red line*) at $f = 20\,\text{Hz}$: (**a**) without the inverse compensator and (**b**) with the inverse compensator

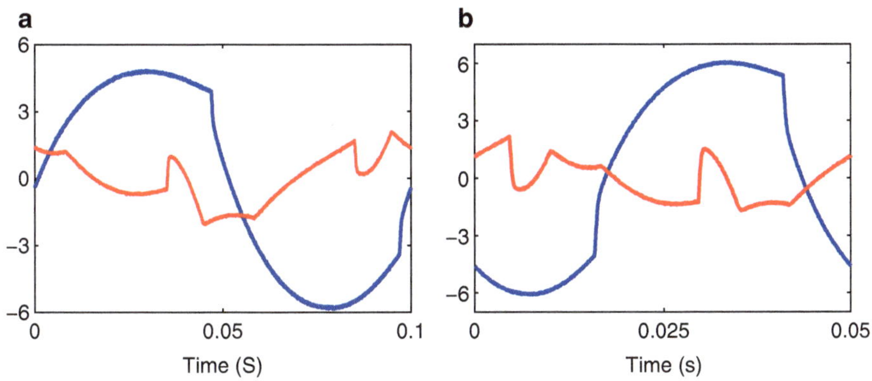

Fig. 9.13 The positioning error of Fig. 9.11 with (*red line*) and without (*blue line*) the inverse compensator: (**a**) $f = 10\,\text{Hz}$ and (**b**) $f = 20\,\text{Hz}$

when a triangular input voltage is applied. Figure 9.12 shows the triangular input voltage and the output displacement at the excitation frequency of 20 Hz with and without the inverse compensator. Figure 9.13 shows the time history of the positioning error with and without the inverse compensator. The results show the effectiveness of the inverse compensator when a triangular input voltage is applied at different excitation frequencies.

9.5.5 Major and Minor Hysteresis Loops

Major and minor hysteresis loops with the inverse rate-dependent Prandtl–Ishlinskii model are tested in this section (Fig. 9.14). Harmonic input voltages of $u(t) = 10\cos(2\pi ft) + 30\cos(4\pi ft)$ are applied to the piezomicropositioning actuator to

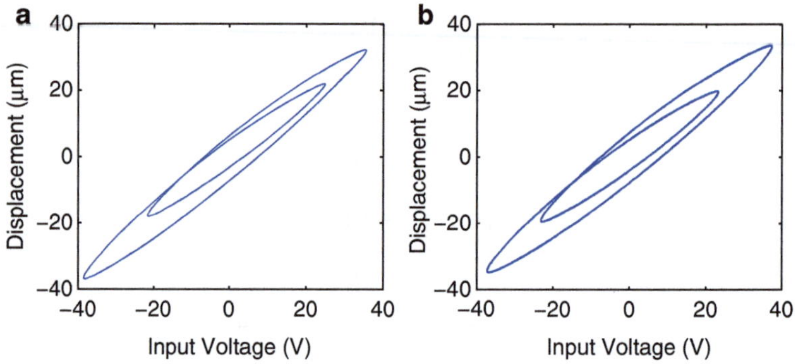

Fig. 9.14 The output of the piezomicropositioning actuator with the input voltage $u(t) = 10\cos(2\pi ft) + 30\cos(4\pi ft)$, where (**a**) $f = 5$ Hz and (**b**) $f = 10$ Hz

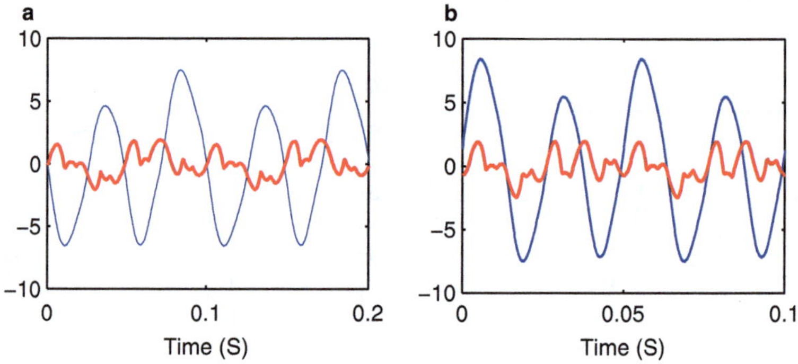

Fig. 9.15 The time history of the positioning error of Fig. 9.16 with the inverse compensator (*red line*) and without the inverse compensator (*blue line*) at: (**a**) $f = 5$ Hz and (**b**) $f = 10$ Hz

show major and minor hysteresis loops at excitation frequencies of $f = 5$ Hz and $f = 10$ Hz. Figure 9.15 shows the output of the inverse compensation. Figure 9.16 shows the time history of the positioning error with and without the inverse compensator.

9.6 Discussions

The above analysis shows that the inverse rate-dependent Prandtl–Ishlinskii model is capable of suppressing the error due to hysteresis, regardless of the excitation frequency of the input voltage, while maintaining consistency in the tracking accuracy (Fig. 9.17). The results manifest the effectiveness of the inverse rate-dependent model in compensating for hysteresis under low- and high-excitation frequencies. However, the inverse compensator shows some deviation in the output, which is attributed to prediction errors attained between the output of the rate-dependent model and the measured displacement of the piezomicropositioning

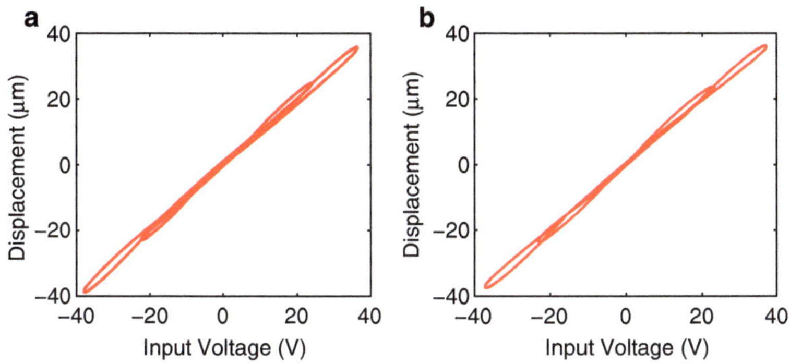

Fig. 9.16 The output of the piezomicropositioning actuator when the inverse rate-dependent Prandtl–Ishlinskii is applied as a feedforward compensator with the input voltage of $u(t) = 10\cos(2\pi ft) + 30\cos(4\pi ft)$, where (**a**) $f = 5$ Hz and (**b**) $f = 10$ Hz

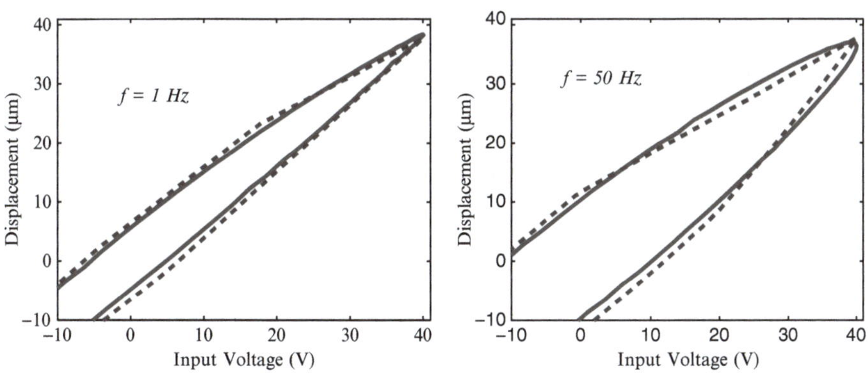

Fig. 9.17 Comparison between the measured displacement (*solid line*) and the output of the rate-dependent Prandtl–Ishlinskii model (*dashed line*) at excitation frequencies of 1 and 50 Hz

actuator. Figure 9.18 shows a comparison between the characterization error of the rate-dependent Prandtl–Ishlinskii model and the positioning error in the displacement output of the piezomicropositioning actuator. The figure shows a similarity between the error in both cases. It can be seen from the experimental results that the characterization error of the rate-dependent Prandtl–Ishlinskii model at the turning points is relatively larger than elsewhere. It should be mentioned that piezomicropositioning actuators also show creep effects during slow-speed actuation. These dynamic effects cause positioning errors in the output displacement.

As shown in the previous section, the inverse rate-dependent Prandtl–Ishlinskii model shows perfect compensation for symmetric hysteresis nonlinearities. However, the inverse model may not show the same performance when applied to compensate for asymmetric rate-dependent hysteresis nonlinearities. In future work, the inverse rate-dependent Prandtl–Ishlinskii model will be developed to compensate for asymmetric rate-dependent hysteresis nonlinearities of smart-material based actuators.

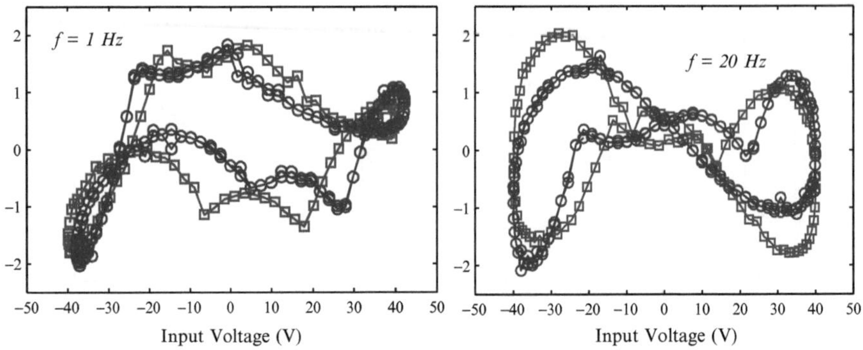

Fig. 9.18 Comparison between the error of the inverse compensation when the inverse rate-dependent Prandtl–Ishlinskii model is applied as a feedforward compensator (*circle*) and the characterization error between the measured displacement and the output of the rate-dependent Prandtl–Ishlinskii model (*square*)

It should be mentioned that different piezomicropositioning actuators exhibit asymmetric rate-dependent hysteresis nonlinearities that increase as the excitation frequencies of the applied input voltage increase. These effects can be accurately compensated for using the method proposed in this chapter. The results presented in this chapter can also be extended to complex hysteresis nonlinearities studied by Kuhnen [28] and Visone and Sjöström [29].

9.7 Conclusions

The inverse rate-dependent PrandtlIshlinskii model is analytical and exact. This makes the inverse PrandtlIshlinskii model attractive for control piezomicropositioning actuators at different excitation frequencies. The proposed compensation algorithm is easy to use and can be applied to compensate for rate-dependent hysteresis nonlinearities in micro/nano-positioning applications where the use of feedback sensors and feedback control techniques are not easy.

References

1. A. Cavallo, C. Natale, S. Pirozzi, C.Visone, A. Formisano, Feedback control systems for micropositioning tasks with hysteresis compensation. IEEE Trans. Magn. **40**(2), 876–879 (2004)
2. M. Rakotondrabe, C. Clevy, P. Lutz, Complete open loop control of hysteretic, creeped, and oscillating piezoelectric cantilevers. IEEE Trans. Autom. Sci. Eng. **7**(3), 440–450 (2010)
3. Y. Li, Q. Xu, A totally decoupled piezo-driven XYZ flexure parallel micropositioning stage for micro/nanomanipulation. IEEE Trans. Autom. Sci. Eng. **8**(2), 265–279 (2011)

4. D. Davinoa, C. Nataleb, S. Pirozzib, C. Visone, A phenomenological dynamic model of a magnetostrictive actuator. Physica B **343**(1–4), 112116 (2004)
5. B. Choi, M. Han, Vibration control of a rotating cantilevered beam using piezoactuators: experimental work. J. Sound Vib. **277**(1–2), 436–442 (2004)
6. B. Agrawal, M. Elshafei, G. Song, Adaptive antenna shape control using piezoelectric actuators. Acta Astronaut. **40**(11), 821–826 (1997)
7. P. Krejčí, M. Al Janaideh, F. Deasy, Inversion of hysteresis and creep operators. Physica B **407**(9), 1354–1356 (2012)
8. J. Park, K. Yoshida, S. Yokoto, Resonantly driven piezoelectric micropump-fabrication of a micropump having high power density. Mechatronics **9**(7), 687–702 (1999)
9. B. Mokaberi, A. Requicha, Compensation of scanner creep and hysteresis for AFM nanomanipulation. IEEE Trans. Autom. Sci. Eng. **5**(2), 197–206 (2008)
10. G. Tao, P. Kokotovic, Adaptive control of plants with unknown hysteresis. IEEE Trans. Automat. Contr. **40**(2), 200–212 (1995)
11. M. Al Janaideh, S. Rakheja, C.-Y. Su, Experimental characterization and modeling of rate-dependent hysteresis of a piezoceramic actuator. Mechatronics **17**(5), 656–670 (2009)
12. S. Viswamurthy, R. Ganguli, Modeling and compensation of piezoceramic actuator hysteresis for helicopter vibration control. Sensor Actuator A: Phys. **135**(2), 801–810 (2007)
13. R. Ben Mrad, H. Hu, A model for voltage-to-displacement dynamics in piezoceramic actuators subject to dynamic-voltage excitations. IEEE/ASME Trans. Mechatron. **7**(4), 479–489 (2002)
14. K. Leang, Q. Zou, S. Devasia, Feedforward control of piezoactuators in atomic force microscope systems: inversion-based compensation for dynamics and hysteresis. IEEE Contr. Syst. Mag. **19**(1), 70–82 (2009)
15. M. Grossard, M. Boukallel, N. Chaillet, C. Rotinat-Libersa, Modeling and robust control strategy for a control-optimized piezoelectric microgripper. IEEE/ASME Trans. Mechatron. **16**(4), 674–683 (2011)
16. P. Ge, M. Jouaneh, Tracking control of a piezoceramic actuator. IEEE Trans. Contr. Syst. Technol. **4**(3), 209–216 (1996)
17. H. Hu, H. Georgiou, R. BenMrad, Enhancement of tracking ability in piezoceramic actuators subject to dynamic excitation conditions. IEEE/ASME Trans. Mechatron. **10**(2), 230–240 (2005)
18. G. Song, J. Zhao, X. Zhou, J. Abreu-Garcia, Tracking control of a piezoceramic actuator with hysteresis compensation using inverse Preisach model. IEEE/ASME Trans. Mechatron. **10**(2), 198–209 (2005)
19. A. Esbrook, X. Tan, H. Khalil, Control of systems with hysteresis via servocompensation and its application to nanopositioning. IEEE Trans. Contr. Syst. Technol. 1–12 (2012). doi:10.1109/TCST.2012.2192734
20. Y. Shan, K. Leang, Repetitive control with Prandtl–Ishlinskii hysteresis inverse for piezo-based nanopositioning, in *Proceedings of the American Control Conference*, St. Louis, MO, 2009, pp. 301–306
21. W. Ang, P. Khosla, C. Riviere, Feedforward controller with inverse rate-dependent model for piezoelectric actuators in trajectory-tracking applications. IEEE/ASME Trans. Mechatron. **12**(2), 134–142 (2007)
22. M. Al Janaideh, P. Krejčí, An inversion formula for a Prandtl–Ishlinskii operator with time dependent thresholds. Physica B **406**(8), 1528–1532 (2011)
23. P. Krejci, K. Kuhnen, Inverse control of systems with hysteresis and creep. IEE Proc. Contr. Theor. Appl. **148**(3), 185–192 (2001)
24. A. Bergqvist, On magnetic hysteresis modeling, Ph.D. thesis, Royal Institute of Technology, Stockholm, Sweden, 1994
25. P. Krejčí, Hysteresis and periodic solutions of semilinear and quasilinear wave equations. Math. Z. **193**, 247–264 (1986)
26. M. Brokate, J. Sprekels, *Hysteresis and Phase Transitions* (Springer, New York, 1996)

27. K. Kuhnen, P. Krejčí, Compensation of complex hysteresis and creep effects in piezoelectrically actuated systems: a new preisach modeling approach. IEEE Trans. Automat. Contr. **54**(3), 537–550 (2009)
28. K. Kuhnen, Modeling, identification and compensation of complex hysteretic nonlinearities—a modified Prandtl–Ishlinskii approach. Eur. J. Contr. **9**(4), 407–418 (2003)
29. C. Visone, M. Sjöström, Exact invertible hysteresis models based on play operators. Physica B **343**(1–4), 148–152 (2004)

Chapter 10
Feedforward Control of Flexible and Nonlinear Piezoelectric Actuators

Micky Rakotondrabe

Abstract In this chapter, the control without sensors, also called feedforward control, of piezoelectric actuators is proposed. Typified by hysteresis and creep nonlinearities and by badly damped vibration, the design of the controller (compensator) is based on precise models and on the inversion of the latter. For that, the hysteresis is first modeled and compensated by using the Prandtl–Ishlinskii technique. Then, the creep is treated. Finally, the badly damped vibration is modeled and controlled. Experimental results along the chapter demonstrate the efficiency of the approach.

10.1 Introduction

Piezoelectric materials are well known for the design and development of microactuators in microsystems, microrobotics, and systems working at the micro/nanoscale. Such recognition is particularly due to their high resolution (better than the micron), their high bandwidth (up to some tens of kiloHertz), their high force density, the fact that they can also be used as sensors, and their ease of *control* (electrical supply)

Among the systems working at the micro/nanoscale that are based on piezoelectric materials, there are microgrippers which are used to manipulate or to assemble small objects (micromanipulation and microassembly) [1]. A piezoelectric microgripper, pictured in Fig. 10.1a, is made up of two piezoelectric actuators with cantilever structure (piezocantilever). Each piezocantilever has a rectangular section and can be supplied and controlled independently. Often, while one cantilever is controlled on deflection for the precise positioning, the second cantilever is controlled on force in order to avoid the destruction of the manipulated object

M. Rakotondrabe (✉)
Automatic Control and Micro-Mechatronic Systems Department, AS2M,
FEMTO-ST Institute, 24 rue Alain Savary, Besançon 25000, France
e-mail: mrakoton@femto-st.fr

M. Rakotondrabe (ed.), *Smart Materials-Based Actuators at the Micro/Nano-Scale: Characterization, Control, and Applications*, DOI 10.1007/978-1-4614-6684-0_10,
© Springer Science+Business Media New York 2013

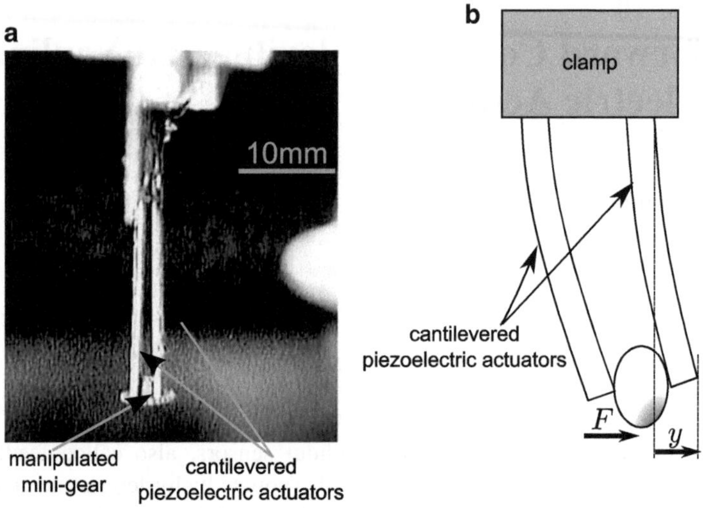

Fig. 10.1 (**a**): A photography of a piezoelectric microgripper developed at FEMTO-ST institute. (**b**): principle of manipulation of a small object using a microgripper

(or conversely to avoid the destruction of the actuator) or in order to estimate the mechanical characteristics of this latter [2] (see Fig. 10.1b). In the figure, y is the displacement and F is the manipulation force. A piezocantilever is generally composed of two or several layers where at least one layer is made of piezoelectric material (piezolayer). The non-piezoelectric layers are called passive layers.

Although piezoelectric actuators offer a very good resolution, they are typified by hysteresis and creep nonlinearities that can drastically decrease the accuracy. Furthermore, cantilever structured piezoelectric actuators are characterized by a badly damped oscillation which not only increases the response time but also affects the stability of the whole system. Closed-loop control techniques (or feedback) have therefore been studied in the past to overcome these nonlinearities and oscillation [2–10]. These techniques offer a high robustness relative to model uncertainties and to external disturbances. However, they are strongly limited by the available sensors. Indeed, sensors that have the required performances at the same time (high accuracy, high bandwidth) are expensive and have bulky dimensions that are not convenient face to the dimensions of the actuators and face to the available space. They include interferometry and optical sensors. On the other hand, embeddable sensors such as strain gages do not often provide the necessary performances (limited range, sensitive to noises, etc.) and are very fragile. Hence, open-loop control techniques (or feedforward) have been emerging. The main advantage of feedforward is that no sensor is required. This makes these techniques very attractive for applications where packaging and compacity are essential. These applications include micromanipulation and microassembly, measurement and scanning at the micro/nanoscale with small systems, intra-body microrobotics, etc. The main principle of feedforward control consists in modeling as precise as possible the

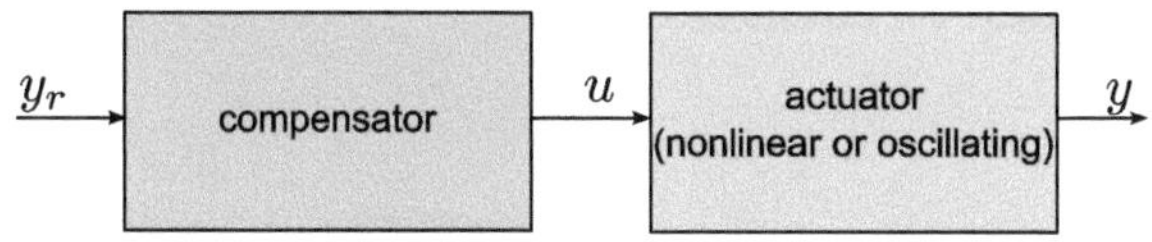

Fig. 10.2 Principle of a feedforward control technique

nonlinearities or the oscillation to be cancelled and then putting in cascade with the actuator a kind of inverse model as in Fig. 10.2. The objective is therefore to find the compensator such that $y = y_r$. In the figure, u represents the input voltage (control signal).

Controlling the hysteresis in piezocantilevers by feedforward has attracted several attentions. Different approaches have been proposed: the Preisach [11, 12], the Prandtl–Ishlinskii [13–17], and the Bouc–Wen [18]. In the two first approaches (Preisach and Prandtl–Ishlinskii), the complex hysteresis model is a superposition of many elementary hysteresis operators. These approaches could be very accurate at a price of a complexity of implementation: increasing the number of the elementary operators increases not only the accuracy but also the model/compensator complexity. The last approach (Bouc–Wen) is based on a set of two nonlinear differential equations that contain very few number of parameters. This approach is very interesting in terms of simplicity of structure. Concerning the creep nonlinearity, the main approaches that have been used are the logarithmic method [19] and the linear dynamic operator [11, 13, 15]. Finally, to compensate the vibration and the badly damped oscillation, Croft et al. [11] and Clayton et al. [20] use once again linear dynamic models and their direct inversion while in [13, 21] an input shaping technique is employed. A complete survey of control of piezocantilevers in general is presented in [22].

This chapter deals with the feedforward control of piezocantilever used in micromanipulation and microassembly. The main challenge remains in the fact that the feedforward controller should account the hysteresis and the creep nonlinearities and the badly damped oscillation at the same time. For that, we propose to compensate first the hysteresis, then the creep, and finally the vibration. The final compensator includes therefore three *sub-compensators* conveniently calculated. For the hysteresis, we propose to employ the Bouc–Wen approach which presents the advantage of structure simplicity. The creep will be modeled with a linear dynamic model and then compensated by using an inverse multiplicative structure. Finally, to efficiently damp the vibration, we propose to use an input shaping technique which is very simple in computation and in implementation. The chapter will treat the displacement y control and the force F shall not be considered.

The experimental setup used for the rest of the paper is pictured in Fig. 10.3. It includes:

- The actuator which is a unimorph piezocantilever. It is composed of one piezoelectric layer based on PZT material (lead zirconate titanate) and of a Nickel layer (passive layer). When a voltage u is applied to the piezolayer, it contracts or expands. Due to the constraint between the two layers, a global flexion y of the cantilever is obtained. The total dimensions of the cantilever

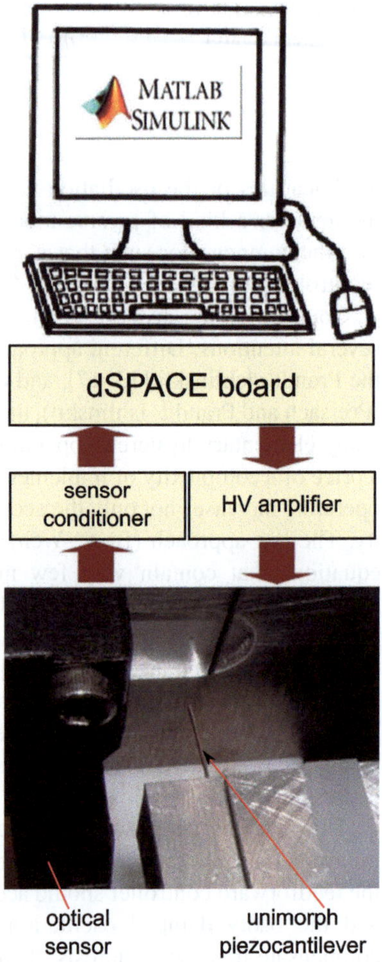

Fig. 10.3 The experimental setup

are 15 mm × 2 mm × 0.3 mm where the thicknesses of the piezolayer and of the Nickel layer are 0.2 mm and 0.1 mm, respectively.

- An optical sensor (*Keyence LC2420*) that is set to have a resolution of 50 nm. This sensor is used to measure y for the identifications of the model parameters and to validate the control technique.
- A computer and a dSPACE board to acquire the measured displacement y, to implement the feedforward controller, and to generate the control signals u. The software Matlab-Simulink is used for the implementation and for the signal management. The sampling time to cadence the computer and the dSPACE board is 0.2 ms. It has been chosen to conveniently account the dynamics of the piezocantilever.
- A high-voltage amplifier (±200 V) that amplifies the signals from the dSPACE board before supplying the actuator.

10.2 Hysteresis Compensation Using the Bouc–Wen Approach and the Inverse Multiplicative Structure

In this section, the hysteresis is first characterized, modeled, and compensated. As we will further characterize and compensate the dynamics, the hysteresis studied here is reduced to a static or rate-dependent hysteresis.

10.2.1 Characterization

To characterize the static hysteresis, a sine or triangular input voltage $u(t)$ is applied to the piezocantilever. The amplitude u_A of the signal should include the maximal range of use. In the applications considered, an amplitude of $u_A = 80\,\mathrm{V}$ is sufficient. The frequency f should be low enough such that the dynamics of the structure will not affect the hysteresis, i.e., to avoid the phase lag. However, it should not be too low in order to avoid the effect of the creep (which is seen at very low rate/frequency) on the hysteresis curve [13]. For the employed piezocantilever, $f = 0.1\,\mathrm{Hz}$ is convenient. After applying the sine input voltage, the output displacement $y(t)$ is recorded. Figure 10.4a, b picture the sine voltage applied to the piezocantilever and the output deflection, respectively. The hysteresis curve is afterwards obtained and can be characterized by plotting the (u,y)-map (see Fig. 10.4c). From this figure, we observe that the hysteresis amplitude is of $\frac{h_h}{H_h} \approx \frac{40\,\mu\mathrm{m}}{160\,\mu\mathrm{m}} = 25\%$ which is non-negligible. It is therefore essential to correctly control such a hysteresis in order to improve the accuracy of the piezocantilever during its utilization. For that, a feedforward control technique based on the Bouc–Wen approach is proposed in the next subsections.

10.2.2 Modeling and Identification

To calculate a feedforward controller able to correctly compensate the hysteresis pictured in Fig. 10.4c, a model is first required. The controller (compensator) can be afterwards derived from this model. In the Bouc–Wen approach [23, 24], the model has the advantage to be a simple structure and to have lower number of parameters to be identified. The following set of two equations describes the Bouc–Wen model adapted to piezoelectric actuators in general [25, 26]:

$$\begin{cases} y(t) = d_p u(t) - h(t),\ y(t_0) = y_0 \\ \frac{\mathrm{d}h}{\mathrm{d}t} = A_{bw}\frac{\mathrm{d}u}{\mathrm{d}t} - B_{bw}\left|\frac{\mathrm{d}u}{\mathrm{d}t}\right| h - \Gamma_{bw}\frac{\mathrm{d}u}{\mathrm{d}t}\left|h\right|,\ h(t_0) = h_0 \end{cases} \tag{10.1}$$

where A_{bw}, B_{bw}, and Γ_{bw} are parameters that determine the amplitude and the shape of the hysteresis and d_p is the piezoelectric coefficient. The signal $h(t)$ is

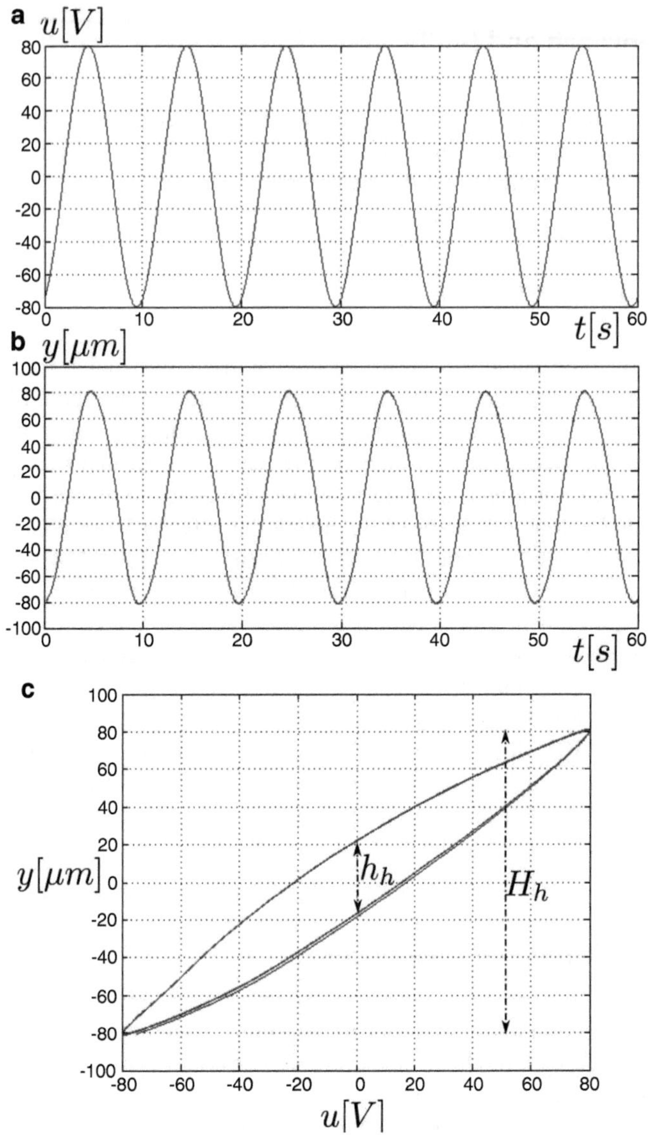

Fig. 10.4 Characterization of the hysteresis. (**a**): the applied input voltage. (**b**): the measured output displacement. (**c**): the obtained hysteresis

an internal state. Remark that if the hysteresis parameters A_{bw}, B_{bw}, and Γ_{bw} are null, the model in (10.1) becomes a linear model: $y(t) = d_p u(t)$. Remind also that this model is a model for static (or rate-independent) hysteresis. The block-scheme corresponding to the Bouc–Wen model is shown in Fig. 10.5. Both the set of equations in (10.1) and this block-scheme can be used to simulate the model.

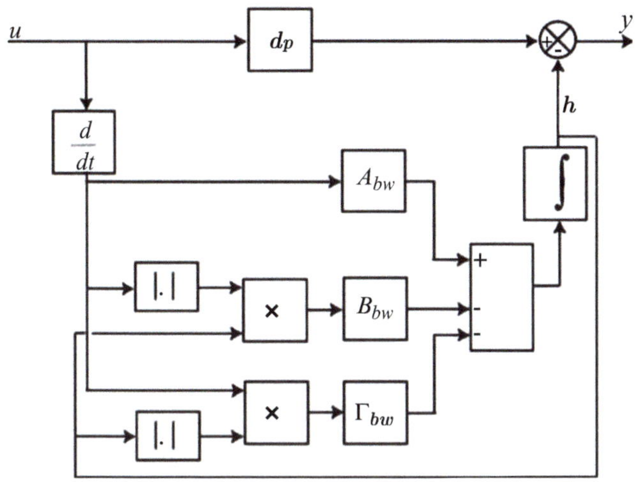

Fig. 10.5 Block diagram of the Bouc–Wen model in (10.1)

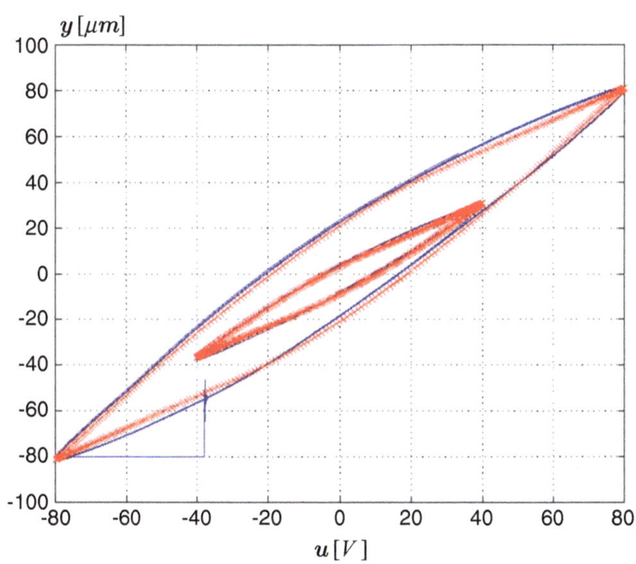

Fig. 10.6 Hysteresis curve: comparison of the model simulation and experimental results

By using a nonlinear filter system identification method [25], we provide approximated values of A_{bw}, B_{bw}, and Γ_{bw}. Afterwards, the identified parameters are validated by comparing the hysteresis curves of the model and of the actuator. During this step, the parameters can be manually refined if required. We finally have $d_p = 1.6\left[\frac{\mu m}{V}\right]$, $A_{bw} = 0.9\left[\frac{\mu m}{V}\right]$, and $B_{bw} = \Gamma_{bw} = 0.008\left[V^{-1}\right]$. Figure 10.6 pictures the comparison of the model simulation with the experimental results. It shows that

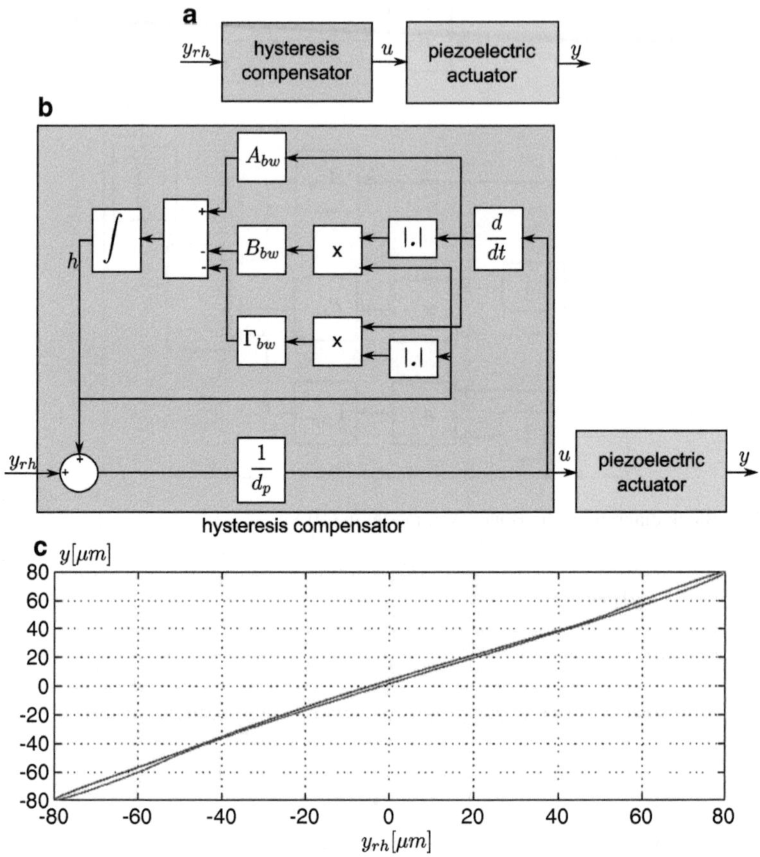

Fig. 10.7 Hysteresis compensation. (**a**) General principle. (**b**) Bouc–Wen model and inverse multiplicative structure as compensator [18]. (**c**) Experimental result

the identified model fits enough to the experiments both for the external loop (with $u = 80\,\text{V}$) and internal loop (here, we use $u = 40\,\text{V}$).

10.2.3 Compensation

To compensate a static hysteresis that has been modeled with the Bouc–Wen model, we proposed to combine the same model with an inverse multiplicative structure for the compensator [18]. The main advantage is that no additional computation is required to have the compensator. This technique is used here to compensate the hysteresis and will further be combined with a creep and a vibration compensator in order to enhance the general performances of the piezocantilever. Let us consider the figure in Fig. 10.7a that corresponds to the piezocantilever with the hysteresis compensator. In the figure, y_{rh} is the input reference. From the first equation of the Bouc–Wen model in (10.1), we derive the necessary control voltage $u(t)$ such that the output $y(t)$ matches the input reference y_{rh}, i.e., such that $y = y_{rh}$. We have

$$u(t) = \frac{1}{d_p}\left(y_{rh}(t) + h(t)\right) \tag{10.2}$$

Thus, the hysteresis compensator is described as follows:

$$\begin{cases} u(t) = \frac{1}{d_p}\left(y_{rh}(t) + h(t)\right) \\ \frac{\mathrm{d}h}{\mathrm{d}t} = A_{bw}\frac{\mathrm{d}u}{\mathrm{d}t} - B_{bw}\left|\frac{\mathrm{d}u}{\mathrm{d}t}\right| h - \Gamma_{bw}\frac{\mathrm{d}u}{\mathrm{d}t}\left|h\right| \end{cases} \tag{10.3}$$

The block diagram that corresponds to the compensator (10.3) is given in Fig. 10.7b. We can see from this diagram that the compensator has an inverse multiplicative structure. The hysteresis compensator has been implemented in the Matlab-Simulink. To check the efficiency of the compensation, a sine reference signal y_{rh} is applied. Figure 10.7c pictures the results. It clearly shows the deletion of the initial hysteresis of 25% (see Fig. 10.4c). Furthermore, not only the new system (piezocantilever with the hysteresis compensator) is linear but also the gain is nearly equal to one, i.e., a high accuracy is obtained: $y \approx y_{rh}$.

10.3 Creep Compensation Using a Linear Model and the Inverse Multiplicative Structure

10.3.1 Characterization

The previous section deals with the hysteresis compensation. The new system (piezocantilever with the hysteresis compensator) becomes linear and quite accurate since $y \approx y_{rh}$. Unfortunately, when a step input reference $y_{rh} = 20\,\mu\text{m}$ is applied and when we report the output displacement y for a long period of time, we will observe a very slow drift as in Fig. 10.8. This drift is called creep and acts at very low frequency. This makes again the previous linearized system non-accurate and requires therefore a control. Once again, to feedforward control this behavior, an accurate model is required first.

10.3.2 Modeling and Identification

Regarding Fig. 10.8, the output displacement y can be seen as the summation of two signals:

- The evolution $y^{\text{transient}}$ from $0\,\mu\text{m}$ to a *final value* before the drift start
- The evolution y^{creep} that tracks the drift

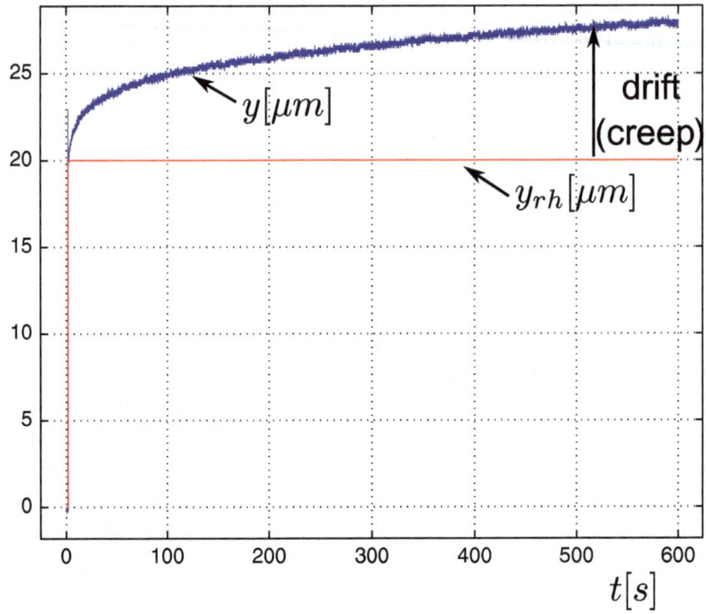

Fig. 10.8 Characterization of the creep

Then, in the time domain, we can write

$$y(t) = y^{\text{transient}}(t) + y^{\text{creep}}(t) \tag{10.4}$$

which, in the Laplace domain, is equivalent to

$$y(s) = y^{\text{transient}}(s) + y^{\text{creep}}(s) \tag{10.5}$$

where s is the Laplace variable.

Under the assumption that we do not regard the dynamics of the piezocantilever, which is valuable since we study the behavior at low and very low frequency so far, the ratio between $y^{\text{transient}}(s)$ and $y_{rh}(s)$ is a constant, denoted K, that corresponds to the slope of the curve in Fig. 10.7c. We have

$$\frac{y^{\text{transient}}(s)}{y_{rh}(s)} = K \tag{10.6}$$

Although the creep behavior is classified as a nonlinear behavior, it can be approximated by a linear model, for instance by a transfer function. Denote $C(s)$ the transfer function that approximates the ratio between $y^{\text{creep}}(s)$ and $y_{rh}(s)$ such that

$$\frac{y^{\text{creep}}(s)}{y_{rh}(s)} = C(s) \tag{10.7}$$

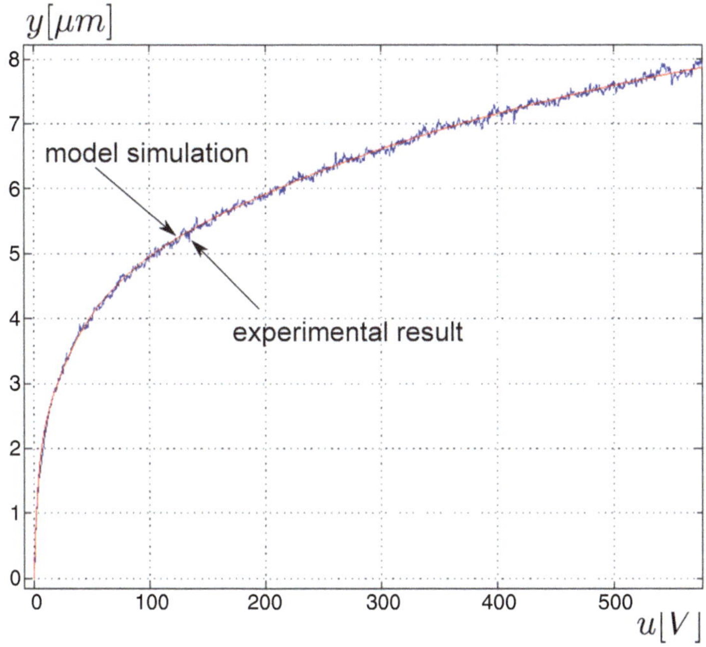

Fig. 10.9 Creep evolution: experimental result and model simulation

From (10.4), (10.6), and (10.7), we derive

$$\frac{y(s)}{y_{rh}(s)} = (K + C(s)) \tag{10.8}$$

K can be identified from Fig. 10.7c, which provide $K \approx 1$. To identify $C(s)$, the drift in Fig. 10.8 needs to be separated from the whole curve. Then applying a system identification technique (such as ARMAX-Auto Regressive Moving Average with eXternal inputs) in the Matlab software [27], we obtain

$$C(s) = 0.465 \cdot \frac{(16 \cdot s + 1) \cdot (203 \cdot s + 1)}{(3 \cdot s + 1) \cdot (34 \cdot s + 1) \cdot (444 \cdot s + 1)} \tag{10.9}$$

Figure 10.9 presents the simulation of the creep model in (10.9) and the experimental result extracted from Fig. 10.7a. It shows the accuracy of the identified model used.

10.3.3 Compensation

Let Fig. 10.10a picture the general block diagram of the new system (linearized system) with the creep compensator where y_{rc} is the new reference. The aim is

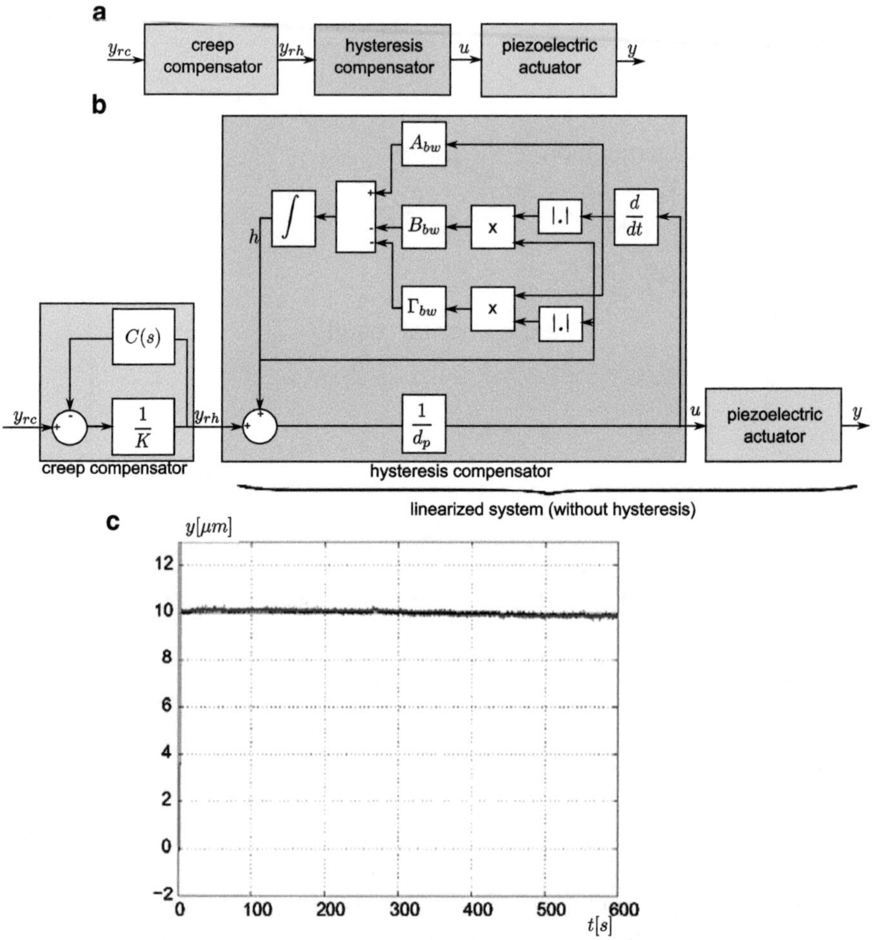

Fig. 10.10 Compensation of the creep. (**a**) General block diagram. (**b**) Detailed diagram of the compensators. (**c**) Experimental results

now to compute this creep compensator. To control the creep, we start with the model (10.8) and derive the necessary input y_{rh} such that the output y meets the new reference y_{rc}, i.e., such that $y_{rc} = y$. We obtain

$$\frac{y_{rh}(s)}{y_{rc}(s)} = \frac{1}{(K + C(s))} \tag{10.10}$$

which provides

$$y_{rh}(s) = \frac{1}{K}\left(y_{rc}(s) - C(s)y_{rh}(s)\right) \tag{10.11}$$

where K is ensured to be invertible since it corresponds to the static gain between the input y_{rh} and y ($K \neq 0$). Equation (10.11) is the equation of the creep compensator and its block diagram, when applied to the linearized system in Fig. 10.7b, is pictured in Fig. 10.10b. The main advantage of the controller in (10.11) is that no direct dynamics inversion is required. Hence, no particular conditions are required for the creep model $C(s)$ [13].

The experimental results effectuated with the piezocantilever are pictured in Fig. 10.10c. It presents the response y of the actuator when a step reference $y_{rc} = 10\,\mu\text{m}$ is applied. It clearly shows that the creep has been completely removed.

10.4 Vibration Compensation Using an Input Shaping Technique

10.4.1 Characterization

Although the hysteresis and the creep have been removed, the piezocantilever has a badly damped vibration behavior. This can be seen by zooming the transient part in the step response pictured in Fig. 10.10c. Figure 10.11a pictures this transient part (step response) from which we observe an overshoot of $\frac{d}{D} \approx 33\%$. In Fig. 10.11b, we show the frequency response of the system. This clearly shows the high peak of resonance. In many applications such as micromanipulation and microassembly, a badly damped vibration is unwanted because the corresponding high overshoot may imply the destruction of the manipulated object. In addition, the settling time is greatly increased due to the long time necessary to damp and stabilize the piezocantilever. The control of this vibration is therefore as essential as the control of the nonlinearities previously presented.

10.4.2 Modeling and Identification

The nonlinear parts having been compensated, the remaining behavior is now linear. The dynamics studied here is therefore linear. The system to be considered is pictured in Fig. 10.10b and has as input the reference y_{rc} and as output the displacement y. A general model of this system is

$$\frac{y(s)}{y_{rc}(s)} = \frac{\sum_{j=0}^{m} b_j s^j}{\sum_{i=0}^{n} a_i s^i} \tag{10.12}$$

where a_i and b_j are the coefficients to be identified and where $m \leq n$ for the causality of the system.

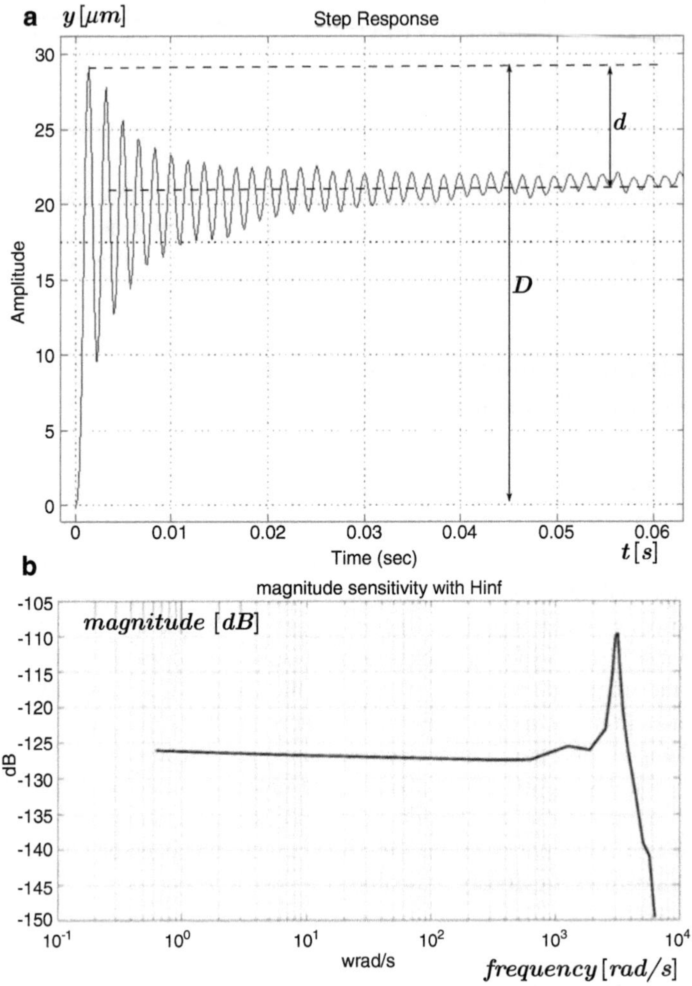

Fig. 10.11 (**a**) Response of the piezocantilever to a step $y_{rc} = 21\,\mu\text{m}$. (**b**) Frequency response of the piezocantilever

The higher the order n is, more accurate will be the identified model. However this will increase the complexity of the model and therefore the complexity of the vibration compensator. This is the case for the zero magnitude error tracking control (ZMETC), the zero phase error tracking control (ZPETC), and the direct dynamics inversion compensation techniques for which the compensators orders are at least equal to the model's orders [22]. Some techniques such as the input shaping and the Posicast techniques can however use lower model (second-order model) by still maintaining their efficiency to strongly damp the vibration. We will present here the input shaping technique. Hence, we impose the dynamics that will be used as a model of Fig. 10.11 as a second-order model. The ARMAX method and the Matalb software are again used to identify the parameters. We obtain

$$\frac{y(s)}{y_{rc}(s)} = \frac{1}{\left(\frac{1}{\omega_n}\right)^2 s^2 + \frac{2\xi}{\omega_n} s + 1} \tag{10.13}$$

where the natural frequency is $\omega_n = 3,092$ rad/s and the damping ratio is $\xi = 0.029$.

10.4.3 Compensation by Using an Input Shaping Technique

The input shaping approach is a simple approach to damp vibration in oscillatory structures. It consists in shaping the input control and generating the right signal such that the steady state is obtained without having an oscillating transient part. The way to shape the input control is by using a sequence of impulses convolved with a reference input. There are several kinds of input shaping techniques as surveyed in [28]. Among them, the zero vibration input shaping technique (ZVIS) [29] is the simplest one that can minimize the vibration and that we will use in this chapter. First, we remind its principle.

When an impulse is applied to an oscillating system, for instance the system in (10.13), a vibration appears. When a second impulse is applied at time $t_2 = T_p/2$, with $T_p = \frac{2.\pi}{\omega_n.\sqrt{1-\xi^2}}$, the vibration caused by the second impulse can cancel the one caused by the first impulse if the amplitudes of both are judiciously chosen (Fig. 10.12). For any reference input y_r, the precedent sequence of impulses, also called shaper, is convolved with it to obtain a new signal control that will cause no vibration. As example, if the reference input is a step, the resulting signal control will be a staircase with two steps.

The shaper is calculated as follows. Consider A_i and t_i the amplitudes of the impulses and their application times. Consider $K = e^{-\frac{\xi.\pi}{\sqrt{1-\xi^2}}}$. Then:

$$\begin{aligned} &\left[A_1 = \tfrac{1}{1+K},\, A_2 = \tfrac{K}{1+K}\right] \\ &\left[t_1 = 0,\, t_2 = \tfrac{T_p}{2}\right] \end{aligned} \tag{10.14}$$

Sometimes, it is hard to precisely identify the parameters ω_n and ξ. The model uncertainty will therefore generate residual vibration when applying the calculated shaper. In such a case, it is recognized to use a higher number of impulses in the shaper. Then, when a step reference input y_r is applied, the amplitudes of the

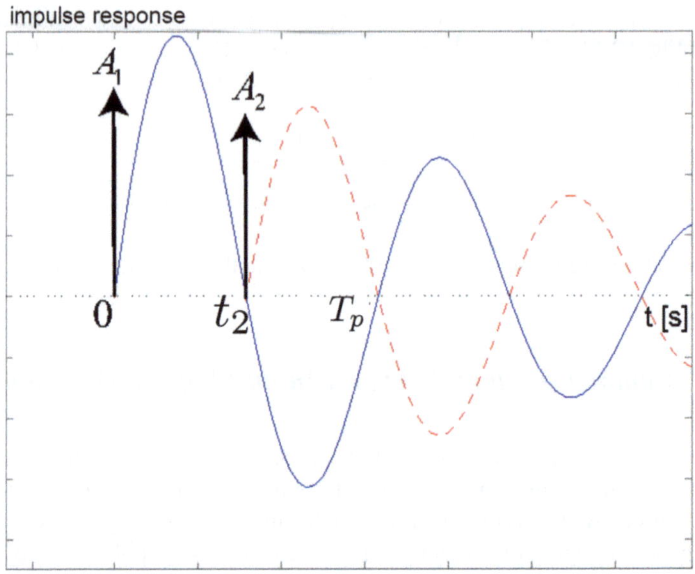

Fig. 10.12 Response of the piezocantilever to two delayed impulses

different impulses will be lower and the residual vibration will be reduced. For a shaper with k impulses, we have

$$\left[A_1 = \frac{a_1}{(1+K)^{m-1}},\, A_2 = \frac{a_2}{(1+K)^{m-1}},\right.$$
$$\left.\ldots,\, A_m = \frac{a_m}{(1+K)^{m-1}}\right]$$
$$\left[t_1 = 0,\, t_2 = \frac{T_p}{2},\, \ldots,\, t_m = (m-1)\,\frac{T_p}{2}\right] \tag{10.15}$$

with a_i the ith coefficient of the polynomial from $(1+K)^{m-1}$. We have $a_1 = 1$ and $a_m = K^{m-1}$. Using the dynamic model given by (10.13) and the identified parameters, a shaper has been computed. It has been implemented in cascade with the linearized system as pictured in Fig. 10.13.

First, a step response characterization was carried out. The step reference applied is $y_r = \pm 15\,\mu\text{m}$. Different shapers with different number of impulses were tested. The step responses indicated that the performances did not increase substantially when using number of impulses greater than four. The results are pictured in Fig. 10.14. This figure includes the response of the vibration compensated system with a shaper with 1, 2, 3, and 4 impulses. The response of the system without vibration compensation is also pictured in the same figure. Table 10.1 summarizes the performances. They point out that the overshoot has been reduced from 50.8 to

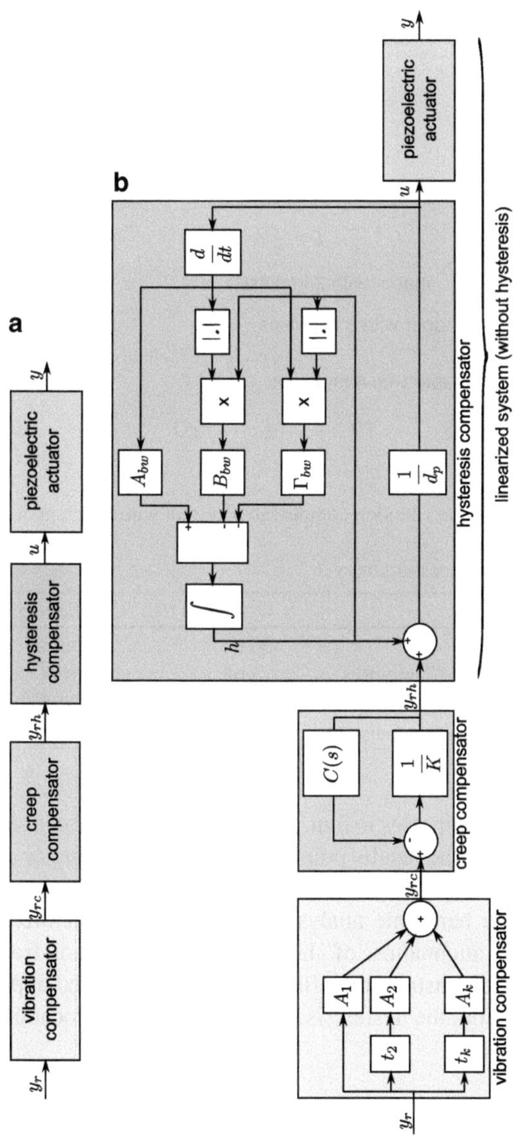

Fig. 10.13 Block diagram of the vibration compensation. (**a**) General block diagram. (**b**) Detailed diagram of the compensators

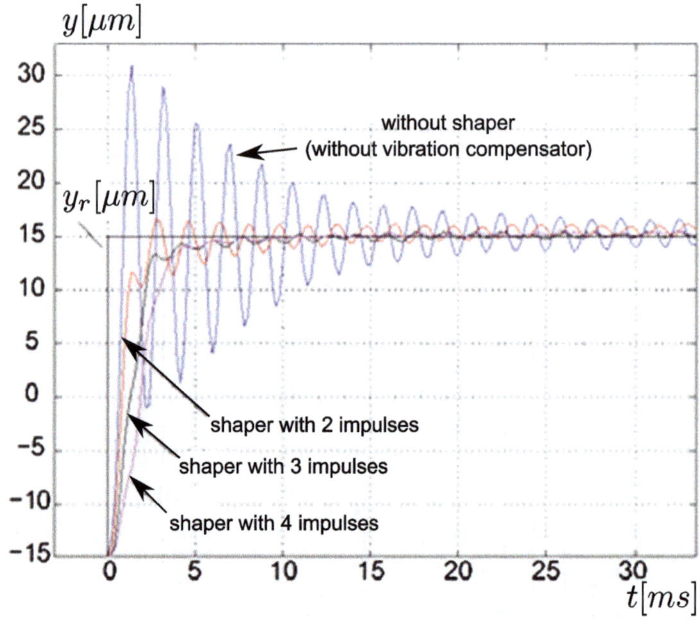

Fig. 10.14 Step response of the vibration compensated system with different order of shaper

Table 10.1 Performances summary

	Overshoot (%)	Setting time (ms)
Uncompensated vibration	50.8	>28.5
Compensated vibration (2 impulses)	3.65	10.7
Compensated vibration (3 impulses)	0	3.95
Compensated vibration (4 impulses)	0	≈4

0% when the number of impulses is four. In addition, a high reduction of the settling time is obtained. Finally, the results point out that there is no major amelioration of the performances when using more than 3 impulses.

Next, we perform a harmonic analysis. The results are pictured in Fig. 10.15. They show the great attenuation of the peak at the resonant frequency. These experimental results demonstrate the efficiency of the proposed approach to open-loop control the vibration, the hysteresis, and the creep in piezocantilevers.

10.5 Conclusion

Piezoelectric materials are well recognized for the design and development of microactuators and systems for micro and nanopositioning. This recognition is due to their high resolution, good bandwidth, high force density, and the ease

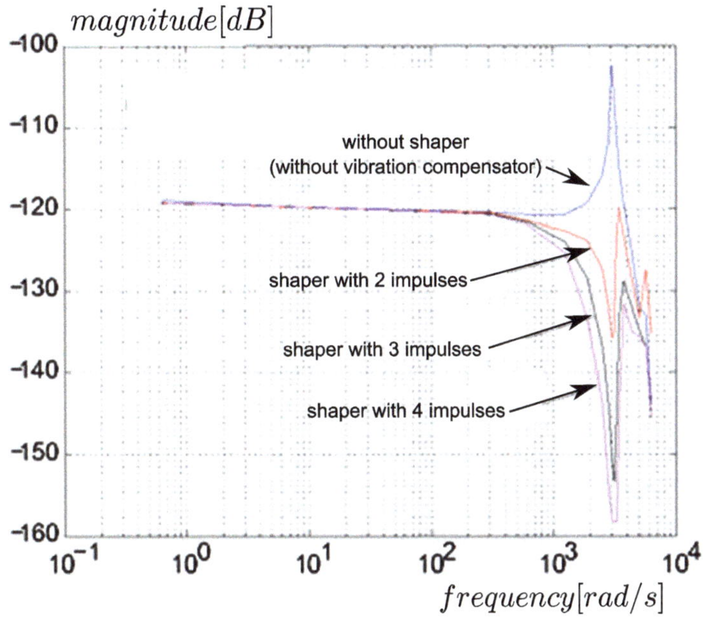

Fig. 10.15 Harmonic responses of the vibration compensated system with different order of shaper

of supply (electrical). In particular, many systems used in micromanipulation or microassembly are based on piezoelectric actuators with cantilever structure, for instance piezoelectric microgrippers. These actuators allow the positioning of the manipulated object with a very high resolution and with a high dynamics. Unfortunately, piezoelectric material-based actuators are typified by hysteresis and creep nonlinearities that make them lose the final accuracy. In addition, cantilever structured actuators are often characterized by a badly damped vibration which increases the settling time of the process.

This chapter presented the feedforward control of the hysteresis, the creep, and the badly damped oscillation in piezoelectric cantilever actuators used in microgrippers. The main advantage of the feedforward control relative to feedback is the non-necessity of sensors which make them very attractive for applications where the packageability is essential. The approach presented in this chapter consisted to compensate first the hysteresis by using the Bouc–Wen approach and the inverse multiplicative structure. Then the creep was compensated by using a linear approximate model combined with the inverse multiplicative structure again. The interest of using the inverse multiplicative structure was the nonnecessity to compute the compensator: as soon as the model is identified, the compensator is derived by the structure. Finally the vibration was compensated using the input shaping technique. This technique is simple in implementation and could be robust enough even in the presence of uncertain model parameters. Experimental results in a unimorph piezocantilever were carried out and confirmed the efficiency of the proposed approach.

Acknowledgements This work is supported by the national ANR-JCJC C-MUMS-project (National young investigator project ANR-12-JS03007.01: Control of Multivariable Piezoelectric Microsystems with Minimization of Sensors).

References

1. J. Agnus, N. Chaillet, C. Clévy, S. Dembélé, M. Gauthier, Y. Haddab, G. Laurent, P. Lutz, N. Piat, K. Rabenorosoa, M. Rakotondrabe, B. Tamadazte, Robotic microassembly and micromanipulation at FEMTO-ST. J. Micro. Bio. Robot. (JMBR), **8**(2), 91–106 (2013)
2. M. Rakotondrabe, C. Clévy, P. Lutz, Modelling and robust position/force control of a piezoelectric microgripper, in *IEEE - International Conference on Automation Science and Engineering (CASE)*, Scottsdale, AZ, USA, 2007, pp. 39–44
3. M. Rakotondrabe, Y. Haddab, P. Lutz, Quadrilateral modelling and robust control of a nonlinear piezoelectric cantilever. IEEE - Trans. Contr. Syst. Technol. (T-CST) **17**(3), 528–539 (2009)
4. M. Rakotondrabe, K. Rabenorosoa, J. Agnus, N. Chaillet, Robust feedforward-feedback control of a nonlinear and oscillating 2-dof piezocantilever. IEEE - Trans. Autom. Sci. Eng. (T-ASE) **8**(3), 506–519 (2011)
5. S. Khadraoui, M. Rakotondrabe, P. Lutz, Interval modeling and robust control of piezoelectric microactuators. IEEE - Trans. Contr. Syst. Technol. (T-CST) **20**(2), 486–494 (2012)
6. Y. Shan, K.K. Leang, Accounting for hysteresis in repetitive control design. Automatica **48**(8), 1751–1758 (2012)
7. A. Sebastian, A. Gannepalli, M.V. Salapaka, A review of the systems approach to teh analysis of dynamic-mode atomic force microscopy. IEEE Trans. Contr. Syst. Technol. **15**(5), 952–959 (2007)
8. Q. Xu, Y. Li, Model predictive discrete-time sliding mode control of a nanopositioning piezostage without modeling hysteresis. IEEE Trans. Contr. Syst. Technol. **20**(4), 983–994 (2012)
9. A. Bazaei, Y.K. Yong, S.O.R. Moheimani, A. Sebastian, Tracking of triangular references using signal transformation for control of a novel AFM scanner stage. IEEE Trans. Contr. Syst. Technol. **20**(2), 453–464 (2012)
10. S. Devasia, E.E. Eleftheriou, R. Moheimani, A survey of control issues in nanopositioning. IEEE Trans. Contr. Syst. Technol. **15**(5), 802–823 (2007)
11. D. Croft, G. Shed, S. Devasia, Creep, hysteresis and vibration compensation for piezoactuators: atomic force microscopy application. ASME J. Dyn. Syst. Meas. Contr. **123**(1), 35–43 (2001)
12. A. Dubra, J. Massa, C.l. Paterson, Preisach classical and nonlinear modeling of hysteresis in piezoceramic deformable mirrors. Optic. Express **13**(22), 9062–9070 (2005)
13. M. Rakotondrabe, C. Clévy, P. Lutz, Complete open loop control of hysteretic, creeped and oscillating piezoelectric cantilever. IEEE Trans. Autom. Sci. Eng. (TASE) **7**(3), 440–450 (2010)
14. W.T. Ang, P.K. Kholsa, C.N. Riviere, Feedforward controller with inverse rate-dependent model for piezoelectric actuators in trajectory-tracking applications. IEEE/ASME Trans. Mechatron. **12**(2), 134–142 (2007)
15. B. Mokaberi, A.A.G. Requicha, Compensation of scanner creep and hysteresis for AFM nanomanipulation. IEEE Trans. ASE **5**(2), 197–0208 (2008)

16. M. Al Janaideh, P. Krejci, Inverse rate-dependent Prandtl–Ishlinskii model for feedforward compensation of hysteresis in a piezomicropositioning actuator. IEEE/ASME Trans. Mechatron. (2012). doi:10.1109/TMECH.2012.2205265
17. M. Rakotondrabe, Classical Prandtl–Ishlinskii modeling and inverse multiplicative structure to compensate hysteresis in piezoactuators, in *ACC (American Control Conference)*, Montréal, Canada, June 2012, pp. 1646–1651
18. M. Rakotondrabe, Bouc–Wen modeling and inverse multiplicative structure to compensate hysteresis nonlinearity in piezoelectric actuators. IEEE Trans. ASE **8**(2), 428–431 (2011)
19. H. Jung, J.Y. Shim, D. Gweon, New open-loop actuating method of piezoelectric actuators for removing hysteresis and creep. Rev. Sci. Instrum. **71**(9), 3436–3440 (2000)
20. G.M. Clayton, S. Tien, S. Devasia, A.J. Fleming, S.O.R. Moheimani, Inverse-feedforward of charge-controlled piezopositioners. Mechatronics **18**, 273–281 (2008)
21. M. Rakotondrabe, C. Clévy, P. Lutz, Hysteresis and vibration compensation in a nonlinear unimorph piezocantilever, in *IEEE/RSJ - IROS, (International Conference on Intelligent Robots and Systems)*, Nice, France, Sept 2008, pp. 558–563
22. M. Rakotondrabe, *Piezoelectric Cantilevered Structures: Modeling Control and Measurement/Estimation Aspects* (Springer, Berlin, 2013)
23. R. Bouc, Forced vibration of mechanical systems with hysteresis, in *Conference on Nonlinear Oscillation*, Prague, 1967
24. Y.K. Wen, Method for random vibration of hysteresis systems. J. Eng. Mech. Div. **102**(2), 249–263 (1976)
25. M. Jouaneh, H. Tian, Accuracy enhancement of a piezoelectric actuators with hysteresis, in *ASME Japan/USA Symp. Flexible Automation*, Proceedings of the Japan-USA Symposium on Flexible Automation, A Pacific Rim Conference, San Francisco, California, USA, ASME/ISCIE, ISBN 0-7918-0765-8, 1992, pp. 631–637
26. T.S. Low, W. Guo, Modeling of a three-layer piezoelectric bimorph beam with hysteresis. J. Microelectromech. Syst. **4**(4), 230–237 (1995)
27. L. Ljung, System identification toolbox, for use with Matlab. The Matworks (1995)
28. T. Singh, W. Singhose, Tutorial on input shaping/time delay control of maneuvering flexible structures, in *American Control Conference*, Proceedings of the American Control Conference, Anchorage Alaska USA, 2002, pp. 1717–1731
29. N.C. Singer, W.P. Seering, K.A. Pasch, Shaping command inputs to minimize unwanted dynamics. Patent No. US-4.916.635, 1990

Chapter 11
Micro/Nanorobotic Manufacturing of Thin-Film NEMS Force Sensor

Gilgueng Hwang and Hideki Hashimoto

Abstract This chapter presents the fabrication and characterization of piezoresistive force sensors based on helical nanobelts. The three-dimensional helical nanobelts are self-formed from 27-nm-thick n-type InGaAs/GaAs bilayers using rolled-up techniques and assembled onto electrodes on a micropipette using nanorobotic manipulation. Patterned gold electrodes were fabricated using thermal evaporation or fountain-pen-based gold nanoink deposition. Nanomanipulation inside a scanning electron microscope was conducted to locate small metal pads of helical nanobelts to be connected to the fabricated pipette-type electrodes. Gold nanoink was deposited under optical micrograph using the fountain-pen method. Nanomanipulation inside a scanning electron microscope using a calibrated atomic force microscope cantilever was conducted to calibrate the assembled force sensors, and the values were compared with finite-element-method simulation results. With their strong piezoresistive response, low stiffness, large-displacement capability, and good fatigue resistance, these force sensors are well suited to function as sensing elements for high-resolution and large-range electromechanical sensors.

11.1 Introduction

In recent decades, various micro-/nanoelectromechanical systems (MEMS/NEMS) have been used for many applications. Much effort has been devoted to the innovative process of synthesizing micro-/nanostructures as the building blocks

G. Hwang (✉)
Laboratory for Photonics and Nanostructures, Centre National de la Recherche Scientifique, Marcoussis 91460, France
e-mail: gilgueng.hwang@lpn.cnrs.fr

H. Hashimoto
Department of EECE, Chuo University 1-13-27 Kasuga, Bunkyo-ku, Tokyo, Japan

M. Rakotondrabe (ed.), *Smart Materials-Based Actuators at the Micro/Nano-Scale: Characterization, Control, and Applications*, DOI 10.1007/978-1-4614-6684-0_11,

© Springer Science+Business Media New York 2013

for creating such MEMS or NEMS [1, 2]. Carbon nanotubes (CNTs) [3–6], nanowires (NWs) [7], and Nanohelixes [8–13] are the most widely synthesized and considered the promising elements in NEMS and nanoelectronics. For their successful applications, the packaging process to construct devices from these new building blocks and the characterization of the devices are among the most critical steps to proving their quality for working in application environments. For example, with field emitting displays, the major packaging challenge is to achieve directionally controlled growing of NWs and CNTs. For this purpose, the mechanical properties of these nanostructures should be understood precisely. Electrical property characterizations is another practical issue along with device packaging. Regarding packaging, a bottom-up approach like self-assembly is the most widely used. A top-down approach using micro-/nanolithography and etching processes is another process. In addition to the top-down approach, micro-/nanoassembly could be an alternative approach to creating devices or prototypes [14–16]. This approach is based on micro-/nanorobotic manipulation systems with precise MEMS/NEMS sensors and actuators installed inside nanoscale imaging devices such as scanning/transmission electron microscopes (SEM/TEM). In particular, wide-range mechanical pressure/force sensors are among the most important devices that should be integrated into robotic manipulations to characterize the mechanical/electrical and even electromechanical properties of various nanostructures. The characterized physical properties are essential to establish a precise model of nanostructures. Physical models of single nanostructures are essential for the device engineering and performance optimization of the NEMS sensors/actuators constructed from them. Force sensing is the most widely used tool to characterize the mechanical properties of these structures.

In this chapter, we demonstrate the micro-/nanorobotic manufacture of thin-film NEMS force sensors to have a large force-sensing range. For example, force sensing probes based on the piezoresistivity of InGaAs/GaAs helical nanobelts (HNBs) is introduced. HNBs can serve as a mechanism to transduce force to displacement. The deformation is detected through a piezoresistive effect to measure the corresponding force after calibration. A major challenge in the development of the proposed force sensor is the lack of manufacturing processes. Therefore, this chapter describes the details of the micro-/nanorobotic manufacturing process of such three-dimensional thin-film nanodevices.

11.2 Helical Nanobelt Force Sensors

11.2.1 Large-Range Force Sensors

Force sensing with high enough precision but large bandwidth is essential to in particular small-scale robotics applications (mechanical characterizations of nanostructures, robotic drug delivery, single-molecule detection from a whole bunch of molecules, robotic injection, etc.) [17, 18]. Kinking and buckling force mea-

Table 11.1 Required specification of force sensor in nanomanipulation

Dimension	~a few tens of microns
Force-sensing range	nN ~ a few hundred nanonewtons
Displacement range	1 nm to a few microns
Force-sensing direction	Arbitrary direction
Force-sensing mechanism	Self-sensing mechanism

surements are shown to be very important in the understanding of the mechanical properties of newly synthesized nanomaterials for determining their competitive NEMS applications [19, 20]. To fulfill these nanomanipulation tasks, it is highly expected that NEMS-based force sensors will be developed.

The required features of force sensors to fulfill these tasks are summarized in Table 11.1. These requirements were based on empirically obtained knowledge from nanomanipulations of nanostuctures such as CNTs, NWs, etc. They were also obtained from published works on nanomanipulations [19, 20]. Meanwhile, an increasing number of applications in nanorobotic manipulations require an nanonewton range force sensing [21]. Conventionally, mechanical transducers have been developed. A scanning force microscopy (SFM) cantilever is used mostly for sensing forces in a range of 10 pN–100 nN [22]. Microneedles have been used to measure the force of a single actin filament [23]. Photon-field-based optical tweezers [24] have been used for force sensing in a range of 0.1–100 pN [22]. This laser-based sensing can heat biological samples, and so its application is limited. A magnetic field can measure below 10 pN by manipulating an attached magnetic bead [25]. However, it has also a drawback in that it requires indirect measurement of the magnetic force. Flow fields in a laminar flow can measure 0.1 pN–1 nN [22]. To reach this sensing resolution in a more systematic way, many NEMS force-sensing devices have been demonstrated in several different types such as in-plane devices and out-of-plane probes. As an in-plane force and pressure sensor, an individual single-walled carbon nanotube (SWNT) was bridged between two electrodes using the characterized piezoresistivity [26, 27]. For out-of-plane device transduction, CNTs were attached to an atomic force microscope (AFM) cantilever [3–5]. However, there are still no built-in sensing elements have been demonstrated for such cantilevers due to their nanometer sizes, whereas similar MEMS force sensors, such as piezoresistive cantilevers [28, 29] and capacitance sensors [30], have been fabricated. On the other hand, three-dimensional (3D) helical structures with micro- and nanofeatures have been synthesized from various materials. Typical examples include microcoils based on amorphous carbon [8], nanocoils based on CNTs [9], and zinc oxide HNBs [10, 11]. Because of their interesting morphology, as well as mechanical [12, 13], electrical, and electromagnetic properties, these micro-/nanostructures can be used as components for MEMS and NEMS such as springs, inductors, sensors, and actuators. Recently, the electrical and mechanical properties of SiGe/Si/Cr and SiGe/Si HNBs were characterized separately through experiments and simulations [12]. The fabrication and mechanical characterization of InGaAs/GaAs HNBs have been also described [13]. Their excellent flexibility provides a new avenue for fabricating ultra small force sensors with high resolution.

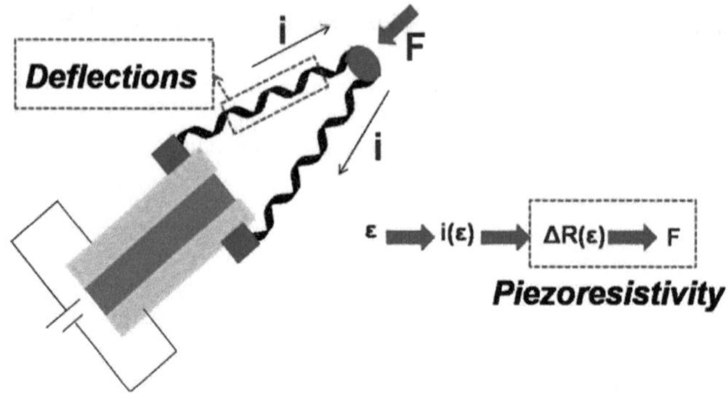

Fig. 11.1 Schematic of working principle of HNB force sensors

11.2.2 Piezoresistive Helical Nanobelt Force Sensors

The helical morphology benefits the ultraflexible mechanical property compared to beam-type cantilevers. Therefore, helical nanostructures such as HNBs have a large displacement range and high resolution. By determining the piezoresistivity in HNBs, arbitrary directional force sensing and easy integration with nanomanipulators from self-sensing are achievable. Therefore, our approach is to build a force sensor with an ultraflexible structure from a helical morphology. As a sensing mechanism, the piezoresistivity of InGaAs/GaAs HNBs for force sensing is characterized first. Since the proposed design undergoes both axial and bending forces, the corresponding piezoresistivities in multiple axes were also characterized to estimate the piezoresistivity effect after the force sensor assembly. Nanorobotic assembly processes were proposed mainly for the field-assisted alignment of HNBs onto a pipette electrode and for electrical soldering to assure an ohmic conductivity. The assembled HNB force sensor was characterized using an as-calibrated AFM cantilever. The force sensor was able to measure the applied force by reading the resistance change from the HNBs' piezoresistivity. HNBs represent a new material and are much more flexible and fit our out-of-plane devices. HNBs have very nice features: for example, they are ultrasoft and uniform in geometry and have easy band-gap tuning. Their drawbacks include mainly that they have a high surface-to-volume ratio, making soldering difficult, and their ultra flexibility with an ultra-thin film can lead to misalignment or uncontrolled assembly. Most of the assembly technologies that have been investigated so far do not work well with 3D HNBs. To solve these problems, we have proposed better alignment technologies of HNBs using an external field assist. Furthermore, we propose robust soldering technologies which can be applied to various scales (gold nanoink deposition, in situ extension of gold nanoink soldering, and chemical-free resistance spot welding) for 3D structures.

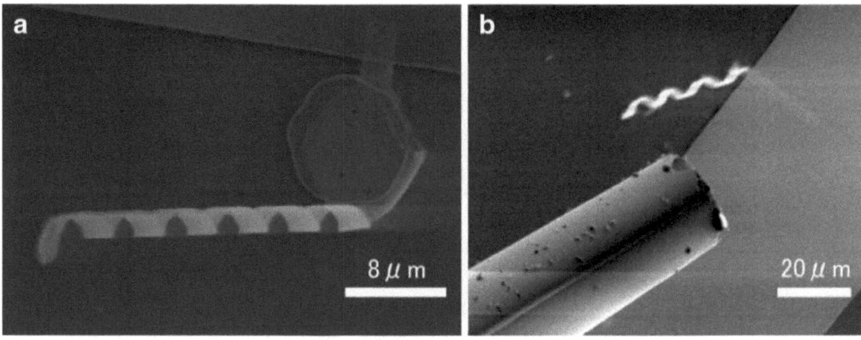

Fig. 11.2 As-fabricated (**a**) InGaAs/GaAs HNBs with metal pads and (**b**) electrodes on tapered micropipettes

11.2.3 Working Principle of HNB Force Sensors

As shown in Fig. 11.1, HNB force sensors measure the force by piezoresistivity. When a certain amount of force is applied to a HNB force sensor, the deflection of the HNB changes the resistance of the assembled structure. Figure 11.1 shows a schematic view of the proposed HNB force sensors. Two HNBs with metal connectors are assembled on an independently patterned micropipette electrode. Then each electrode passing through two HNBs is interfaced outside to determine the strain (ε)-induced current change given a constant voltage input.

11.3 Force Sensor Assembly

11.3.1 Interconnection Layer Fabrication

NEMS using HNBs include two typical configurations [31], i.e., a HNB bridging two electrodes horizontally or standing vertically on electrodes. An as-fabricated HNB is shown in Fig. 11.2. To obtain a better interconnection conductivity, HNBs were fabricated with metal connectors (Cr/Ni/Au 20/200/25 nm) on both ends [32], which is different from the standard design [31]. Microtapered pipette-type electrodes were prepared [33]. A ferromagnetic Ni layer was evaporated at the end of the HNB for electromagnetic actuation. Figure 11.2 shows the as-fabricated pipette electrodes used to assemble HNBs. The electrode pattern was generated by thin-film evaporation. Our objective is to assemble suspended HNBs on the as-fabricated pipette electrodes (Fig. 11.2b), Cr/Ni/Au deposited independent electrodes, for precise location of HNBs with metal deposited connectors. First, the borosilicate capillary was pulled to make tapered micropipettes. The dimensions of the pipette opening were controlled in a reproducible way using a micropipette puller (DMZ Universal Puller, Zeitz Instruments, Germany). Pipettes with 1 and

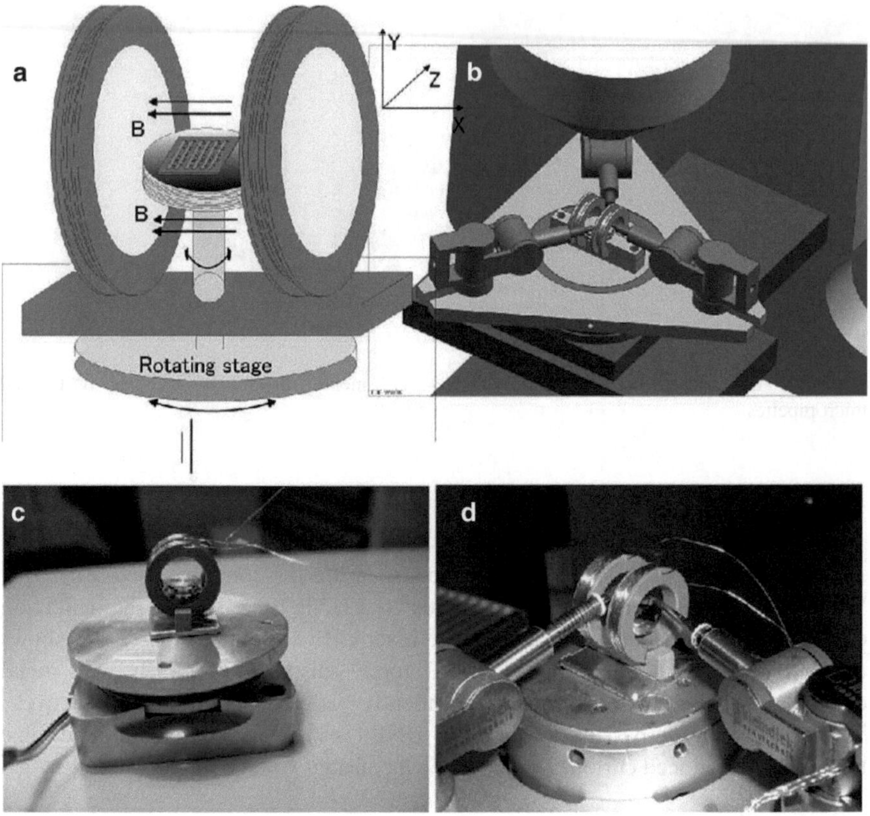

Fig. 11.3 Experimental setup: (**a**) Helmholtz coil on piezoelectric rotational stage, (**b**) CAD model of integration with manipulators, (**c**), (**d**) manipulators and Helmholtz coil installed inside SEM

15 μm openings were fabricated. Then, independent Cr/Ni/Au metal layers were evaporated on both sides of the pipette by changing the exposure to the target electron-beam-heated metal source. A homemade pipette holder with wiring was used to mount to a nanomanipulator and connect to the power supply.

11.3.2 External Force-Generating System

The nanorobotic manipulation system shown in Fig. 11.3 was used for the manipulation of the as-fabricated HNBs inside a SEM (Carl Zeiss DSM 962). Three nanorobotic manipulators (Kleindiek, MM3A) were installed inside the SEM; each had three degrees of freedom and 5-, 3.5-, and 0.25-nm resolution in the X-, Y-, and Z-directions at the tip. A metal probe (Picoprobe, T-4-10-1 mm, tip radius: 100 nm) was mounted on the nanomanipulator. The same manipulation setup was used for both the assembly and the characterizations.

Table 11.2 Experiment specifications

Gap *r* [μm]	5
Length [μm]	36.3
R_{ext} [μm]	1.05
R_{int} [μm]	1.023
Electric constant e_0 [C^2/Nm^2]	$8.85 \cdot 10^{-12}$
Voltage range *V* [V]	0-1;Step:0.1 V
E-module [N/m^2]	$8.0215 \cdot 10^{10}$

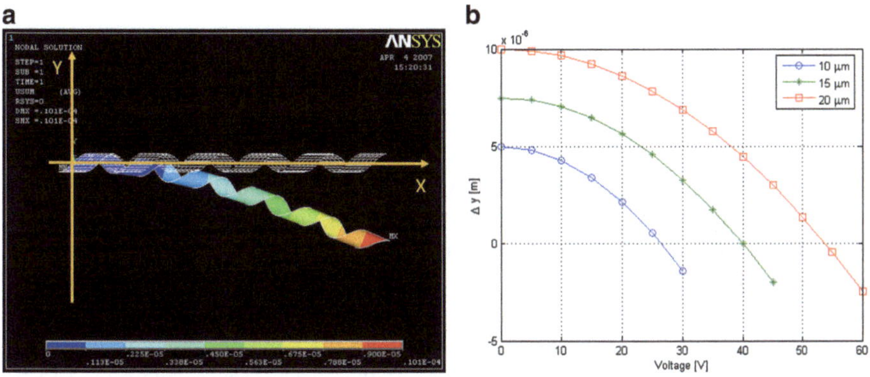

Fig. 11.4 FEM simulation of HNB by ANSYS. (**a**) Bending force simulation for closing HNBs; (**b**) voltage as a function of HNB deflection for different gaps (10, 15, and 20 μm)

Additionally, we designed a Helmholtz coil to generate the required external magnetic force to assemble the HNB (Fig. 11.3). For use inside the SEM chamber, we should consider the working distance of an electron beam and sample stage. Rings 21 mm in diameter were used to wind coils, and these coils were grounded onto the sample stage to prevent charging from the electron beam.

A single SEM sample holder was located between two coils. Samples were placed onto the sample holder between two coils. With this coil configuration, we measured a 2-mT magnetic field at 2.3 V, 0.254 A, which was required to deflect the magnetic pads on both ends of the HNB. A sample holder was also coiled so as to have a vertical axis magnetic field that achieved 1.3 mT at 2 V, 0.554 A. This coil was mounted onto a piezoactuated rotating nanostage, as shown in Fig. 11.3. Two nanomanipulators were installed through the coils to work over the sample chip. We experienced SEM imaging distortion over 5.5 V, which caused heating and melted the plastic part of coils.

11.3.3 External Field-Assisted Assembly

Finite-element-method (FEM) simulation was used to estimate the applied force on the HNBs in the experiments. The dimensions of the HNBs used in the simulation were the same as in the experiments, as summarized in Table 11.3. The simulation

Table 11.3 Specifications of HNBs used for simulation

Length [μm]	36.3
Radius [μm]	1.05
Pitch [μm]	6.6
Width [μm]	3
♯ of turns	5.5
Stiffness [N/m]	0.0001
Force [pN]	0.001–1

result was previously validated with experimental results for similar structures [13]. Simulation was carried out in an linear elastic range (small displacements). The values of the material properties in the model were taken from [13] with the rule of mixture applied for the InGaAs layer. Both ends of the helix were constrained from rotation around all three axes. Moreover, on one end of the helix was constrained from all translational movements, and on the other end it was constrained from translational movement perpendicular to the axis. On this end, a force in the axial (X-axis) or bending (Y-axis) direction was applied to compute the displacement.

In Fig. 11.4, a plot of the displacement along the bending direction is shown. From the simulation, the bending stiffness of the structure was determined to be 0.0001 N/m, as summarized in Table 11.3. The first thing to do was the preparation of the sample and the installation of the manipulators. In fact, if the pipette were touched with bare hands, without protection, the electrostatic discharge (ESD) could have broken the thin part of the pipette. For this reason, a bracelet and special gloves were used to ground it during the installation.

In Fig. 11.5, we see that the probe is in contact with and forms an electric circuit with the suspended HNB. The HNB plays the role of a switch. In Fig. 11.5c, the circuit is closed, whereas in Fig. 11.5d it is open. The pipette had to be as close as possible to the HNB until it touched the HNB (the circuit was closed). At this point, an SEM image was grabbed for the initial state. Then the pipette was moved from its present position on the y-axis until the contact between the pipette and the HNB was broken and another new image was grabbed. In the end two images (Fig. 11.5) were compared to determine the extent to which the HNB was deflected (Δd). This procedure was repeated with different voltages. When all the results were analyzed (voltage or current versus Δd), we finally obtained the curve shown in Fig. 11.6a. It shows the linear relation between the voltage and the deflection, except for the drop at 8 V, which was caused by an unequal contact configuration.

The next experiment to be discussed is similar to previous tests. This time, however, we used Helmholtz coils to generate a uniform EM field (Fig. 11.5). Between the coils, a sample with HNBs was mounted. The experiment consisted in moving the pipette until it was in contact with the HNB. This was the initial state; then the pipette was moved from this position until the pipette–HNB contact was released. As was described in the previous experiment, images at each time were grabbed for the deflection measurement.

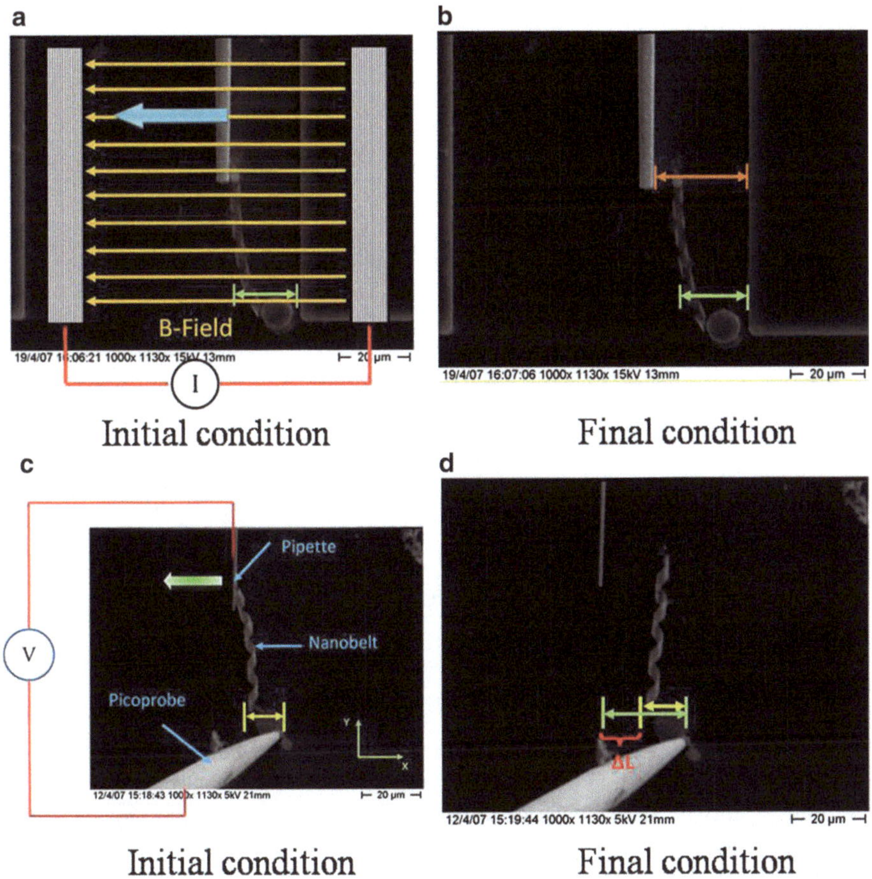

Fig. 11.5 Experiment: deflection of HNB by EM force (a), (b), deflection of HNB by ES force (c), (d)

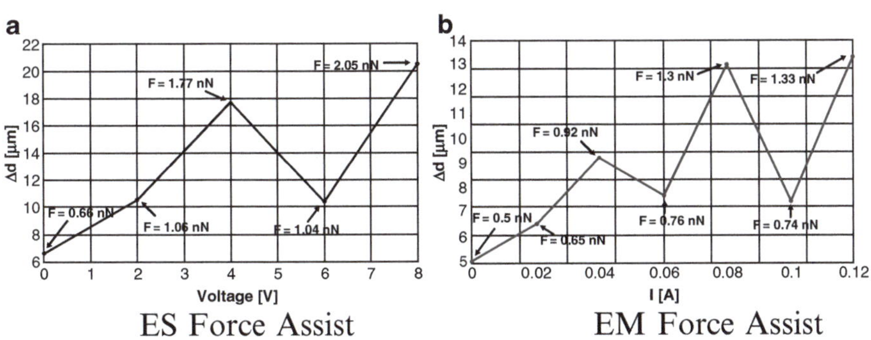

Fig. 11.6 Experiments with ES force (a) and EM force assistance (b): deflection [μm]

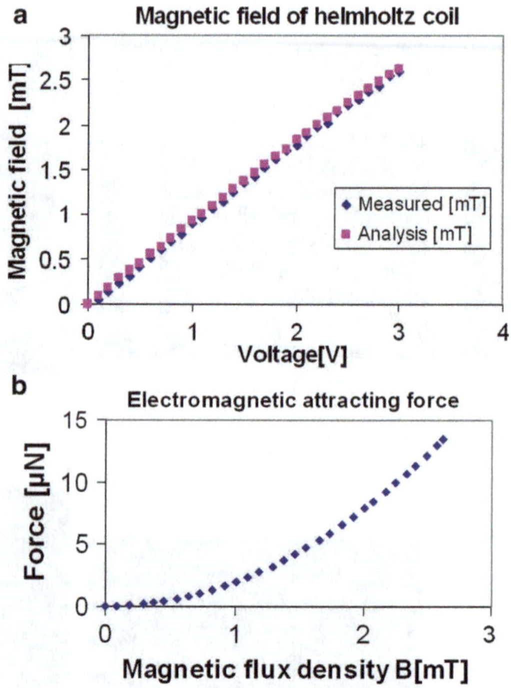

Fig. 11.7 Electromechanical characterization of magnetic HNB. (**a**) magnetic field measurement of Helmholtz coil. (**b**) Magnetic attracting force between probe and Ni pad

Figure 11.6b plots deflection versus the current flowing through the Helmholtz coil. It should be noted that there is a limit on the current, and a current that exceeds the limit results in SEM image distortion. In fact, the curve of the plot increases linearly. It was improved by several attempts with an ESD. A few problems remain with respect to decoupling the EM force itself from ES or van der Waals forces. However, the current result is sufficient to show that the EM field contributed the magnetization of the metal pads of the HNBs.

Given the coil setup, we measured a 2-mT magnetic field at 2.3 V, 0.254 A. Then the resistance was calculated using Ohm's law with $R = 2.3[\mathrm{V}]/0.254\,[\mathrm{A}] = 9.055\ [\Omega]$. In Fig. 11.6b, we use the current $I = 0.12\,[\mathrm{A}]$ to measure the voltage: $V = R/I = 1.086\,[\mathrm{V}]$. For a voltage of 1.086 [V] we obtained a B-field of 0.9 [mT] from Fig. 11.7a. We obtained an attracting force between the probe and Ni pad of 0.2 [μN] from a magnetic field of 0.9 [mT]. Then we were able to compare the estimated force with the experimental result in Fig. 11.6b. It showed a force of 1.33 [nN] for a current of $I = 0.12\,[\mathrm{A}]$. This very large difference can be explained by the fact that the analytical result was an ideal case with surface-to-surface contact between metal pad and pipette. However, as shown in Fig. 11.5, we could only on the side of the pad, which reduced the adhesive force. We also needed to consider that the HNB was fixed on one side; this means that the torsion force played a bigger role.

$$q_{\text{elec}} = \frac{\pi\varepsilon_0 V^2}{R\sqrt{\frac{r(r+2R)}{R^2}}\log^2(1+\frac{r}{R}+\sqrt{r(r+2R)}R^2)} \tag{11.1}$$

In theory, we have explained the equation of the ES force (11.1), and along with that we have conducted a few simulations to understand the behavior of the voltage as a function of the deflection. A MATLAB script to calculate the force with different HNBs was prepared, which was useful for the iterative simulations. We used different gaps between HNBs. The gap distances were set at 10, 15, and 20 μm. It should be noted that the pull-in voltage or the voltage necessary to collapse HNB to the electrode was found in the middle of the gap. This means that two HNBs were attached together at this position. In (11.1), we must insert the gap (r), the voltage (V), and the radius of the HNB turn (R_{ext}) (Table 11.2).

The calculated force using the MATLAB script was used in another MATLAB script to create a HNB model for simulation in ANSYS. We had to change step by step the force data in the file and compile and start in ANSYS the simulation in accordance with the determined deflection. Finally, we were able to obtain the relation between the voltage and the deflection of the HNBs. The results of this simulation are shown in Fig. 11.4b. The pull-in voltage of 27 V was obtained in the case of a 10-μm gap. We should not consider the result with a negative value in the graph because the two HNBs, when the distance 0 μm was reached, were attached together, so the HNBs were not able to exceed this distance. We had these negative data from the ANSYS simulation because we had used a range of voltages (for the force) without considering the limit. In the second case (15-μm gap), the voltage was 40 V, and in the third (20-μm gap) it was 54 V. An entire assembly procedure inside the SEM with the assistance of an external field was conducted and is introduced at the end of this section. A piezoresistive HNB force-sensing probe was assembled using the proposed method. It was conducted by serial nanorobotic assembly with an external electrostatic and electromagnetic force assist. The force-sensing probe showed piezoresistivity by deflection and was calibrated with an as-calibrated atomic force microscope cantilever.

The HNB force-sensing probe was assembled using an external-force-assisted nanorobotic assembly. Both the ES and EM forces were characterized quantitatively to show their contribution to the whole assembly process. The ES force is a relatively stronger force than the EM force in a SEM environment constraint. However, the hybrid approach of using both fields might be useful for a variety of future assembly tasks that will requires a certain amount of assembly force, such as, for example, soldering. The work is expected to be applied to real assembly tasks and steps toward future autonomous nanorobotic manufacturing. Table 11.4 summarizes the comparison of assembly steps in coarse and fine motions. The field-assist assembly reduces the assembly steps and improves the success rate and completion time.

Table 11.4 Summary of assembly process comparison: nanorobotic manipulation only and field assist

	NR only	Field assist
Coarse steps	9	6
Fine steps	9	4
Success rate	Rarely successful	Almost successful
Completion time	No limit	~10 min

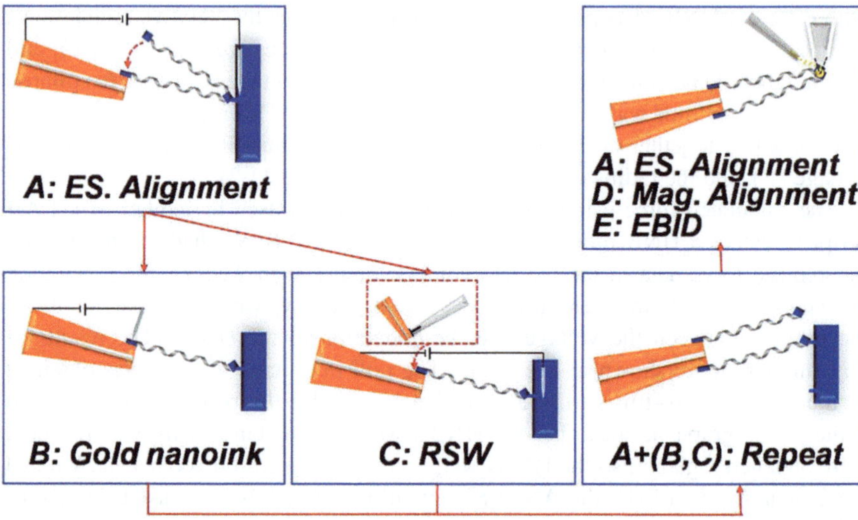

Fig. 11.8 Basic sensor assembly and calibration process sequence: (*A*) fabricated HNBs with metal connectors are aligned using electrostatic force on the two independently fabricated electrodes; (*B* and *C*) gold nanoink or RSW is used to solder the aligned HNBs for electrical measurement; (*D*) electromagnetic force is used to assist the assembly; (*E*) EBID is used to solder the aligned HNBs [[34] (©AIP 2012), reprinted with permission]

11.3.4 Force Sensor Assembly

Figures 11.8 and 11.9 depict the fabrication process. After the previous experiments, to test if the resistance spot welding (RSW) is an optimal choice to fix the HNB over the pipette, we concluded that this was the best way to create our final sensor prototype. In other words, RSW is a good choice because the generated contact is strong but is also a good conductor. After these observations, we decided to create a sensor using RSW. The procedure is similar to the previous one (assembly with glue and nanoink). This means that a pipette with double conductive layers, two picoprobes, chips with HNBs, and the necessary welding equipment was required. To create the sensor, first, HNBs were attached over the surface of the pipette. Figure 11.9 shows a coil attached over the pipette using RSW. As shown in Fig. 11.9, we broke the contact between the surface of the chip and the pad. When we were done with this side, we began with the other. The process was the same, and in the

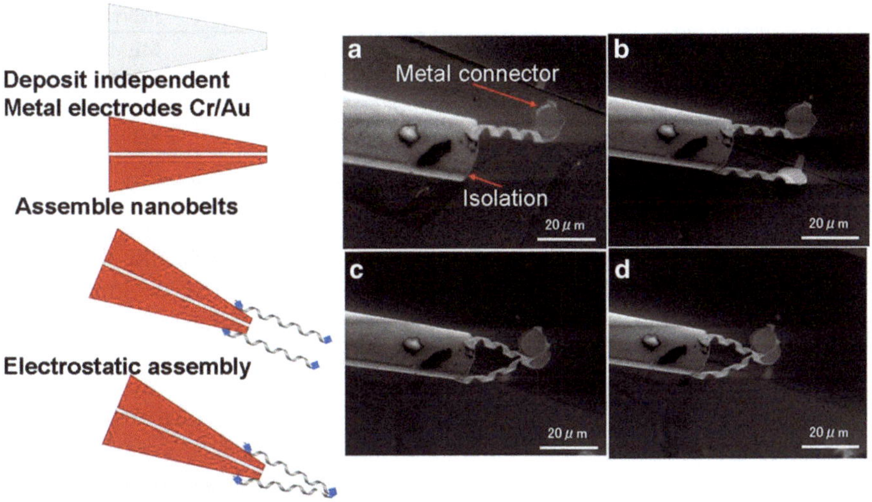

Fig. 11.9 Assembly procedure of HNBs on electrodes with pipette: (**a**) a side of an HNB is selected; (**b**) second HNB connection; (**c**) assembly by electrostatic force; (**d**) assembled device is moved [[34] (©AIP 2012), reprinted with permission]

end we obtained a device like that in the Fig. 11.9. To ensure contact, we decided to perform an extra process, electron-beam-induced deposition (EBID) (Fig. 11.9).

11.4 Characterizations

11.4.1 Giant Piezoresistivity of InGaAs/GaAs HNBs

For the electromechanical characterization experiments, two nanomanipulators (Kleindiek, MM3A), each with two metal probes (Picoprobe, T-4-10-1 mm) with a tip radius of 100 nm attached, were installed inside an SEM (Zeiss, DSM 962). The experimental procedure was explained in [35]. One manipulator was used to break and pick up a HNB on one side. For this purpose the HNBs were fabricated with a small length between the support and the first metal pad. The other manipulator was used to make contact with the other side. To achieve good electrical contacts on both sides of the HNBs, EBID with $W(CO)6$ precursor was used. In this way, a voltage could be applied to both sides of the HNBs and the current could be measured with a low-current electrometer (Keithley 6517A). After a HNB was attached as described, a tensile force was applied to it by moving one probe away from the other in the axial direction. Continuous frames of images were taken to analyze the deformation, and $I-V$ curves were recorded for the different positions. The characterization was carried out for three different HNBs. It was verified from

Table 11.5 Piezoresistivity and fabrication methods [21,26,35]

	Piezo. coef. $\pi_l^\sigma [10^{-10}\,\mathrm{Pa}^{-1}]$	Fabrication
Si bulk	1.7–9.4	MEMS
Bn-Si	4	MEMS
SiNW	3.5–355	Self-assembly
CNT	400	Self-assembly
HNB	996–3,560	Robotic assembly

the images that the boundary conditions did not change significantly during the experiments. The SEM images were analyzed to extract the HNB deformation for a certain $I-V$ measurement. This negative piezoresistivity behavior was observed in many different experiments and used as the force transduction of the proposed force sensors. The piezoresistivity of the structures can further be increased if Al is incorporated in the bilayer. Further details on the piezoresistive HNBs can be found in [35].

The material property of piezoresistivity in several piezoresistors is compared. The piezoresistance coefficients (π_l^σ) and fabrication methods of several piezoresistors are summarized in Table 11.5. The piezoresistance coefficients of HNBs were measured from our work (35). And Bn Si show that $|\pi_l^\sigma|$ is less than 10 $[10^{-10}\,\mathrm{Pa}^{-1}]$ [26]. But their fabrication was controlled with MEMS-compatible processes. It was recently reported that very large piezore- sistivities in SiNW and CNTs showed that a piezoresistance coefficient of $|\pi_l^\sigma|$ is 3.5–400 $[10^{-10}\,\mathrm{Pa}^{-1}]$ [21]. The high response was explained by the size effect. From this work, HNBs were found to be much higher ($|\pi_l^\sigma|$ was 996–3,560 $[10^{-10}\,\mathrm{Pa}^{-1}]$ [35]) than other piezoresistors. They are considered to be 249–890 times higher than Si piezoresistors. Therefore, HNBs are promising piezoresistors that will prove useful in high-resolution force sensors.

11.4.2 Force Transduction of Assembled HNB Force Sensor

The second experiment was conducted to characterize the assembled HNB force sensor. We aimed to measure the change in resistance when a force was applied and we wanted to find the parameters of the HNB for the force calibration. As we characterized the piezoresistive HNB, we expected to observe a change in resistance when a force was applied to it. Before we started this force transduction experiment, we needed to calibrate an AFM cantilever. For the mechanical characterization experiments, a nanomanipulator (Kleindiek, MM3A) and an AFM cantilever (Mikromasch, CSC38/Al BS, nominal stiffness 0.03 N/m) were installed inside the SEM (Zeiss, DSM 962). The AFM cantilever was calibrated using the method shown by Sader et al. [36], and the stiffness was found to be 0.132 N/m. Table 11.6 shows the proprieties of the cantilever and data items such as stiffness after calibration. The calibration procedure is shown in Fig. 11.10. The sensor pressed against the top of the cantilever, and so we measured the deflection of the

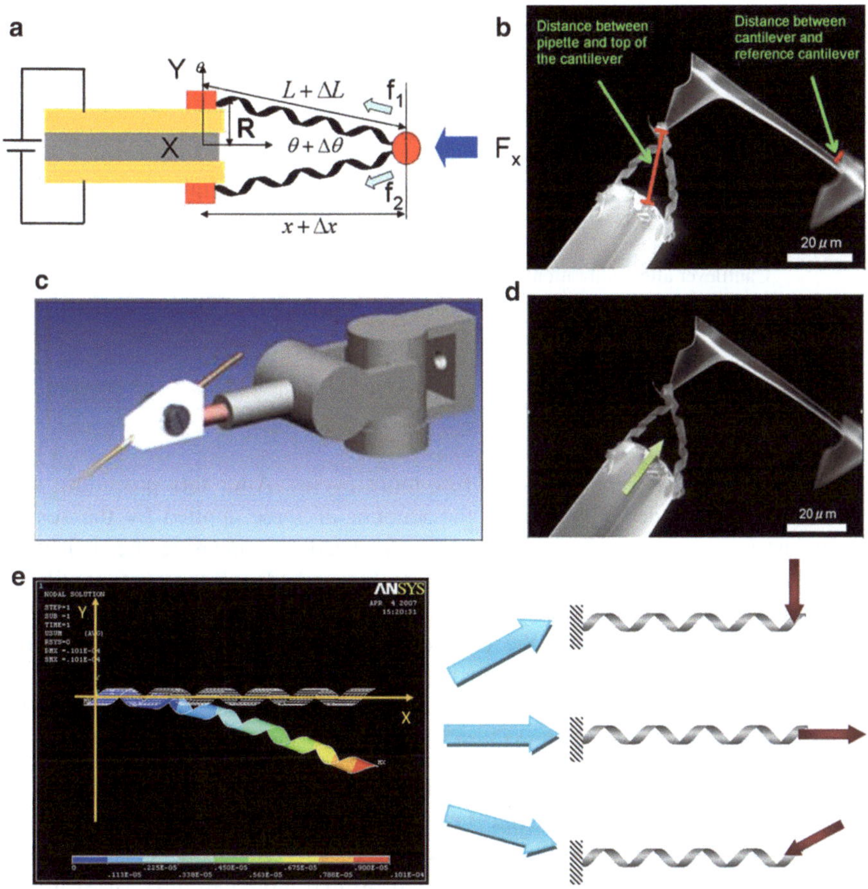

Fig. 11.10 Force measurement setup: (**a**) sensor geometry and force diagram; (**b**) SEM photo of calibration with AFM cantilever; (**c**) sensor mounted on manipulator; (**d**) SEM photo calibration after application of compressive force; (**e**) ANSYS simulation of HNB [[34] (©AIP 2012), reprinted with permission]

cantilever and the compression of the sensor. During these measurements, the data of the resistance change were saved. The process was as follows.

1. We took an initial picture of the experiment, made a voltage sweep from 0 to 1 [V], for a step of 0.1 [V], and recorded the measured current *I* [A] at each step. It was also important to measure the distance between the pipette and the top of the cantilever (for the compression of the sensor) and the distance between the cantilever used and the reference cantilever (see Fig. 11.10 for the deflection of the cantilever).
2. The pipette was moved forward. The sensor began to compress itself, and the cantilever started to deflect. As before, we swept the voltage from 0 to 1 [V] and measured the current *I* [A], and so we needed new photo frames for the calculation of the distances.

Table 11.6 Specification of cantilevers

Cantilever properties:	
Type	B
Length	350 ± 5 [μm]
Width	±3 [μm]
Thickness	min = 0.7, typical = 1, max = 1.3 [μm]
Resonant frequency	min = 7, typical = 10, max = 14 [kHz]
Force constant	min = 0.01, typical = 0.03, max = 0.08 [N/m]
Cantilever after calibration:	
Stiffness	$K = 0.132$ [N/m]
Q-factor	$Q = 59.9$
Fluid density	1.18 [kg/m^3]
Fluid viscosity	1.86 $\times 10^{-5}$ [kg·m/s]

3. The last step was repeated for a few more cycles. After this procedure was finished, we needed to calculate the amount of force applied by the sensor against the cantilever. To calculate this, we needed to measure the deflection of the cantilever between the cantilever itself (touched) and the other cantilever (reference, Fig. 11.10). To calculate the force, we used Hooke's law ($F = k \cdot x$), where k is the stiffness of the cantilever. Now we knew the force for each compression of the sensor. The next step was to determine whether there was a relationship between the force and the change in resistance of the sensor. The resistance was calculated using Ohm's law at a constant voltage (1 V).

Measurement results are summarized in Figs. 11.11 and 11.12. We applied 0–80 nN in five different steps (trial 1) and applied 0–154 nN in a wider force range (trial 2) in Fig. 11.11. Better curves were observed in the second trial. In a smaller force range, deflection should be measured more carefully. SEM image frames were grabbed by the image acquisition software (DISS-5, Point Electronic GmbH) and analyzed using image processing software (DIPS, Point Electronic) to examine the cantilever deflections. The image analysis resolution of SEM was 100 nm, which could have minimum detectable forces of cantilever and HNB force sensing resolution of 13.2 nN and 2.42 nN respectively. The resistance changed at the second and third points of trial 1 while the same force was measured but the same 13.2 nN was recorded as in Fig. 11.12. Both trials show quite good repeatability. To further estimate the piezoresistivity behavior of HNBs, we decided to analyze HNB deflections that had much lower stiffness than that of the AFM cantilever. However, we needed to make the model estimate the force that could be eligible for experimental data. In the FEM simulation (axial direction) using ANSYS, we estimated a single HNB's axial stiffness as 0.0121 N/m. As the sensor had two angled HNBs connected at the end, a simple model to estimate stiffness of parallel aligned HNBs caused considerable error. In particular, the stiffness of the sensor varied by the angle change between the two HNBs when a force was applied. Finally, we wanted to determine the relation between the axial force on the sensor and the single HNB in the arm. Based on the geometric information from Fig. 11.10,

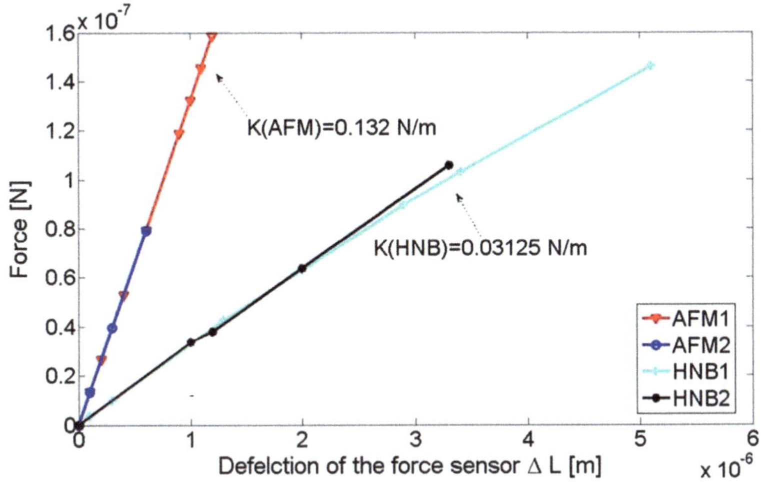

Fig. 11.11 Stiffness calibration of HNB force sensor using as-calibrated AFM cantilever [[34] (©AIP 2012), reprinted with permission]

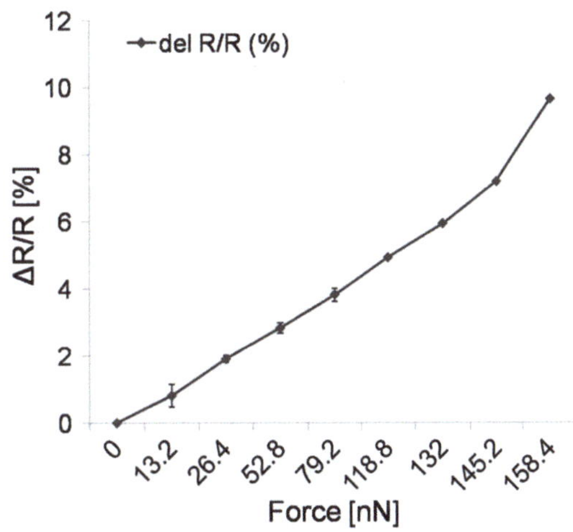

Fig. 11.12 Electromechanical measurements on a piezoresistive HNB force sensor: change in percentage of resistance as a function of axial force [nN] [[34] (©AIP 2012), reprinted with permission]

a trigonometric method was used to derive (11.2), which describes the mentioned relation. A detailed derivation is not included here:

$$K = \frac{2k_1 \Delta L}{\Delta X} \cos\frac{\theta}{2}. \tag{11.2}$$

where K is the stiffness of the sensor, k_1 is the stiffness of a single HNB, ΔX is the deflection of a single HNB arm, ΔL is the axial deflection of the sensor, and θ is the angle between two HNB arms. As the stiffness varied while the sensor underwent deflection, the model simulation of a sensor with constant stiffness resulted in a nonuniform error during deflection. Therefore applying (11.2) to estimate the stiffness at each deflection yielded a better fitted calibration curve of the sensor model and real experiment, as shown in Fig. 11.11. It shows the stiffness difference between the AFM cantilever and the sensor, which varied by the deflection. Figure 11.11 shows the stiffness calibration of the HNB force sensor using an as-calibrated AFM cantilever. From the measurement, the stiffness of the HNB force sensor varied slightly by the deflection but was almost constant in the measured region. The stiffness of the measured HNB force sensor was approximately 0.03125 [N/m] compared with the value (0.132 [N/m]) of the as-calibrated AFM cantilever. Furthermore, the optimized design of the HNB force sensor could be predicted. From (11.2), the higher range of the force sensor could be designed by decreasing the angle (θ) and increasing the given design of individual HNBs. In addition to the assembly design parameters, the individual HNBs could also be tuned their stiffness. When the two HNBs are assembled serially (θ is 180°), the highest resolution but minimum range force sensing is possible. For a wider range of force sensing compensating for resolution, two HNBs should be assembled in parallel (θ is 180°).

Figure 11.12 shows the response of the piezoresistive HNB force sensor. The stiffness of HNB force sensor is close to the stiffness of the cantilever. It should be noted that these results are considered for an ideal simulation, where the device is symmetric and the HNBs have the same properties. The piezoresistance coefficient (π_l^{σ}) of the assembled sensor was calculated to be $515 \cdot 10^{-10}$ [Pa^{-1}] from the measurement and from (11.3):

$$\pi_l^{\sigma} = \frac{1}{X}\frac{\Delta\sigma}{\sigma_0}, \tag{11.3}$$

where σ_0 is the conductivity under zero stress and X is the stress applied. This is close to the individual HNB that was measured in a range of 996–3,560$\cdot 10^{-10}$ [Pa^{-1}]. It should be noted that the stiffness of the assembled structure was almost doubled; thus the piezoresistance coefficient was decreased by half. It could explain the fact that the design of triangular shape of the HNB force sensor did not sacrifice the force sensitivity compared to the single HNB. The stiffness of the HNB force sensors can also be tuned by the assembly geometry using constant stiffness of individual HNBs. It allows for the control of the force-sensing range and resolution by editing both the assembly and individual HNBs' design parameters. The calibration experiment and simulation based on the assembled force sensor model shown here were not ideal for measuring the sensitivity of force sensing. It should be considered that the calibration was performed in a full force-sensing range to verify how the piezoresistivity of HNBs contributed to measuring the force in the range. To measure to more accurately measure the sensing resolution, we should apply smaller force steps (displacement) under magnified SEM image. Another consideration is on the shape of the sensor (Fig. 11.10). The minimum detectable force resolution using an individual HNB was estimated to be 0.91 nN by considering the standard deviation of the measured noise (0.03 nA), which is within the range of the high-resistance electrometer (Keithley 6517 A, measurable up to 1 fA).

11.5 Conclusions

Large bandwidth force sensing based on thin-film three-dimensional nanostructures could be a useful tool for various micro-/nanomanipulation applications. For example, this chapter described a manufacturing challenge and the proposed assembly process of thin-film piezoresistive HNB force sensors assembled by nanorobotic manipulations. The proposed process consists of assembly, characterizations, and calibrations done in an in situ manner inside a SEM. The assembled HNB force sensors showed a large displacement range, high-resolution force sensing, self-sensing, and low weight as a result of the unusually high piezoresistivity, low stiffness, and high-strain capability of HNBs. There are open applications from electronics (electric contact probing and testing for microelectronic circuits), biology (wide range mechanical characterizations of tissues, fibers), and MEMS/NEMS (mechanical characterizations of CNTs, NWs). Moreover, this alternative technology to the conventional micro-/nanomanufacturing process holds great potential for MEMS/NEMS.

Acknowledgements This work was sponsored by the Japan Society for the Promotion of Science (JSPS) research fellow program and the Ministry of Education, Science and Culture, Japan. I would like to thank Bradley J. Nelson, Dominik J. Bell, Lixin Dong, and Li Zhang at IRIS ETHZ for my research visit and valuable discussions.

References

1. R. Smith, D.R. Sparks, D. Riley, D. Najafi, A MEMS-based coriolis mass flow sensor for industrial applications. IEEE Trans. Ind. Electron. **56**(4), 1066–1071 (2009)
2. R. Saeidpourazar, N. Jalili, Microcantilever-based force tracking with applications to high-resolution imaging and nanomanipulation. IEEE Trans. Ind. Electron. **55**(11), 3935–3943 (2008)
3. H.J. Dai, J.H. Hafner, A.G. Rinzler, D.T. Colbert, R.E. Smalley, Nanotubes as nanoprobes in scanning probe microscopy. Nature **384**(6605), 147–150 (1996)
4. J.H. Hafner, C.L. Cheung, T.H. Oosterkamp, C.M. Lieber, High-yield assembly of individual single-walled carbon nanotube tips for scanning probe microscopies. J. Phys. Chem. B **105**(4), 743–746 (2001)
5. J. Li, A.M. Cassell, H.J. Dai, Carbon nanotubes as AFM tips: measuring DNA molecules at the liquid/solid interface. Surf. Int. Anal. **28**(1), 8–11 (1999)
6. S. Iijima, Helical microtubules of graphitic carbon. Nature **354**, 56–58 (1991)
7. Y. Cui, C.M. Lieber, Functional nanoscale electronic devices assembled using silicon nanowire building blocks. Science **291**(5505), 851–853 (2001)
8. S. Motojima, M. Kawaguchi, K. Nozaki, H. Iwanaga, Growth of regularly coiled carbon filaments by Ni catalyzed pyrolysis of acetylene, and their morphology and extension characteristics. Appl. Phys. Lett. **56**(4), 321–323 (1990)
9. X.B. Zhang, X.F. Zhang, D. Bernaerts, G.T. Vantendeloo, S. Amelinckx, J. Vanlanduyt, V. Ivanov, J.B. Nagy, P. Lambin, A.A. Lucas, The texture of catalytically grown coil-shaped carbon nanotubules. Europhys. Lett. **27**(2), 141–146 (1994)
10. X.Y. Kong, Z.L. Wang, Spontaneous polarization-induced nanohelixes, nanosprings, and nanorings of piezoelectric nanobelts. Nano Lett. **3**(12), 1625–1631 (2003)

11. P.X. Gao, Y. Ding, W.J. Mai, W.L. Hughes, C.S. Lao, Z.L. Wang, Conversion of zinc oxide nanobelts into superlattice-structured nanohelices. Science **309**(5741), 1700–1704 (2005)
12. D.J. Bell, Y. Sun, L. Zhang, L.X. Dong, B.J. Nelson, D. Grutzmacher, Three-dimensional nanosprings for electromechanical sensors. Sens. Act. A-Phys. **130**, 54–61 (2006)
13. D.J. Bell, L.X. Dong, B.J. Nelson, M. Golling, L. Zhang, D. Grutzmacher, Fabrication and characterization of three-dimensional InGaAs/GaAs nanosprings. Nano Lett. **6**(4), 725–729 (2006)
14. G. Yang, J.A. Gaines, B.J. Nelson, Optomechatronic design of microassembly systems for manufacturing hybrid microsystems. IEEE Trans. Ind. Electron. **52**(4), 1013–1023 (2005)
15. E. Enikov, L. Minkov, S. Clark, Microassembly experiments with transparent electrostatic gripper under optical and vision-based control. IEEE Trans. Ind. Electron. **52**(4), 1005–1012 (2005)
16. B. Borovic, A. Liu, D. Popa, H. Cai, F. Lewis, Light-intensity-feedback-waveform generator based on MEMS variable optical attenuator. IEEE Trans. Ind. Electron. **55**(1), 417–426 (2008)
17. J.J. Abbott, Z. Nagy, F. Beyeler, B.J. Nelson, Robotics in the small, part I: microrobotics. IEEE Rob. Auto. Mag. **14**(2), 92–103 (2007)
18. L.X. Dong, B.J. Nelson, Robotics in the small, part II: nanorobotics. IEEE Rob. Auto. Mag. **14**(3), 111–121 (2007)
19. D. Golberg, P. Costa, O. Lourie, M. Mitome, C. Tang, C. Zhi, K. Kurashima, Y. Bando, Direct force measurements and kinking under elastic deformation of individual multiwalled boron nitride nanotubes. Nano Lett. **7**(7), 2146–2151 (2007)
20. L. Zhang, L. Dong, B.J. Nelson, Bending and buckling of rolled-up SiGe/Si microtubes using nanorobotic manipulation. Appl. Phys. Lett. **92**, 243102 (2008)
21. C. Bustamante, Z. Bryant, S.B. Smith, Ten years of tension: single-molecule DNA mechanics. Nature **421**, 423–427 (2003)
22. C. Bustamante, J.C. Macosko, G.J.L. Wuite, Grabbing the cat by the tail: manipulating molecules one by one. Nature Rev. Mol. Cell. Biol. **1**, 130–136 (2000)
23. A. Kishino, T. Yanagida, Force measurements by micromanipulation of a single actin filament by glass needles. Nature **334**, 74–76 (1988)
24. A. Ashkin, J.M. Dziedzic, Optical trapping and manipulation of viruses and bacteria. Science **235**(4975), 1517–1520 (1987)
25. S.B. Smith, L. Finzi, C. Bustamante, Direct mechanical measurements of the elasticity of single DNA-molecules by using magnetic beads. Science **258**(5085), 1122–1126 (1992)
26. C. Stampfer, T. Helbling, D. Obergfell, B. Schoberle, M.K. Tripp, A. Jungen, S. Roth, V.M. Bright, C. Hierold, Fabrication of single-walled carbon-nanotube-based pressure sensors. Nano Lett. **6**(2), 233–237 (2006)
27. C. Stampfer, A. Jungen, R. Linderman, D. Obergfell, S. Roth, C. Hierold, Nano-electromechanical displacement sensing based on single-walled carbon nanotubes. Nano Lett. **6**, 1449–1453 (2006)
28. T.C. Duc, J.F. Creemer, P.M. Sarro, Lateral nano-Newton force-sensing piezoresistive cantilever for microparticle handling. J. Micromech. Microeng. **16**(6), S102–S106 (2006)
29. B.L. Pruitt, W.T. Park, T.W. Kenny, Measurement system for low force and small displacement contacts. J. Microelectromech. Syst. **13**(2), 220–229 (2004)
30. F. Beyeler, A. Neild, S. Oberti, D.J. Bell, Y. Sun, J. Dual, B.J. Nelson, Monolithically fabricated micro-gripper with integrated force sensor for manipulating micro-objects and biological cells aligned in an ultrasonic field. J. Microelectromech. Syst. **16**(1), 7–15 (2007)
31. L. X. Dong, L. Zhang, D.J. Bell, B.J. Nelson, D. Grutzmacher, Hybrid nanorobotic approaches for fabricating NEMS from 3D helical nanostructures. Proc. IEEE ICRA, Orlando, Florida, U.S.A., 1396–1401 (2006)
32. G. Hwang, C. Dockendorf, D.J. Bell, L.X. Dong, H. Hashimoto, D. Poulikakos, B.J. Nelson, 3-D InGaAs/GaAs helical nanobelts for optoelectronic devices. Int. J. Optomech. **2**, 88–103 (2008)
33. P. Kim, C.M. Lieber, Nanotube nanotweezers. Science **286**, 2148–2150 (1999)
34. G. Hwang, H. Hashimoto, Helical nanobelt force sensors. AIP Rev. Sci. Inst. **83**, 126102 (2012)

35. G. Hwang, H. Hashimoto, D.J. Bell, L.X. Dong, B.J. Nelson, S. Schon, Piezoresistive InGaAs/GaAs nanosprings with metal connectors. Nano Lett. **9**(2), 554–561 (2009)
36. J.E. Sader, J.W.M. Chon, P. Mulvaney, Calibration of rectangular atomic force microscope cantilevers. AIP Rev. Sci. Inst. **70**, 3967–3969 (1999)

Chapter 12
Human Sperm Tracking, Analysis, and Manipulation

Jun Liu, Clement Leung, Zhe Lu, and Yu Sun

Abstract Sperm analysis and manipulation play a significant role in biology research and reproductive medicine (assisted reproductive technologies). This chapter reviews computer vision-based sperm tracking methods, sperm analysis techniques, and automated sperm manipulation. Based on computer vision tracking of sperm head and sperm tail, sperm motility can be quantified by calculating the sperm's straight line velocity, curvilinear velocity, moving path linearity, and the sperm tail beating amplitude. Conventional computer-assisted sperm analysis (CASA) systems are capable of performing some of these tasks. Recent progress in this field provides additional, enhanced capabilities to biologists and clinical embryologists. This chapter also introduces recent progress in automating sperm manipulation procedures, including sperm immobilization, aspiration, and positioning inside a micropipette.

12.1 Introduction

A sperm is a male reproductive cell consisting of an ellipsoidal or spherical head, a short midpiece, and a thin motile tail. The sperm head contains the nucleus with genetic materials, surrounded anteriorly by a cap-like acrosome that contains digestive enzymes. These enzymes can break down the outer membrane of ovum during the fertilization process, allowing the haploid nucleus in the sperm cell to join with the haploid nucleus in the ovum. The midpiece of the sperm cell has a central filamentous core with many mitochondria spiralled around it, generating energy for the sperm's motion. The sperm tail executes the swing movement that

J. Liu • C. Leung • Z. Lu • Y. Sun (✉)
Department of Mechanical and Industrial Engineering, University of Toronto,
5 King's College Road, Toronto, ON, Canada M5S 3G8
e-mail: ljun@mie.utoronto.ca; clement.leung@utoronto.ca; zhe.lu@utoronto.ca;
sun@mie.utoronto.ca

M. Rakotondrabe (ed.), *Smart Materials-Based Actuators at the Micro/Nano-Scale: Characterization, Control, and Applications*, DOI 10.1007/978-1-4614-6684-0_12,
© Springer Science+Business Media New York 2013

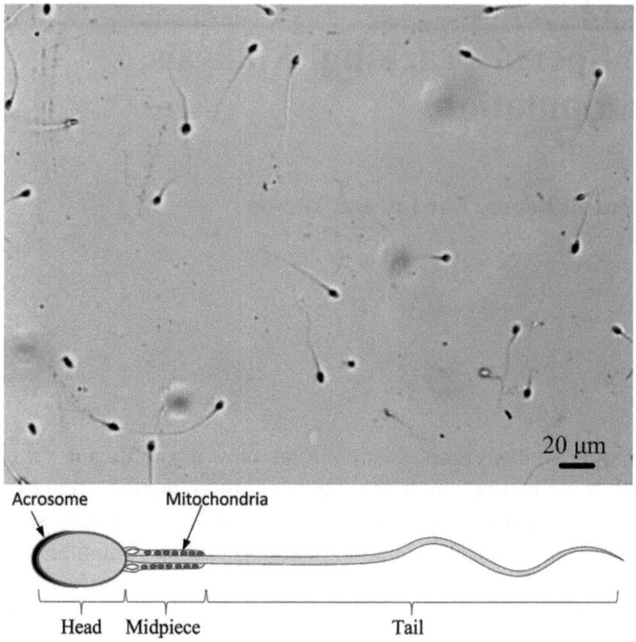

Fig. 12.1 Human sperm cells: (**a**) Image taken under 20× magnification. (**b**) A diagram of human sperm cell

propel the sperm to move. In humans, the average total length of sperm tail and the short midpiece is approximately 50 μm, and the sperm head dimensions are 4.4×2.8 μm [1]. Figure 12.1 is an image of sperms on a Petri dish taken under 20× magnification and a schematic diagram of a human sperm cell.

In natural conception, a healthy sperm overcomes the physiological and biological selection barriers, actively seeks out and fertilizes an egg. Sperm selection occurs naturally in this procedure. However, for couples having infertility issues, assisted reproduction technologies are required to address their reproductive needs. For example, in intracytoplasmic sperm injection (ICSI), an embryologist selects a single sperm cell and injects it directly into an oocyte (i.e., egg cell) to overcome issues such as male infertility [2]. These assisted reproduction technologies bypass the natural sperm selection barriers and demands the operator to select high-quality sperms. The criteria for sperm quality assessment provided by the World Health Organization are vitality, morphology, and motility [3]. A widely used method for sperm selection is motile sperm organelle morphology examination (MSOME) [4–7]. Sperm motility is also a commonly used criterion for sperm quality assessment. A motility grade is often used as a specified measure and classified into four grades:

Grade 1: Sperm with fast progressive movements
Grade 2: Sperm with slow progressive movements
Grade 3: Sperm with slow non-progressive movements (i.e., with curved motion)

Grade 4: Sperm are immotile and fail to move at all

Besides sperm assessment based on morphology and motility, another method for selecting a healthy sperm is based on the analysis of sperm DNA integrity. Some of the DNA analysis methods assess sperm DNA quality directly, such as the TUNEL assay and the sperm chromatin structure assay. Some other systems indirectly measure sperm DNA quality. For instance, Huszar's group recently proposed a hyaluronic acid (HA) assay [8]. Among a consecutive series of studies on HA-based sperm assay, Huszar's group indicated that the HA assay permits the selection of healthy sperm with no DNA damage [9, 10]. In their studies, sperms that bind to HA microdots are proven to have a higher level of DNA integrity compared to those unbound sperm. In clinical HA-based sperm selection, a number of healthy sperm's head bind to the HA microdot and lose their progressive movement despite vigorous tail beating. In this case, the sperm tail beating movement becomes the only indicator to differentiate the HA bound sperm from each other.

The past few decades have witnessed the development of computer-assisted sperm analysis (CASA) methods for measuring both sperm morphology and motility [11]. CASA utilizes an automated system to digitize successive images of sperm, process, and analyze the information and provide the accurate and objective value for individual sperm cell. Since 1970s, many algorithms have been developed to track sperm trajectories, measure sperm velocities, and analyze sperm morphology. Shi et al. reported a robust single-sperm tracking algorithm based on a four-class thresholding method to extract a single sperm in a region of interest (ROI) [12]. The nearest neighbor method is complemented with a speed-check feature to aid tracking in the presence of additional sperm or other particles. In another study, Nafisi et al. demonstrated a template matching algorithm for sperm tracking. The algorithm is insensitive to image acquisition conditions [13]. Existing algorithms for sperm tracking are largely limited to sperm head tracking. The small size (≤ 1 μm in thickness) and low contrast of sperm tails under optical microscopy make sperm tail tracking challenging. In a recent study [14], a maximum intensity region algorithm was developed for sperm tail detection and tracking.

In this chapter, we first discuss methods for tracking sperm head and tail via real-time image processing. We then discuss sperm analysis on the basis of sperm motility, morphology, and DNA quality. Recent developments in automated sperm manipulation systems will then be introduced. We finally discuss future research directions.

12.2 Sperm Tracking

12.2.1 Sperm Head Tracking

Compared with sperm tail, sperm head has clearer contrast under microscopy. Therefore, most CASA systems were designed for tracking sperm head only. To

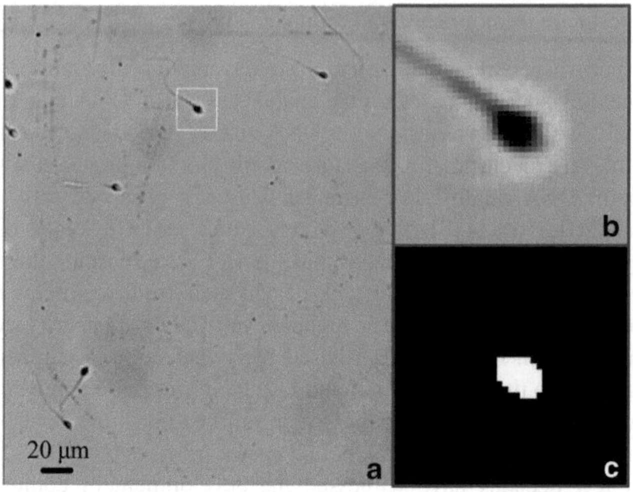

Fig. 12.2 Single sperm head tracking: (**a**) Sperm in the original image. (**b**) The sperm head ROI image is extracted. (**c**) The sperm head is found through adaptive thresholding

successfully track sperm heads, the algorithms should take into account sperms' three-dimensional movements, which can cause targets to be out of focus. A target sperm can also be occluded by other sperms or by foreign particles present in the Petri dish.

Sperm head tracking often begins with filtering noise and image enhancements. Filters (e.g., median filters and Gaussian filters) are used in most of the sperm tracking systems. For image enhancements, Nafisi et al. proposed a two-step enhancement method for noise reduction [13]. The first step is to remove completely stationary objects via frame subtraction. The second step is to remove or reduce the effect of those objects with vague boundaries (due to out-of-focus) by using wavelet transform.

After noise reduction, sperm head detection is performed by using template matching, fitting ellipse, and thresholding approach. Template matching is robust to different imaging modes (e.g., bright field, phase contrast, and DIC). However, the high computing cost could make this method unsuitable for real-time sperm tracking tasks. Besides for sperm head tracking, fitting ellipse can also be used for sperm tail tracking. Through analyzing the shape of sperm head, this method can provide the direction of sperm tail for sperm tail detection. It was reported that sperm tail can be found along the major axis of the ellipse [15]. However, this method is less effective for detecting sperm heads that do not have regular elliptical shapes.

The most commonly used method for sperm detection is the thresholding approach [14, 16, 17]. This approach applies thresholding inside a ROI. The ROI for sperm head can be initiated by a human operator who selects a desired sperm head to track via computer mouse click on the sperm head (Fig. 12.2). The ROI image can be binarized by applying Otsu's adaptive thresholding algorithm (Fig. 12.2c). Within the binarized ROI image, the contour of the sperm head is computed. The sperm head position is obtained by calculating the moments of the contour. The ROI is then updated to be another ROI centered at the sperm head's centroid. For subsequent frames, a similar process is performed to track the sperm head position.

When a sperm is occluded by other sperms or particles, the sperm tracking algorithms should be able to differentiate between the sperm of interest and interfering sperms/particles that are present in the ROI. The swimming direction vector of the sperm of interest can be used as a unique identifier to discriminate the sperm of interest from other sperms/particles. In the situation where only one sperm is present in the ROI, the sperm's current direction vector $D(i)$ in the current frame, represented by (12.1), is found by subtracting the sperm centroid position in the previous frame $P(i-1)$ from the sperm centroid position in the current frame $P(i)$.

$$D(i) = P(i) - P(i-1) \tag{12.1}$$

When more than one sperm or object is present in the sperm head ROI (SHROI), (12.1) is extended to

$$D(i,s) = P(i,s) - P(i-1, s_{\text{prev}}) \tag{12.2}$$

where s represents each sperm in the SHROI and s_{prev} is the sperm of interest in the previous frame. The candidate sperm s that produces the minimum Euclidean distance value is considered the sperm of interest s_{soi}

$$s_{\text{soi}} = \min_{s \in [1,N]} ||D(i,s) - D(i-1, s_{\text{prev}})|| \tag{12.3}$$

where N is the total number of sperm and objects inside the SHROI. Specifically, the nearest neighbor approach is applied to determine the sperm of interest s_{soi} at frame i, with the knowledge of the sperm of interest s_{prev} at frame $i-1$. The s_{soi} is updated for every frame using this nearest neighbor computation.

12.2.2 Sperm Tail Tracking

Tracking sperm tail can also be useful, such as for the manipulation of sperm and for selecting sperms that are bound to HA dots in the HA assay. After the sperm head position is detected, a sperm tail region of interest (STROI) is extracted by using the sperm head position and the average direction vector of its movement, as shown in Fig. 12.3b. The average direction vector $\overline{D}$ is used instead of the direction vector $D(i)$ because the sperm may exhibit abrupt changes in movement direction between two consecutive frames. By averaging the direction vectors of the sperm across a number of frames (e.g., 30 frames), the effect of abrupt changes in the sperm moving direction between frames is mitigated and the extraction of STROI becomes more robust.

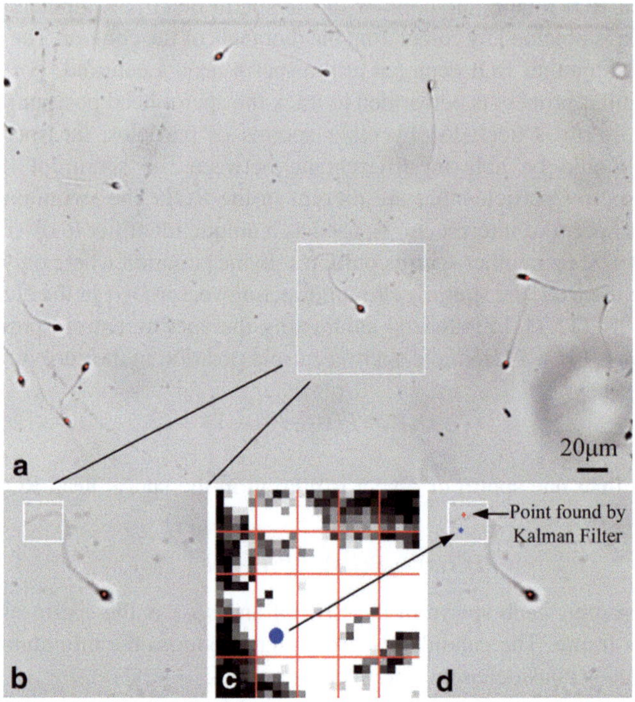

Fig. 12.3 Sperm tail tracking. (**a**) Sperm head position is found. (**b**) STROI (sperm tail region of interest) is determined. (**c**) 5×5 windows are scanned to locate the section with the highest intensity sum in the flicker image. The center point (*blue dot* in figure) of the section is considered the tail location. (**d**) Based on the *blue dot* position found in (**c**), Kalman filter is applied to improve the accuracy of located sperm tail position (©IEEE 2013), reprinted with permission

The STROI's center position in the i frame, $T(i)$, is determined by subtracting a scaled value of the direction vector from the sperm head's centroid

$$T(i) = P(i) - a \cdot \frac{\overline{D}}{\|\overline{D}\|} \tag{12.4}$$

where a is a scalar value determined by the human sperm length. Under the $20\times$ magnification, the average length of human sperms is approximately 90 pixels (i.e., $a = 90$). After the center position is found, a 25×25 ROI is taken as the STROI. The size of 25×25 provides a sufficient tail search area that takes into consideration a range of sperm tail length variations and sperm tail beating amplitudes.

After finding the STROI, the algorithm verifies that a tail is present in the STROI. The fundamental feature of flicker is extracted by taking the absolute difference between six consecutive inverted grayscale image frames.

$$f(i) = \sum_{k=0}^{5} |I(i-k) - I(i-k-1)| \tag{12.5}$$

where $f(i)$ is the flicker image extracted at frame i and I represents the grayscale images containing the sperm of interest in frame i to frame $i-5$. Each pixel in the flicker image is squared to enhance the pixel values of areas in which the tail is present. The sum of pixel value in the STROI of the $f(i)$ image is used as a measure to determine the presence of a sperm tail. If the pixel sum is above a specified threshold value, a tail is considered present. The threshold value was found experimentally by comparing the pixel sum values of STROI images in which a tail exists against cases where no tail exists. An example flicker image is shown in Fig. 12.3c. If the pixel sum is below a threshold value, no tail is found inside the STROI. This situation can occur when the sperm of interest moves out of focus, resulting in the disappearance of the sperm tail.

Once the sperm tail is determined to exist within the STROI, the maximum intensity region algorithm will locate a point on the sperm tail by using the flicker image. This flicker image approach overcomes the challenges that arise from the low-contrast image of the sperm tail in a single frame. The algorithm first finds the location of maximum intensity within the STROI of the flicker image. This is accomplished by evaluating the sum of the intensity values inside a 5×5 window at a spatial sampling interval of 5 pixels in both the x and the y coordinates of the STROI flicker image. The center position of the 5×5 window with the highest intensity is considered the tail location (i.e., a point on the sperm tail). In order to obtain an accurate sperm tail position, a Kalman filter can be applied to optimize the tracked results. Figure 12.3d shows the sperm tail tracking result and Kalman filter optimized result.

12.3 Sperm Analysis

12.3.1 Sperm Motility Analysis

Sperm motility is a basic criterion for assessing sperm quality. The most commonly used criteria for sperm motility analysis are sperm's curvilinear velocity (VCL), straight line velocity (VSL), and movement linearity (LIN). Additionally, sperm tail beating amplitude is also a motility value that reflects the sperm's locomotive behavior. These criteria can be calculated from the sperm tracking results.

Assume a sperm enters the field of view at frame i and swims out of the field of view at frame $i+N$. With the sperm position detected in each frame, the travel distance of the sperm between two consecutive frames can be determined from its direction vector, $D(i)$. The VCL, which is the average velocity of the sperm head along its actual curvilinear path, is

$$\mathrm{VCL} = \frac{1}{N}\sum_{k=0}^{N-1} D(i) \tag{12.6}$$

The VSL, which is the average velocity of the sperm head along the straight line between its first and last detected position, is

$$\mathrm{VSL} = \frac{P(i+N) - P(i)}{N} \tag{12.7}$$

where $P(i)$ represents the position of the sperm at frame i. The linearity (LIN) of the sperm's curvilinear path is

$$\mathrm{LIN} = \frac{\mathrm{VSL}}{\mathrm{VCL}} \tag{12.8}$$

where LIN is the linearity measure ($0 \leq \mathrm{LIN} \leq 1$). A higher LIN value means that the sperm's moving path is more linear. In sperm selection, healthy energetic sperms with progressive/linear movement are desired (vs. those traveling in circles for instance).

With the sperm tail's position, the sperm tail beating amplitude inside the STROI is computed. The relative position inside the STROI in frame i is denoted by $PT(i)$. The sperm tail beating amplitude is

$$A = \frac{1}{N}\sum_{i=1}^{N} \|PT(i) - \overline{PT}\| \tag{12.9}$$

where $\overline{PT}$ is the sperm tail's average position inside the STROI and N is the number of frames until when the sperm tail is successfully detected. Generally speaking, a higher sperm tail's beating amplitude will result in a faster sperm head movement. Thus, sperm tail's beating amplitude can possibly be used as an indicator to reflect sperm's motility when the sperms lose their head motion (e.g., in HA-based sperm selection).

12.3.2 Sperm Morphology Analysis

Sperm morphology has been recognized as a powerful predictor of the outcome of natural conception, intrauterine insemination, and conventional IVF therapies [18, 19]. For sperm morphology analysis, a method of unstained, real-time, high-magnification motile sperm organellar morphology examination (MSOME) was developed [20, 21].

In MSOME, motile sperms are transferred to an observation microdroplet of sperm medium containing PVP solution. PVP solution is added to slow down the sperm moving speed. To reduce the toxicity of PVP, a low concentration of the PVP is typically used (<8%). In order to assess the morphological state of the sperm nucleus, the motile sperm cell must be kept inside the field of view for at least 20 s. According to Berkovitz's study [20], the success of MSOME is dependent on:

(1) optical resolution, which depends on microscope optics and on the light source of the microscope; (2) image contrast, which is enhanced by Nomarski differential interference contrast optics; (3) maximal optical magnification, which is commonly higher than 100×; and (4) magnification of the video system, which is equal to the ratio of TV monitor diagonal dimension to the CCD chip diagonal dimension.

A sperm is considered morphologically normal when it exhibits a normal nucleus as well as acrosome, post-acrosomal lamina, neck, tail, and it does not possess a cytoplasmic droplet or cytoplasm around the head [21]. A normal nucleus reveals a smooth, symmetric, and oval shape. The normal length and width of the nucleus are estimated as $4.75 \pm 2.8\,\mu m$ and $3.28 \pm 0.2\,\mu m$, respectively.

12.3.3 Sperm DNA Integrity Analysis

A number of studies have indicated that infertile men have a higher level of DNA strand breaks and other types of DNA damage than fertile sperm donors [22–24]. In order to detect DNA damage, several sperm analysis methods have been developed, including TUNEL [25], comet [26], in situ nick translation [27], DNA breakage detection fluorescence in situ hybridization [28], sperm chromatin dispersion test [29], and sperm chromatin structure assay [30]. Some of these sperm analysis systems assess DNA quality directly, such as TUNEL or comet at neutral pH while others measure DNA damage levels after denaturation steps, such as the sperm chromatin structure assay, sperm chromatin dispersion test, and comet at acid or alkaline pH.

Indirect methods measure DNA susceptibility to denaturation after exposure to acid conditions [31]. However, it has been reported that these methods can only evaluate acid-labile sites and would not have a significant impact on the formation of male pronucleus because the intracellular pH of the oocyte is approximately 7.0 [32]. Another indirect approach checks sperm's DNA integrity by using the hyaluronic acid (HA) coated Petri dishes [8]. HA is a linear polysaccharide in the extracellular matrix of cumulus oophorus around the oocyte that seems to play an important role in natural human fertilization [33]. Huszar et al. proved that sperms that bind the head to the HA have a higher level of DNA integrity than those sperms that cannot bind to HA. After a sperm binds the head to a HA microdot, it loses the head movement but reveals more vigorous tail motion.

12.4 Sperm Manipulation

Sperm manipulation is routinely performed by embryologists in the ICSI procedure in IVF clinics. Sperm manipulation includes immobilizing a sperm, aspirating it into a micropipette, and positioning it inside the micropipette.

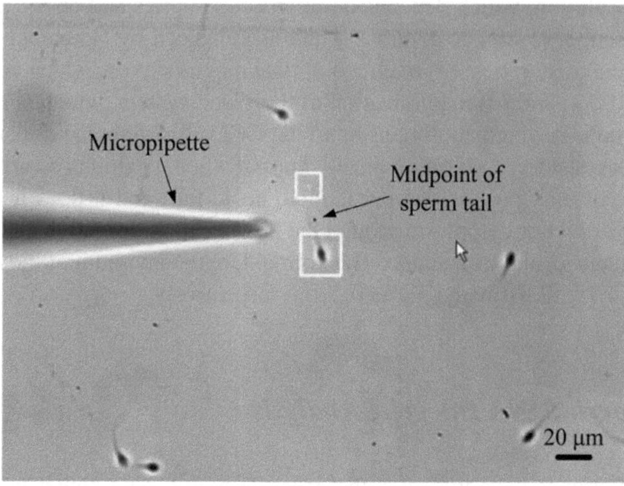

Fig. 12.4 Automated sperm immobilization

12.4.1 Sperm Immobilization

Sperm immobilization must be performed before injecting the sperm into the egg to increase the chance of fertilization since sperm tail movement can cause damage to the intracellular structure of the egg [34]. In sperm immobilization, a micropipette is used to press (tap) the sperm tail against a surface (e.g., the bottom of a Petri dish).

Sperm manipulation requires dexterous operation due to the motile nature of sperms and their small size. Recent study has shown that automated sperm immobilization using a robotic approach is feasible [14]. In automated sperm immobilization, a vision-based contact detection was first performed to determine the vertical depth between the micropipette and the surface of Petri dish. Single sperm head and tail tracking is then conducted to locate the sperm head and tail's positions. The sperm head tracking algorithm enables the system to visually servo a motorized X–Y stage to keep the moving sperm at the center of the field of view. A micropipette is then controlled to tap the midpoint of the sperm tail against the dish bottom to immobilize the sperm (Fig. 12.4). The midpoint of the sperm tail is found by averaging the tracked sperm head and tail position.

Two conditions must be met for tapping the sperm tail: (1) sperm tail needs to be at appropriate orientation and (2) sperm moves within a depth of 25 μm above the dish bottom. In the automated sperm immobilization system, the micropipette is placed on the left side of field of view. Thus, the target sperm should move near-vertically (i.e., within -45 to 45° with respect to the y-axis). For some sperms that do not have the appropriate orientation, a rotational stage can possibly be used to adjust the sperm tail orientation. To immobilize a sperm, the micropipette needs to tap the sperm tail against the dish bottom. If a sperm is too far away from the dish bottom (e.g., $>$25 μm), the system will fail to tap the sperm tail. The automated immobilization system computes the focus measure by using the normalized variance [35] of ROI image of the sperm head. During sperm immobilization, the microscope is focused on the dish bottom. A sperm with higher normalized variance is close to the focal plane and is preferred for immobilization.

Trials of the automated sperm immobilization system on 1,000 sperms indicate the success rate of sperm immobilization is 88.2%. The failure cases include (1) the turbulent flow caused by micropipette movement displacing the sperm's original position; (2) the sperm moving at 25 μm above the dish surface; (3) the sperm tail not staying in the required orientation; (4) sperm tail changing the orientation during the taping step; (5) sperm increasing its speed during the taping step.

12.4.2 Sperm Aspiration and Positioning Inside Micropipette

After the sperm is immobilized on the Petri dish bottom, the next step is to aspirate the sperm into the micropipette and position it to a desired location inside the micropipette. Aspirating a single sperm into a micropipette and precisely controlling the sperm's position within the micropipette is challenging, due to the small volume of a sperm (picoliter) and the nonlinear dynamics involved in the process (e.g., varying mass of culture medium entering the micropipette in cell aspiration).

In manual operation, an operator looks through the eyepieces of a microscope and operates multiple devices (microscope stage, micromanipulator, pump, etc.). When the micropipette approaches a target sperm cell, a small negative pressure is applied to aspirate the sperm into the micropipette. Once the sperm enters the micropipette, which is a rapid event, the operator must quickly apply a positive pressure to stop the sperm movement so that the sperm does not enter too far into the micropipette and disappears. In order to position the sperm to a desired location inside the micropipette, the operator must repeatedly adjust the application of negative and positive pressure skillfully.

In automated sperm aspiration and positioning [36], the sperm aspiration and positioning tasks are achieved via computer vision microscopy and closed-loop motion control. The major components in the micropipette aspiration system are a syringe, a micropipette, and a connecting tube. The syringe and connecting tube are filled with mineral oil. The inner space of micropipette typically consists of three segments: mineral oil, air, and culture medium. A sperm cell moves together with the culture medium segment (no relative motion within the culture medium). Figure 12.5 is a schematic diagram of a sperm moving inside a micropipette.

In the automated system, the micropipette is first controlled to approach the immobilized sperm. The sperm is then aspirated into the micropipette by applying a negative pressure. In this process, the target sperm is detected and tracked. After the sperm is aspired into the micropipette, a robust controller quickly positions the sperm to a desired position in the micropipette. In the study of the sperm positioning in the micropipette, it was reported that the robust controller significantly outperforms PD controllers in terms of efficiency, overshoot, and accuracy.

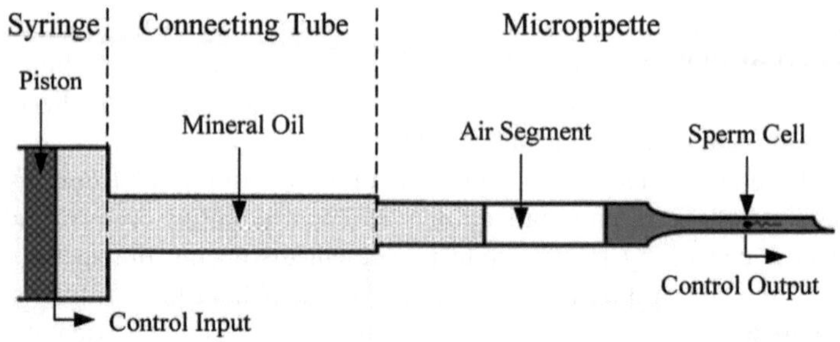

Fig. 12.5 A schematic diagram of a sperm moving inside a micropipette [36] (©IEEE 2012), reprinted with permission

12.5 Conclusion

Sperm analysis, which involves the measurement of sperm's velocity, morphology, and DNA integrity, is necessary for male infertility diagnostics, sperm quality assessment, sperm selection, and other IVF related issues. In the past few decades, CASA systems were developed and used for providing objective and accurate results for sperm analysis. The core of CASA systems is computer vision-based sperm tracking. Most of the sperm tracking algorithms are focused on sperm head tracking. Due to the emerging HA-based sperm assay and the automation of sperm manipulation, sperm tail tracking becomes necessary. This chapter introduced sperm head and tail tracking methods as well as sperm manipulation techniques, including immobilization, aspiration, and positioning in a micropipette.

Further studies using these sperm tracking, analysis, and manipulation technique as well biochemical approaches will answer presently open questions, such as: (1) Does a sperm that moves faster than others have a higher level of DNA integrity? (2) Does a sperm possessing perfect morphologies have less DNA defects? (3) Does a sperm showing faster tail motion, among the HA bound sperms, have higher reproductive quality? The engineering techniques discussed in this chapter will prove instrumental in addressing these questions and will prove useful in both biology research and reproductive medicine.

References

1. D.F. Katz, J.W. Overstreet, S.J. Samuels, P.W. Niswander, T.D. Bloom, E.L. Lewis, Morphometric analysis of spermatozoa in the assessment of human male fertility. J. Androl. **7**, 203–210 (1986)
2. G. Palermo, H. Joris, P. Devroey, A.C. Van Steirteghem, Pregnancies after intracytoplasmic injection of single spermatozoon into an oocyte. Lancet **340**, 17–18 (1992)
3. T.G. Cooper et al., World Health Organization reference values for human semen characteristics. Hum. Reprod. Update **16**, 231–245 (2009)

4. B. Bartoov, A. Berkovitz, F. Eltes, A. Kogosovsky, A. Yagoda, H. Lederman, S. Artzi, M. Gross, Y. Barak, Pregnancy rates are higher with intracytoplasmic morphologically selected sperm injection than with conventional intracytoplasmic injection. Fertil. Steril. **80**, 1413–1419 (2003)
5. H.E. Chemes, Sperm pathology: a step beyond descriptive morphology. Origin, characterization and fertility potential of abnormal sperm phenotypes in infertile men. Hum. Reprod. Update **9**, 405–428 (2003)
6. J. Tesarik, C. Mendoza, E. Greco, Paternal effects acting during the first cell cycle of human preimplantation development after ICSI. Hum. Reprod. (Oxford, England) **17**, 184–189 (2002)
7. A. De Vos, H. Van De Velde, H. Joris, G. Verheyen, P. Devroey, A. Van Steirteghem, Influence of individual sperm morphology on fertilization, embryo morphology, and pregnancy outcome of intracytoplasmic sperm injection. Fertil. Steril. **79**, 42–48 (2003)
8. G. Huszar, C.C. Ozenci, S. Cayli, Z. Zavaczki, E. Hansch, L. Vigue, Hyaluronic acid binding by human sperm indicates cellular maturity, viability, and unreacted acrosomal status. Fertil. Steril. **79**, 1616–1624 (2003)
9. A. Jakab, D. Sakkas, E. Delpiano, S. Cayli, E. Kovanci, D. Ward, A. Ravelli, G. Huszar, Intracytoplasmic sperm injection: a novel selection method for sperm with normal frequency of chromosomal aneuploidies. Fertil. Steril. **84**, 1665–1673 (2005)
10. A. Yagci, W. Murk, J. Stronk, G. Huszar, Spermatozoa bound to solid state hyaluronic acid show chromatin structure with high DNA chain integrity: an acridine orange fluorescence study. J. Androl. **31**, 566–572 (2010)
11. R.P. Amann, D.F. Katz, Reflections on CASA after 25 years. J. Androl. **25**, 317–325 (2004)
12. L.Z. Shi, J.M. Nascimento, M.W. Berns, E.L. Botvinick, Computer-based tracking of single sperm. J. Biomed. Optic. **11**, 054009 (2006)
13. V.R. Nafisi, M.H. Moradi, M.H. Nasr-Esfahani, A template matching algorithm for sperm tracking and classification. Physiol. Meas. **26**, 639–651 (2005)
14. C. Leung, Z. Lu, N. Esfandiari, R.F. Casper, Y. Sun, Automated sperm immobilization for intracytoplasmic sperm injection. IEEE Trans. Bio-med. Eng. **58**, 935–942 (2011)
15. V.R. Nafisi, M.H. Moradi, M.H. Nasr-esfahani, Sperm identification using elliptic model and tail detection. World Acad. Sci. Eng. Technol. **6**, 205–208 (2005)
16. A.M. Groenewald, E. Botha, Preprocessing and tracking algorithms for automatic sperm analysis, in *COMSIG 1991 Proceedings: South African Symposium on Communications and Signal Processing* (IEEE, New York, 1991), pp. 64–68
17. L.Z. Shi, J. Nascimento, C. Chandsawangbhuwana, M.W. Berns, E.L. Botvinick, Real-time automated tracking and trapping system for sperm. Microsc. Res. Tech. **69**, 894–902 (2006)
18. S. Oehninger, T. Kruger, The diagnosis of male infertility by semen quality. Clinical significance of sperm morphology assessment. Hum. Reprod. (Oxford, England) **10**, 1037–1038 (1995)
19. Z.P. Nagy, J. Liu, H. Joris, G. Verheyen, H. Tournaye, M. Camus, M.C. Derde, P. Devroey, A.C. Van Steirteghem, The result of intracytoplasmic sperm injection is not related to any of the three basic sperm parameters. Hum. Reprod. (Oxford, England) **10**, 1123–1129 (1995)
20. A. Berkovitz, F. Eltes, S. Yaari, N. Katz, I. Barr, A. Fishman, B. Bartoov, The morphological normalcy of the sperm nucleus and pregnancy rate of intracytoplasmic injection with morphologically selected sperm. Hum. Reprod. (Oxford, England) **20**, 185–190 (2005)
21. B. Bartoov, A. Berkovitz, F. Eltes, A. Kogosowski, Y. Menezo, Y. Barak, Real-time fine morphology of motile human sperm cells is associated with IVF-ICSI outcome. J. Androl. **23**, 1–8 (2002)
22. A. Zini, R. Bielecki, D. Phang, M.T. Zenzes, Correlations between two markers of sperm DNA integrity, DNA denaturation and DNA fragmentation, in fertile and infertile men. Fertil. Steril. **75**, 674–677 (2001)
23. M. Benchaib, V. Braun, J. Lornage, S. Hadj, B. Salle, H. Lejeune, J.F. Guérin, Sperm DNA fragmentation decreases the pregnancy rate in an assisted reproductive technique. Hum. Reprod. (Oxford, England) **18**, 1023–1028 (2003)

24. M. Sergerie, G. Laforest, L. Bujan, F. Bissonnette, G. Bleau, Sperm DNA fragmentation: threshold value in male fertility. Hum. Reprod. (Oxford, England), **20**, 3446–3451 (2005)
25. W. Gorczyca, F. Traganos, H. Jesionowska, Z. Darzynkiewicz, Presence of DNA strand breaks and increased sensitivity of DNA in situ to denaturation in abnormal human sperm cells: analogy to apoptosis of somatic cells. Exp. Cell Res. **207**, 202–205 (1993)
26. C.M. Hughes, S.E. Lewis, V.J. McKelvey-Martin, W. Thompson, A comparison of baseline and induced DNA damage in human spermatozoa from fertile and infertile men, using a modified comet assay. Mol. Hum. Reprod. **2**, 613–619 (1996)
27. M.J. Tomlinson, O. Moffatt, G.C. Manicardi, D. Bizzaro, M. Afnan, D. Sakkas, Interrelationships between seminal parameters and sperm nuclear DNA damage before and after density gradient centrifugation: implications for assisted conception. Hum. Reprod. (Oxford, England) **16**, 2160–2165 (2001)
28. J.L. Fernández, F. Vázquez-Gundín, A. Delgado, V.J. Goyanes, J. Ramiro-Díaz, J. de la Torre, J. Gosálvez, DNA breakage detection-FISH (DBD-FISH) in human spermatozoa: technical variants evidence different structural features. Mutat. Res. **453**, 77–82 (2000)
29. J.L. Fernández, L. Muriel, M.T. Rivero, V. Goyanes, R. Vazquez, J.G. Alvarez, The sperm chromatin dispersion test: a simple method for the determination of sperm DNA fragmentation. J. Androl. **24**, 59–66 (2003)
30. D.P. Evenson, Z. Darzynkiewicz, M.R. Melamed, Relation of mammalian sperm chromatin heterogeneity to fertility. Science (New York, NY) **210**, 1131–1133 (1980)
31. D.P. Evenson, R. Wixon, Comparison of the Halosperm test kit with the sperm chromatin structure assay (SCSA) infertility test in relation to patient diagnosis and prognosis. Fertil. Steril. **84**, 846–849 (2005)
32. D. Sakkas, J.G. Alvarez, Sperm DNA fragmentation: mechanisms of origin, impact on reproductive outcome, and analysis. Fertil. Steril. **93**, 1027–1036 (2010)
33. C.G. Petersen, F.C. Massaro, A.L. Mauri, J.B. Oliveira, R.L. Baruffi, J.G. Franco, Efficacy of hyaluronic acid binding assay in selecting motile spermatozoa with normal morphology at high magnification. Reprod. Biol. Endocrinol.: RB&E **8**, 149 (2010)
34. G.D. Palermo, P.N. Schlegel, L.T. Colombero, N. Zaninovic, F. Moy, Z. Rosenwaks, Aggressive sperm immobilization prior to intracytoplasmic sperm injection with immature spermatozoa improves fertilization and pregnancy rates. Hum. Reprod. (Oxford, England) **11**, 1023–1029 (1996)
35. Y. Sun, S. Duthaler, B.J. Nelson, Autofocusing in computer microscopy: selecting the optimal focus algorithm. Microsc. Res. Tech. **65**, 139–149 (2004)
36. X. Zhang, C. Leung, Z. Lu, N. Esfandiari, R.F. Casper, Y. Sun, Controlled aspiration and positioning of biological cells in a micropipette. IEEE Trans. Bio-med. Eng. **59**, 1032–1040 (2012)

Index

M. Rakotondrabe (ed.), *Smart Materials-Based Actuators at the Micro/Nano-Scale: Characterization, Control, and Applications*, DOI 10.1007/978-1-4614-6684-0,
© Springer Science+Business Media New York 2013

MIX
Papier aus verantwortungsvollen Quellen
Paper from responsible sources
FSC® C105338

If you have any concerns about our products,
you can contact us on
ProductSafety@springernature.com

In case Publisher is established outside the EU,
the EU authorized representative is:
Springer Nature Customer Service Center GmbH
Europaplatz 3, 69115 Heidelberg, Germany

Printed by Libri Plureos GmbH
in Hamburg, Germany